FIXED POINTS AND ECONOMIC EQUILIBRIA

Series on Mathematical Economics and Game Theory

Series Editor: Ezra Einy *(Ben Gurion University)*

Editorial Advisory Board

Tatsuro Ichiishi
Hitotsubashi University

James S. Jordan
The Penn State University

Ehud Kalai
Northwestern University

Semih Koray
Bilkent University

John O. Ledyard
California Institute of Technology

Richard P. McLean
Rutgers University

Dov Monderer
The Technion

Bezalel Peleg
The Hebrew University of Jerusalem

Stanley Reiter
Northwestern University

Dov E. Samet
Tel Aviv University

Timothy Van Zandt
INSEAD

Eyal Winter
The Hebrew University of Jerusalem

Itzhak Zilcha
Tel Aviv University

Published

Vol. 1: Theory of Regular Economies
by Ryo Nagata

Vol. 2: Theory of Conjectural Variations
by C. Figuières, A. Jean-Marie, N. Quérou & M. Tidball

Vol. 3: Cooperative Extensions of the Bayesian Game
by Tatsuro Ichiishi & Akira Yamazaki

Vol. 4: General Equilibrium Analysis of Production and Increasing Returns
by Takashi Suzuki

Vol. 5: Fixed Points and Economic Equilibria
by Ken Urai

Series on Mathematical Economics and Game Theory
Vol. 5

FIXED POINTS AND ECONOMIC EQUILIBRIA

Ken Urai
Osaka University, Japan

NEW JERSEY • LONDON • SINGAPORE • BEIJING • SHANGHAI • HONG KONG • TAIPEI • CHENNAI

Published by
World Scientific Publishing Co. Pte. Ltd.
5 Toh Tuck Link, Singapore 596224
USA office: 27 Warren Street, Suite 401-402, Hackensack, NJ 07601
UK office: 57 Shelton Street, Covent Garden, London WC2H 9HE

British Library Cataloguing-in-Publication Data
A catalogue record for this book is available from the British Library.

Series on Mathematical Economics and Game Theory — Vol. 5
FIXED POINTS AND ECONOMIC EQUILIBRIA
Copyright © 2010 by World Scientific Publishing Co. Pte. Ltd.
All rights reserved. This book, or parts thereof, may not be reproduced in any form or by any means, electronic or mechanical, including photocopying, recording or any information storage and retrieval system now known or to be invented, without written permission from the Publisher.

For photocopying of material in this volume, please pay a copying fee through the Copyright Clearance Center, Inc., 222 Rosewood Drive, Danvers, MA 01923, USA. In this case permission to photocopy is not required from the publisher.

ISBN-13 978-981-283-718-9
ISBN-10 981-283-718-3

Typeset by Stallion Press
Email: enquiries@stallionpress.com

Printed in Singapore.

To My
Father and Mother
Megumi Urai — Masue Urai

Preface

In Spring 2004 I obtained the opportunity to write a book of my researches on mathematical fixed-point theory and economic equilibrium from Professor Tatsuro Ichiishi, the editor (in that period) of this monograph series. I had written several papers on a certain general class of fixed-point theorems (that are treated in Chapter 2 of this book), and he suggested that I develop those arguments into a monograph and submit it to his series. Since then, for five years, I have devoted myself to this special task: the existence of economic equilibria and fixed-point arguments in general topological spaces. Usually, researches in my area (economic theory) are first supplied as relatively short papers, so writing a book based on a certain subject without being bothered by restrictions of pages or contents (except for the minimal ones that I had written in the prospectus), and the development of discussions is an ideally luxurious condition. With this opportunity, I could explore many topics that I had never intended to visit; e.g., almost all equilibrium arguments are given under one of the most general abstract convexity structures without vector space structures (Chapters 3–5); fixed-point arguments are related to the Vietoris–Čech homology theory and the extension theorems for the Lefschetz number and fixed-point indexes are obtained (Chapters 6 and 7); the fixed-point and equilibrium arguments are reconsidered from the methodological viewpoint of social science to give extensional arguments for Gödel's Second Incompleteness Theorem and Tarski's Truth Definition Theorem (Chapter 9), and the final development into the perspective for human rationality as a fixed-point view of the world (Chapter 10). I am very grateful to Professor Ichiishi for giving me this fantastic experience.

1 This Book's Outline

The material included does not comprehensively cover every topic of mathematical economics; instead, I have collected the principal results and issues from (1) the central topics of the foundation of general equilibrium analysis, and (2) what seems highly mathematical and abstract but crucial from a methodological viewpoint in the social sciences. The former includes fixed-point theorems for multivalued mappings (Chapter 2), the existence of Nash and generalized Nash equilibria (Chapter 3), market equilibrium (Gale–Nikaido–Debreu) theorems (Chapter 4), and general equilibrium theory with non-ordered preferences and infinite dimensional commodity spaces (Chapter 5). The latter includes homological (algebraic) methods in fixed-point arguments (Chapter 6), a homological type of index theory (Chapter 7), and axiomatic set theory with mathematical logic as the most fundamental method to describe objects in the social sciences (Chapter 9).

In addition, there is a general introduction for basic tools (Chapter 1), a chapter on miscellaneous arguments about closely related issues (Chapter 8), a concluding discussion (Chapter 10), and appendices for further mathematical study and supplemental theorems and proofs (Appendices I, II, and III).

2 This Book and Current Stream of Economic Theory

Economic theory explains human society, and a theory of human society is difficult to construct on a static or an unchangeable framework. Although the topics in almost all the chapters of this book are taken from traditional ones in mathematical economics, the methods, discussions, and results are aimed to develop new theories and approaches in economics that are firmly based on abstract (algebraic topological) settings and systematic (axiomatic set-theoretical) arguments.

Since the 1950s, rigorous axiomatic approaches to general equilibrium theory have advanced on the basis of the framework of the Arrow–Debreu economy, or, more generally, non-cooperative n-person games. Of course, the six decades since World War II have seen great strides in the formation of the total system of economic theories. Even from this book's particular focus, many influential works exist, including Debreu's *Theory of Value* (1959), Nikaido's *Convex Structures and Economic Theory* (1969), Arrow and Hahn's *General Competitive Analysis* (1971), Scarf's *The Computation of Economic Equilibria* (1973), and Hildenbrand's *Core and Equilibria of a Large Economy* (1974). We have also seen the introduction of game-theoretic methods and tools for economic equilibrium theory, including Ichiishi's *Game Theory for Economic Analysis* (1983). From the late 1970s to the early 1990s, research was particularly vibrant on equilibrium analysis for economies with non-ordered preferences, non-convex preferences and technologies, and infinite dimensional commodity and price spaces. Note also the introduction of structural stability and global analysis methods of using differentiable approaches as treated in Mas-Colell's *The Theory of General Economic Equilibrium* (1985) and Balasko's *Foundations of the Theory of General Equilibrium* (1988).

Such developments may be considered great successes in the description of human society as a total mechanism consisting of rational individuals, for example, the static economic worldview. The main purpose of this book is to give a unified perspective on these arguments through a fixed-point method and equilibrium-existence theorems by clarifying their minimal requirements or their limits in view of social science. Particularly, the use of algebraic settings under homological arguments over the problems and a systematic approach based on the general space settings forms this book's methodologically distinctive features and contributes to this area.

Note also that the developments of the static worldview the past six decades do not seem successfully directed to other important problems in describing human society, i.e., dealing with the dynamic aspects of the mutual interaction between individuals and society as a whole. For example, an axiomatic (Arrow–Debreu type) approach to economic agents clearly fails to deal with changing informational structures or developing knowledge, which causes many serious problems in describing the dynamic world, e.g., firm's objective functions, market viability problems under asymmetric information, survival conditions and defaults, objective demand functions in oligopolistic markets, and many other situations inquiring what our (economic) rationality (based on the model we have confronted) is.

Although completely describing such dynamic aspects in this world would be impossible, it does not seem to be a good idea to restrict or even condense our arguments to a static framework.[1] Indeed, many general equilibrium theorists today are unwilling to interpret their models to approximate (even in an idealized sense) the "whole" world. For example, Balasko

---

[1]One way to treat all dynamic aspects through the static framework is the perfect foresight equilibrium concept.

(1988, p. 2) says, "A widespread but rather unfortunate practice of economic theory is to consider the market as universal, in the sense that it is unique and that every commodity is traded there, an interpretation that cannot be seriously defended." Such a standpoint, however, sacrifices a clear distinction between general and partial equilibrium concepts, and one of the most important implications of general equilibrium is to grasp human society as a whole (or giving an economic worldview).[2]

In standard economic theory, the market is a clearly defined mathematical mechanism that can be treated as a fact for each member of society. A price, however, is in some sense a value that depends on our thoughts, expectations, beliefs, reason, etc., as we can easily see in cases with dynamic or *incomplete market* models. It may be possible to study market mechanisms simply from mathematical or engineering viewpoints, though in such cases, as with *financial theory*, we must take the price process as given. In many cases, we must treat expectations exactly as our preferences (i.e., as merely data given from outside the world being considered), as in cases with the *temporary general equilibrium theory*, or to use the model's consistency as an excuse for accepting the validity of such prices, as in cases with *perfect foresight* or *rational expectation equilibria* for the *incomplete-market general equilibrium theory*. The former is one extreme (eliminating the consideration of dynamic human factors), and the latter is another extreme (considering only economic agents compatible with the theory). It follows that we have to restrict, condense, or even sacrifice the meaning of general equilibrium theory and/or stop thinking about man and dynamism in the human world as long as we base our discussion on the classical static worldview.

Although this book's generalized fixed-point theorems and homological arguments by themselves contain many applied examples, the main purpose of their algebraic and axiomatic settings is not a simple generalization of conditions for such a static framework. In this book, using fixed-point arguments and methods not merely as tools for *describing* the world but as tools of agents for *thinking about* the world, I have attempted to clarify minimal mathematical requirements or restrictions for constructing

---

[2]For example, in *Dougakuteki Keizai Riron* (*Dynamic Economic Theory*, p. 1, 1950), Morishima says that economics and those economists who support general equilibrium theory aim to obtain "ultimate principles of explanation" for "all economic phenomena" through analysis "finally based on non-economic world" described as the data that economists take as given.

Preface   xi

a social theory with equilibrium views of the world and reconstructing standard equilibrium arguments to more properly deal with man and society in economic theory. The concept "fixed point" is used not only as a mathematical term but also as a general notion or method of thinking in ordinary language, like its use for thoughts, language, recognition, and knowledge. From this viewpoint, the book's generalizations on such issues as convexity, continuity, and the reconstruction of commodity/price duality and homological settings in Chapters 2–8 will have different meanings: the dual-system and fixed-point arguments will be considered a general method to capture the world, and the minimal settings for equilibrium arguments under the algebraic structure are crucial requirements for equilibrium theory itself to be coded as objects in mathematical theory. Several interesting consequences of such a unified viewpoint or method are obtained in a rigorous mathematical framework through axiomatic set theory including formal logic and model theory in Chapters 9 and 10. I believe they construct important methodological arguments in economics, and, as a conclusion to this book, provide a new basis for using mathematical arguments in the social sciences.

3 For General Readers

Since this is a research monograph in mathematical economics, the tools and methods presented include highly abstract mathematics. Abstract convexity, homology theory, and mathematical logic under axiomatic set theory may be less familiar to economists (even mathematical economists) than tools in linear algebra, measure theory, differential topology, etc. One purpose of this book is to serve as an introductory text in mathematical economics including these concepts and methods for all researchers and students who are interested in rigorous and new mathematical methods in economics.

Indeed, the basic mathematics in Chapter 1 with Appendices I and II may be used as a preparatory course of mathematics (general topology and topological vector space) for undergraduate mathematical economics. Following it, Chapter 2, Section 2.1 can provide the essence of all fixed-point arguments without using the abstract convexity concept. Almost all the theorems in Chapters 3–5 retain their generality even under such ordinary interpretations. (The abstract convexity will help us, however, incorporate notions in vector spaces with the homological arguments in

Chapters 6 and 7.) Chapter 9, Section 9.1 may also be used independently as an introductory course in mathematical logic and model theory based on axiomatic set theory for economic and game-theoretic treatment of individuals, rationality, social scientific recognition, and society. The necessary tools and concepts for the theorems in Chapter 9 are closed in this chapter. For the main discussions in Chapter 10, one must add the definition of direct limits (in Section 6.1) and one existence theorem of Nash equilibrium in Chapter 3.

I recommend that readers concerned with the homological argument in this book (1) study the ordinary proof of Brouwer's fixed-point theorem through Sperner's lemma (see, e.g., Nikaido (1968), Ichiishi (1983)), and (2) then study Section 6.1, where one may skip the proof of Theorem 6.1.3, (3) and directly go to subsection, Analogue of Sperner's Lemma, in Section 6.3, where one should omit the concept of Vietoris–Begle barycentric subdivision and take Theorems 6.2.2 and 6.3.3 (The Vietoris–Begle Mapping Theorem) as granted. One may read to the end of Chapter 6. Then consider their proofs as the final purpose for Chapter 6.

I also hope this book can serve as an introduction to economic equilibrium theory for mathematicians and researchers in all areas who are familiar with the type of reasoning used in mathematics and who are interested in the use of general equilibrium theory as an economic way to view the world.

For updated information about this book, see the following URL:

http://math.econ.osaka-u.ac.jp/LABORATORYe.html

4 Relation to Other Works

Beside the books mentioned above, most works in mathematical economics since the 1980s have important connections to at least one chapter in this book. Looking at the theorems presented in Chapters 2–5, this book is nothing but an extension of the work presented in books published in the 1980s and 1990s, such as Border's *Fixed Point Theorems with Applications to Economics and Game Theory* (1985), Aliprantis–Brown–Burkinshaw's *Existence and Optimality of Competitive Equilibria* (1989), *Equilibrium Theory in Infinite Dimensional Spaces* (1991) edited by Kahn and Yannelis, and Aliprantis's *Problems in Equilibrium Theory* (1996).

The new perspective on algebraic topological methods for equilibrium analysis (Chapters 6 and 7) may provide a unified viewpoint on classical convexity approaches (Debreu (1959), Hildenbrand (1974), etc.) and differentiable approaches (Mas-Colell (1985), Balasko (1988), etc.) based on generalized convexity and duality structure. The mathematical arguments for economic rationality and knowledge based on axiomatic set theory with logic and model theory (Chapters 9 and 10) are closely related to recent arguments on the foundation of game theory (player's rationality, common knowledge, etc.) or game logic. I have based, however, my arguments on an axiomatic set theory that is strong enough to code itself into its objects (sets). At least for describing human society (rationality or knowledge) and discussing the methodology of social science, I believe that this recursive feature is essential as a minimal requirement that the basic theory must have.[3]

5 Acknowledgments

I wish to express gratitude to all my teachers, colleagues, research assistants, and collaborators who so generously supported me during the years of preparation for this manuscript. I am much indebted to Professor Kiyoshi Kuga for his many lectures and for all the mathematical bases in this book, including the algebraic, the topological and axiomatic set-theoretic ones. I also appreciate the instructions of Professor Toshihiko Hayashi who gave me my basic standpoint as a theorist in social science. Professor Hiroaki Nagatani, the chief referee of my doctoral dissertation, and Professors Masamitsu Ohnishi and Takuo Dome, also gave me lots of important instructions and comments on the preparatory draft of this monograph. During these preparation years, I obtained much from conversations with my colleagues at Osaka University, especially Yoshiyuki Takeuchi and Takuo Dome and their insights into social science beyond economic theory. With respect to philosophical arguments, I owe much to the lectures and books of Professor Takenori Inoki and conversations with Tadashi Shigoka and Yasushi Urai. For my basic knowledge in mathematical logic, I am obliged to Professor Mariko Yasugi when I was a member of the Kyoto-Sangyo University. Takashi Hayashi, Akihiko Yoshimachi, Kousuke Yokota, the coauthors of my earlier papers, Professors Jun Iritani, Tomoyuki

---

[3] Discussions that fail to have this recursive feature will be called *arguments from God's eyes* (Chapter 10).

Kamo, Toshiji Miyakawa, the members of the Joint Seminar in Kobe-Osaka University, Kazuya Kamiya, Mamoru Kaneko, Akira Yamazaki, Toru Maruyama, Hidetoshi Komiya, and Wataru Takahashi offered valuable suggestions and inspiration over the years and in various stages of writing, all of which I sincerely appreciate.

A part of this research was supported by the Grant-in-Aid for Scientific Research from the Japan Society for the Promotion of Science in 2009 (No. 21653017).

I must also thank the anonymous referees of this monograph. With their efforts, at least four chapters of this book (Chapters 1, 2, 9, and 10) were drastically improved. Moreover, my thanks are due to Professor Ezra Einy, the editor-in-chief of this monograph series, Kayoko Araki and Akiko Watabe, secretaries of the joint laboratory, Ron Read, Chris Oleson, Atsuko Watanabe, members of Kurdyla and Associates Co., Ltd., and Pauline Chan, Juliet Lee, and Yvonne Tan, editors of World Scientific Publishing Co. Pte. Ltd. for their kind help.

I would like to thank my family (Kai, Momoka, Reiko) for their tolerance of my late-night work. During the last five years, my son has moved from being a toddler to elementary school and my elementary school daughter has become a high school student. Finally, let me again express my special gratitude to Professor Tatsuro Ichiishi. Without his encouragement and inspiration, I could not have conceived or completed this project.

Ken Urai
Osaka
April 2009

Contents

Preface   vii

1 Introduction   1

1.1 Mathematics is Language   1
1.2 Notes on Some Mathematical Tools
    in This Book   2
1.3 Basic Mathematical Concepts and Definitions   12

2 Fixed-Point Theorems   27

2.1 Classical Results and Basic Extensions   27
2.2 Convexity and Duality for General Spaces   39
2.3 Extension of Classical Results to General Spaces   46

3 Nash Equilibrium and Abstract Economy   57

3.1 Multi-Agent Product Settings for Games   57
3.2 Nash Equilibrium   67
3.3 Abstract Economy   75

4 Gale–Nikaido–Debreu's Theorem   87

4.1 Gale–Nikaido–Debreu's Theorem   87
4.2 Market Equilibria in General Vector Spaces   91
4.3 Demand-Supply Coincidence in General Spaces   99

5 General Economic Equilibrium — 101
- 5.1 General Preferences and Basic Existence Theorems ... 101
- 5.2 Pareto Optimal Allocations ... 109
- 5.3 Existence of General Equilibrium ... 111

6 The Čech Type Homology Theory and Fixed Points — 125
- 6.1 Basic Concepts in Algebraic Topology ... 125
- 6.2 Vietoris–Begle Mapping and Local Connectedness ... 142
- 6.3 Nikaido's Analogue of Sperner's Lemma ... 150
- 6.4 Eilenberg–Montgomery's Theorem ... 164

7 Convex Structure and Fixed-Point Index — 167
- 7.1 Lefschetz's Fixed-Point Theorem and Its Extensions ... 167
- 7.2 Cohomology Theory for General Spaces ... 175
- 7.3 Dual-System Structure and Differentiability ... 177
- 7.4 Linear Approximation for Isolated Fixed Points ... 177
- 7.5 Indices for Compact Set of Fixed Points ... 180

8 Applications to Related Topics — 183
- 8.1 KKM, KKMS, and Core Existence ... 183
- 8.2 Eaves' Theorem ... 187
- 8.3 Fan–Browder's Coincidence Theorem ... 189
- 8.4 L-majorized Mappings ... 190
- 8.5 Variational Inequality Problem ... 191
- 8.6 Equilibrium with Cooperative Concepts ... 192
- 8.7 System of Inequalities and Affine Transformations ... 195

9 Mathematics and Social Science — 201
- 9.1 Basic Concepts in Axiomatic Set Theory ... 201
- 9.2 Individuals and Rationality ... 218
- 9.3 Society and Values ... 232

10 Concluding Discussions — 243
- 10.1 Fixed Points and Economic Equilibria ... 243
- 10.2 Rationality and Fixed-Point Views of the World ... 250

Contents xvii

Mathematical Appendix I 257

Mathematical Appendix II 263

Mathematical Appendix III 273

References 277

Index 285

Chapter 1
Introduction

1.1 Mathematics is Language

For the role of mathematics in the economic theory, Paul A. Samuelson's front page motto in his landmark, *Foundation of Economic Analysis* (Samuelson 1947), "Mathematics is a language," is famous among economists. The statement was taken from J. Willard Gibbs. Later, Samuelson (1952) shortened it to "Mathematics is language," to emphasize its fundamental role of mathematics as a methodological and analytical prerogative tool, not only for communication but also for thinking and constructing economic theory and economic problems.[1] The statement declares that economics is an exact science based on logical and precise mathematical language.

Mathematical economics may generally be considered a wide-sense application of various mathematical methods to many problems posed in economics. In this book, however, I base my arguments on the above standpoint: "Mathematics is language." In other words, this book shows how a special mathematical method (a tool for thinking) can be utilized for constructing or developing part of an economic theory. This is the main justification for my restriction of arguments on a single mathematical method, a "fixed point," and a special topic, "economic equilibria."

In economics, the mathematical theory of a fixed point is closely related to the classical theorem on the existence of competitive equilibrium.[2] Moreover, it is also related to many equilibrium arguments and existence results in economics, such as the Nash equilibrium, core allocation existence,

---

[1] Dixit wrote: "...'Mathematics is a language.' Paul improved this to 'Mathematics is language.' Viewed thus, it should be a tool for thinking as well as for communication. The dichotomy that many of us make between economics or intuition on the one hand and mathematics on the other is just as artificial... Ideally, mathematics and intuition should fuse into one overall Weltanschauung about economics...." (Dixit 2005).

[2] It is even possible to recognize the existence of competitive equilibrium as a mathematical problem that is equivalent to Brouwer's fixed-point theorem (Uzawa 1962).

the rational expectation equilibrium, maximal element existence, and maximal balanced growth.[3]

Many of the theorems in this book are technical extensions of such mathematical fixed-point arguments and methods for economic equilibrium results. The main concern of this book, however, is not only to show abundant ways to apply such extensions, but also to list the *minimal* logical, set-theoretical, or algebraic *requirements* for the construction of an economic equilibrium theory. Simplification by such abstraction is essential for further generalization and theory construction. To extract an indispensable framework in the construction of economic arguments, we expect theory development from the basic level of language (i.e., a necessary development in mathematics for economic theory).

Accordingly, in this book I use many highly abstract settings (e.g., fixed-point arguments based on algebraic settings, preference or demand actions without continuity conditions and/or convexity conditions, spaces without linear structures, and axiomatic set theory with mathematical logic) while basing my arguments on topics that are quite orthodox. Among others, the concept of *convex combination*, the *dual system* of spaces, *algebraic approaches* in topology (*general homology theories*), and methods based on *mathematical logic* form the distinguishing features of this book's mathematical arguments. The next section briefly introduces these subjects in an informal discussion preceding rigorous treatments in later chapters with comments on their necessity as vocabulary for further construction of an economic equilibrium theory.

1.2 Notes on Some Mathematical Tools in This Book

1.2.1 *Convexity*

Convexity theory and topology have been the central tools for the rigorous axiomatic treatment of economic theory since the 1950s. The notion of convexity is used to describe ideas within a mixture of alternative choices, a moderate view among extremes, and especially to ensure the existence

---

[3]The paper of von Neumann (1937) should also be listed as one major predecessor of work on the existence of competitive equilibrium in the fifties. According to McKenzie (1981), he first used a fixed-point theorem for an existence argument in economics.

1: Introduction

of equilibrium depending on such stable actions as a fixed point for a mathematical model of society.

Strictly speaking, convexity, in the ordinary sense, is not a mathematical *structure* but a property for subsets in a space with a vector-space (linear) structure.[4] In this book, however, we often treat the concept as an independent mathematical structure, as an *abstract convex structure*, without referring to linear structure in the space. Of course, such abstract convexity can always be replaced with conventional convexity as long as the space has a linear structure. (Although some theorems might lose generality, most general results in this book do not depend on convexity but on the more general concept of acyclicity.[5]) As seen in the classical fixed-point theorem of Eilenberg and Montgomery (1946), the vector structure for topological space is not a necessary setting for fixed-point arguments. Even for classical equilibrium theorems for non-cooperative games, it is recognized that the vector structure is superfluous and that the necessary setting is "a compact... set in which the convex linear combination of finitely many points depends continuously on its coefficients" (Nikaido 1959, p. 362, Main Theorem). The main reason we use the concept of abstract convexity is to utilize intuitive images (like concepts in vector-space fixed-point and game-theoretic equilibrium arguments in Chapters 2–5) even for general spaces without linear structures in Chapters 6 and 7.

The mathematical structure of abstract convexity is given by axiomatizing the concept of convex combination among finite points in a space. Briefly, for each non-empty finite subset B of topological space X, a *set-B-dependent weighted sum* among points in B is defined as the value of continuous function $f_B : \Delta^B \to X$, where $\Delta^B = \{e | e : B \to R, \forall a \in B, e(a) \geq 0, \sum_{a \in B} e(a) = 1\}$.[6] Axioms define for each non-empty finite

---

[4]The word structure, which has a special meaning in mathematics, is a rigorously defined mathematical object in axiomatic set theory, e.g., an order structure (an order relation), a group structure (a group operator), and a topological structure (the family of all open sets). Names like "order," "topology," "group," and "vector space" are all used to represent a *species* of such structures. (See, e.g., Bourbaki (1939).) This concept will be explained more fully in Chapters 6 and 9 as needed. Until then, interpret structure as used in ordinary language.

[5] The ordinary definition of convexity with linear (vector) space structure is given in the next section (Section 1.3.3) as a basic mathematical concept. Acyclicity is described in the first section of Chapter 6.

[6]Throughout this book, function f on set U to V is denoted by $f : U \to V$, and R denotes the set of real numbers. Basic mathematical notation and concepts are restated in the next section. A complete definition of convexity is given in Chapter 2 (Section 2.2).

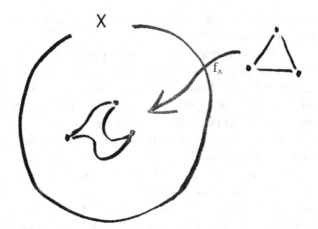

Figure 1: Convex combination of finite points

subset A of X how the *set of all convex combinations among points in* A, $C(A) \subset X$, is characterized by set-dependent continuous functions. For example, we may request $C(A)$ to include all possible set A'-dependent weighted sums among points in A such that $A' \supset A$.[7] The list of such continuous functions and sets,

$$\{(f_A, C(A)) | A \text{ is a non-empty finite subset of } X\}, \quad (1.1)$$

forms a *convex structure* on X. Set $Z \subset X$ is said to be *convex* if for each finite subset B of Z, we have $C(B) \subset Z$. The set-dependent notion of weighted sums among finite points is intuitively represented by the continuous image of the abstract simplex formed by those points (Figure 1). Although the image varies from finite set to finite set to which it belongs, these tools with some axioms are sufficient for the essential part of convexity arguments, especially for fixed-point (and the existence of economic equilibrium) problems.

Throughout this book, all economic equilibrium arguments are based on algebraic or purely set-theoretic methods for fixed-point arguments. For the fixed-point results in Chapters 6 and 7, we do not presuppose any vector space structures on the domain of mappings. The abstract convexity and duality structure (discussed below) of general spaces is useful for relating

---

[7]The convexity concept obtained from this condition is called L-convexity (Ben-El-Mechaiekh *et al.* 1998).

1: Introduction

our ordinary methods and intuitions in vector-space convexity theory and general equilibrium analysis to more advanced contexts.

1.2.2 Duality

In Gerard Debreu's celebrated work on the modern axiomatic analysis of economic equilibrium theory, the *Theory of Value*, the duality between facts and values, the *commodity/price duality*, is one crucial framework for characterizing economic equilibrium theory. General equilibrium theory is nothing but an attempt to describe the total system defining social value (prices), based on our individual judgments of the facts (preferences, technologies, and rules like price-taking behaviors). Such fact/value classification (not by its rigid and total dichotomy or the dualism on its objects but merely as a method for analysis) establishes the essential features of economic equilibrium theory.

When commodity space is taken to be finite dimensional vector space E, prices may usually be taken in set E' of all real valued linear functions on the commodity space (the *dual vector space*). Mathematically, the triplet of E, E', and the evaluation function (the *canonical bilinear form*) $f : E \times E' \ni (x, p) \mapsto p(x) \in R$ is called the *dual system*. More generally, for two topological vector spaces, X and Y, if there is bilinear form $f : X \times Y \to R$, we may identify each element $y \in Y$ with linear function $f(\cdot, y) : X \to R$ and $x \in X$ with $f(x, \cdot) : Y \to R$. If for each $x \neq 0$ and $y \neq 0$, there exist y' and x' such that $f(x, y') \neq 0$ and $f(x', y) \neq 0$, the triplet (X, Y, f) is called a *dual system* or *a system of duality* over real field R.

Linear functions have important meanings and properties based on linear (vector-space) structure. For our purpose, however, such a totality of structure on the base space may not be necessary or, in some cases, may even be restrictive to describe the world. For example, when X is a commodity space and P is a price space, each $p \in P$ gives value $p(x)$ for each point $x \in X$ as well as the value necessary to change the position from x to $x' \in X$, $p(x' - x)$, which inherently equals $p(x') - p(x)$ under a linear structure (Figure 2a). In view of equilibrium theory, the latter concept (value necessary to change the position) must be used to describe the constraints for individual actions under such a social mechanism as the market. In other words, we require the concept of the value of commodity bundle x' *relative to* initial holding x or, more generally, the set of all *available alternatives for* x under a certain social value system. (See Figure 2b, where the value of x' depends on the initial holding. If the initial

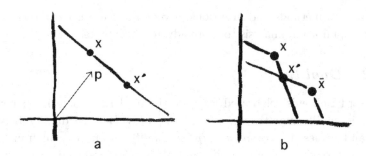

Figure 2: Constraints for individual choices

holding is x, then the value of x' equals the value of x. If the initial holding is $\bar{x}$, the value of x' also equals the value of $\bar{x}$. The value of x, however, is greater than $\bar{x}$ if the initial holding is $\bar{x}$. On the other hand, the value of $\bar{x}$ is also greater than x if the initial holding is x. We often experience such a situation in reality when we confront nonlinear prices, transaction costs, bid-ask spreads, etc.)

If we want to describe the above situation while preserving our (methodological) fact/value classification, we should extend the notion of duality between two spaces, X and Y, as a structure for recognizing each $y \in Y$ as a mechanism to define a subset of X for each $x \in X$, for example, the *set of available alternatives for x*. For this purpose, consider that given two topological spaces, X and Y, function g on $X \times X \times Y \times Y$ to real field R gives a *generalized duality (dual-system) structure* or, more directly, we identify each point $y \in Y$ with a correspondence $V(\cdot, y) : X \to X$ (e.g., $V(x, y) = \{z \in X \mid g(x, z, y, y) > 0\}$) defining a subset of X for each $x \in X$, and (X, Y, V) forms a generalized dual-system structure on X and Y. Here, we do not assume any extra property of g or V, although in later chapters we assume additional conditions on them such as continuity, closedness for their graphs, and (abstract) convexity or acyclicity for their sections.

The generalized system of duality in this sense is intended to be a direct generalization of ordinary commodity/price duality, so we can directly utilize it in many social equilibrium settings (as extensions of ordinary commodity/price or fact/value settings) without referring to any linear (vector-space) structures. Later, we obtain one of the most general types of Gale–Nikaido–Debreu's lemma (for market demand/supply coincidence) and the existence of a competitive equilibrium theorem as a direct application of our concept of duality for spaces with or without linear structures (Chapters 4 and 5).

Mathematically, in combination with the concept of convexity, the duality concept may be used to intuitively describe a certain kind of direction in a space, in the same way that we might characterize the normal vector of a hyperplane or the gradient vector at each point for a real valued mapping on the space. This concept also enables us to extend several important fixed-point theorems, including the theorems of Fan-Browder and Kakutani, to more general spaces without locally convex vector-space structures (Chapters 2 and 3). Moreover, by applying such methods to fixed-point arguments in general homology theories (where the concept of convexity is mainly replaced by the more essential property of being acyclic), we can obtain further results, including an extension of *Lefschetz's fixed-point theorem* (Chapter 6) and arguments on the *fixed-point index* (Chapter 7) for non-continuous functions and multivalued mappings. Of course, these results may also be utilized for further developments in economic theory.

1.2.3   *Algebraic methods*

The most significant feature of this book's integration of convexity, duality, and differentiability is the algebraic methods provided in fixed-point and equilibrium arguments. In particular, the *general homology theories* of the Čech type play essential roles in Chapters 6 and 7.

We use algebraic methods because basing the theory on more elementary tools is preferable than those in standard calculus, convex analysis, differential topology, and so on. Each mathematical theory is associated with a different way of analyzing the world. Since algebra as well as the theory of sets is one of the most fundamental tools for any mathematical argument, a crucial difference exists between, for example, a differentiable approach (research based on differential calculus) and an approach based directly on set-theoretical and algebraic methods. The former mainly consists of *analytic* works that result from seeing the world as a differentiable object, while the latter is mainly a *synthetic* attempt or method to construct models that are more appropriate for describing our real world. They should be reexamined under more primitive concepts, like finiteness, sequences, or limits under set-theoretical or algebraic methods.[8] In this sense, we must

---

[8]The 'limit' is listed here not as standing for the topological one, but for more primitive concepts such as inverse (projective) and direct (inductive) limits.

always use more primitive or fundamental mathematical concepts with more general mathematical methods.

One may ask why some theories of sets and elementary algebra are specially classified as fundamental tools for all mathematical arguments. Of course, what is called fundamental or elementary may also vary as our knowledge or common sense changes. Therefore, some of today's theories of sets and algebra might be replaced by more desirable ones in the future. In this sense, perhaps we cannot obtain what is ultimately fundamental or elementary. Even so, I am convinced that the *linguistic feature* of mathematics itself will never change in our development of knowledge. With the theory of sets, algebraic theory provides a structure to describe our words, sentences, and even our logic used in mathematics or ordinary mathematical arguments.

As long as our knowledge is represented by language, it can be *coded* into algebraic or at least elementary set-theoretical objects. Therefore, the set-theoretic and algebraic methods used in this book must provide a basic framework for arguments that depend not on well-behaved (continuous and/or differentiable) mappings, but on the well-founded minimum requirements for mappings on a primitive finite set of points. They provide a possible framework for *coding* themselves as knowledge to be used in well-founded theories constructed by themselves.

Homology theory is the algebraic study of the connectivity characteristics of a space. Čech-type theory begins this study by approximating the space by sufficiently refined open coverings, thus reducing the connectivity problem to the intersection property among open sets. In Figure 3a, 1-dimensional space X is covered by open covering $\mathscr{M} = \{M_0, M_1, M_2, M_3\}$. In this case, X is approximated by the set of abstract points and lines represented in Figure 3b. Each abstract point (vertex or 0-dimensional simplex) is associated with the name of an open set in the covering, and each line (1-dimensional simplex) indicates that two open sets related to the two vertices of the line intersect. The totality of such abstract simplices (the abstract *complex*) is called the *nerve of covering $\mathscr{M}$*. By taking refinement $\mathscr{N} = \{N_0, N_1, N_2, N_3, N_4\}$ of covering $\mathscr{M}$ (Figure 4a), we obtain the nerve of covering $\mathscr{N}$ as a better approximation for space X (Figure 4b).

A careful reader might think that even if a covering refinement gives a better approximation for the connectivity of space, it may also cause a problem: The dimensions of approximating simplices become too high. In Figure 5a, the nerve of the covering, two open sets, offers a sufficiently good approximation for space X. If we take further refinement for the

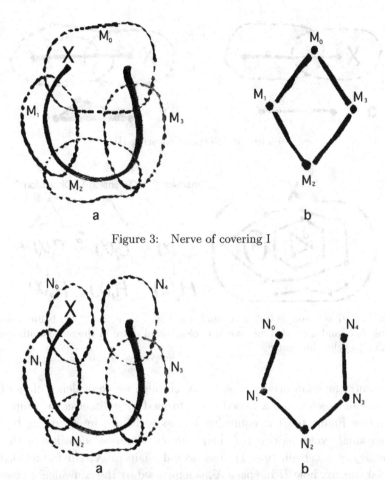

Figure 3: Nerve of covering I

Figure 4: Nerve of covering II

covering, as shown in Figure 5b, the dimensions of simplices approximating X increase, which apparently cannot be reduced under any process of taking refinements. How can we argue that 5b is a better approximation than 5a?

The answer precisely illustrates the homological argument. In homology theory, the difference between the shapes in Figures 5a and 5b is not important. Both sets are called *acyclic*, which is essentially identified with a *single point* under homological arguments. Homology theory associates topological space X with set $H_q(X)$, (q-th homology group of X) with an algebraic structure (e.g., groups, modules, and vector spaces) for each dimension $q = 0, 1, 2, \ldots$. Intuitively, the q-th homology group, $H_q(X)$,

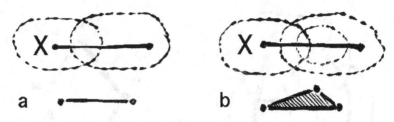

Figure 5: Nerve of covering III

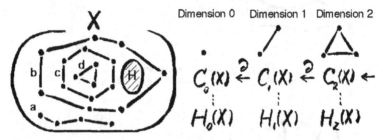

Figure 6: H is a hole in 2-dimensional space X. a, b, c, and d are 1-dimensional chains. b, c, and d are 1-dimensional cycles. c and d are in the same equivalence class of 1-cycles, but not b

represents the connectivity of space X through an equivalence class of q-dimensional *cycles*, i.e., a closed *chain* formed by q-dimensional simplexes in X (see Figure 6). The equivalence class is defined by regarding two q-dimensional cycles as equal if their difference can be identified with the *boundary* of a certain $(q + 1)$-dimensional chain in X. (In Figure 6, the special feature, hole H in space X, is expressed by the equivalence class of 1-cycle b.) For the present, suppose that such $H_q(X)$'s, $q = 0, 1, 2, \ldots$, are vector spaces over real field R and the algebraic structure on each of them successfully stands for the above intuitive discussion about chain formation. The series of such a *graded* vector space

$$\cdots \to H_{q+1}(X) \to H_q(X) \to H_{q-1} \to \cdots \to H_1(X) \to H_0(X) \to 0$$
(1.2)

describes all of the necessary features of space X, where arrows represent the canonical linear functions determined by the concept to take the *boundary* of the chains and 0 denotes the vector space of $\{0\}$. If X is a single point, each q-th homology group is 0 except for $H_0(X) \simeq R$. The acyclic sets are those whose homology groups are exactly the same as those of a single point.

Since there is essentially no difference between acyclic sets and single points, it is not surprising that under homology theory we obtain fixed-point theorems for acyclic valued mappings. *Eilenberg–Montgomery's fixed-point theorem*, one of the most basic results for such multivalued mappings, is far more general than theorems for convex valued mappings, as long as we permit conditions on the space that enable us to use such homological arguments as Čech theory (e.g., polyhedron, absolute neighborhood retracts, and the *local connectedness condition*). Such an identification of sets and points also plays an essential role in relating our convex and topological arguments to algebraic ones (e.g., there is a standard method for constructing associated algebraic mappings, the *method of acyclic models*) and in presenting basic theorems for the settings of Čech-type homology theory (e.g., the *Vietoris mapping theorem*).

The methods and approaches in Chapters 2–5 may be summarized as the replacement of continuity and/or convexity conditions for equilibrium and fixed-point theorems by weaker conditions using the directions of points defined by the dual-system structure. Chapters 6 and 7 show their relationship to the algebraic properties described under homological theory. Moreover, one can see how the underlying ideas in earlier chapters are also utilized for further progress in the arguments, even in algebraic homological settings.

As noted before, the use of algebraic topology rather than differential topology is related to the main purpose of this book: to list *minimal* logical, set-theoretic, and algebraic *requirements* for economic equilibrium and its closely related arguments. The point will also be summarized in Chapters 9 and 10, where all basic features and discussions before Chapter 9 are reexamined through finitistic and recursive methods in an axiomatic set theory.

1.2.4 *Axiomatic set theory*

Chapter 9 refers to the foundation of mathematics as a basic tool or a language for describing a theory of social sciences. Since one important purpose of social science is to describe human society as a well-founded and -defined entity, or a *model*, a formal treatment of our basic tool of thought (mathematics) itself is critical to formalize the foundation of our knowledge.

Mathematically, all arguments in this book are based on *ZF*, the Zermelo–Fraenkel set theory (the most standard axiomatic set theory) written by first-order predicate logic (one of the most popular formal languages). Such a framework, as our basic standpoint, is necessary because our basic theory must be sufficiently strong to incorporate not only our ordinary mathematical arguments but also all necessary procedures in describing the theory itself as formal objects. Indeed, the list of axioms in Zermelo–Fraenkel set theory, which can be used to develop almost all of our ordinary mathematics, is also simple enough to be characterized by standard finitistic or recursive methods that are obviously incorporated in *ZF*.

Of course, until Chapter 9, readers need not be concerned about what axioms our basic theory depends on. The basic mathematical concepts and methods in this book (introduced in the next section) presume a merely natural and naive interpretation of ordinary language. Note, however, that such set-theoretic axioms and their finitistic or recursive methods are not special concepts for a certain field of mathematics, but rather relate to the one thing that never changes in our development of knowledge: the linguistic feature of mathematics.

1.3 Basic Mathematical Concepts and Definitions

The mathematical concepts and definitions that are necessary but not immediately connected with this book's fixed-point or economic equilibrium arguments are gathered into three parts: this section for general fundamental notions, the first section of Chapter 6 for an introduction to algebraic topology, and the first section of Chapter 9 for concepts in axiomatic set theory and mathematical logic. The main purpose of these sections is merely to give definitions of mathematical terms.

In principle, all the concepts and theorems in this book can be explained without any presupposed notion in mathematics and are completely supported in the book. Consequently, all the mathematical topics could be arranged from the basic to the advanced ones so that no theorem is used to prove other results before its own proof is presented. Such an attempt, however, would almost certainly force readers to study several boring mathematical textbooks before reaching the special topics of this book that are not necessarily based on mathematical details and proofs. Therefore, throughout this book, several mathematical theorems and properties are treated (at first) as given, and their proofs are given in later chapters.

Moreover, to facilitate the descriptions of ordinary notions in mathematical economics in Chapters 2–5, readers are expected at least to have a basic knowledge of Euclidean spaces equivalent to college freshman calculus and linear algebra.

In this section, with the definitions of mathematical terms in elementary topology (Subsection 1.3.1), I will introduce two important basic theorems: the partition of unity theorem (Subsection 1.3.2) and the separation hyperplane theorem (Subsection 1.3.3).[9]

1.3.1 Sets, topologies, and notational conventions

All the mathematical arguments in this book are based on Zermelo–Fraenkel set theory with Axiom of Choice, written by first-order predicate calculus. As stated before, these comprise one of the most common pairs of an axiomatic theory of sets and a basic formal language. I merely note here that the following chapters are based on a very standard foundation of mathematics. The formal treatments of the axiomatic set theory and formal language are given in Chapter 9.[10] We also use the notions of Bourbaki (1939) (e.g., structures and inverse and direct limits) in later chapters and Kelley (1955) (many definitions in topology), as long as the underlying set-theoretic differences are not significant.

Sets

Theory of sets is a theory that has only two predicates, $\in$ and $=$, elementhood and equality. We often denote a *set* by the form $\{x|P(x)\}$, where $P(x)$ denotes a property of x described under our formal language (first-order predicate calculus). Notation $\{x|P(x)\}$ represents the class of objects having property P which, in some cases, may not be treated as a proper mathematical object or a *set*. The axioms of the theory of sets (e.g., *ZF* with Axiom of Choice) give rules for a property under which class $\{x|P(x)\}$ may be called a *set*. (A careless use of such properties may cause

---

[9] In this book, adding to these two theorems, Brouwer's fixed-point theorem (in Chapter 2) will be introduced and repeatedly used before its proof is presented. The proof of Brouwer's fixed-point theorem is given in Chapter 6. Proofs for other theorems are given in Mathematical Appendices I and II.
[10] For references, see also Fraenkel *et al.* (1973), Kunen (1980), Jech (1997), etc.

problems like the well-known Russell Paradox.[11]) For example, the class of all natural numbers, $N = \{0, 1, 2, \ldots\}$, the family of two sets x and y, $\{x, y\}$, the *ordered pair* of two sets x and y, (x, y), the class of all subsets of set A (the *power set* of A), $\mathscr{P}(A) = \{X \mid X \subset A\}$, and unions and products for the family of sets (see below) are assured to be sets under the axioms of ZF.

Family of sets

For family (set of sets) $\mathscr{U}$, we denote by $\bigcup \mathscr{U}$ the *union* of elements of $\mathscr{U}$. If the elements of family $\mathscr{U}$ are indexed by set I as $\mathscr{U} = \{U_i \mid i \in I\}$, we often write $\bigcup_{i \in I} U_i$ instead of $\bigcup \mathscr{U}$. If family $\mathscr{U} = \{U_i \mid i \in I\}$ is not empty, we denote by $\bigcap \mathscr{U}$ or $\bigcap_{i \in I} U_i$ the *intersection* of elements of $\mathscr{U}$. Denote by $A \setminus B$ the *set-theoretic difference* between two sets A and B, i.e., $A \setminus B = \{x \mid x \in A \text{ and } x \notin B\}$. Given set X and non-empty family $\{U_i \mid i \in I\}$, the following important relations hold among unions, intersections, and differences: $X \setminus \bigcup_{i \in I} U_i = \bigcap_{i \in I} (X \setminus U_i)$ and $X \setminus \bigcap_{i \in I} U_i = \bigcup_{i \in I} (X \setminus U_i)$ (De Morgan's laws).

Cartesian products and relations

Given two sets, X and Y, the *Cartesian product* (or *direct product*) $X \times Y$ is the set of all ordered pairs (x, y) such that $x \in X$ and $y \in Y$. A *relation* is a set of ordered pairs. A subset of Cartesian product $X \times Y$ of X and Y is called a *relation on X to Y*. For relation φ, the *domain* of φ is the set $\text{dom}(\varphi) = \{x \mid \exists y, (x, y) \in \varphi\}$, and the *range* of φ is the set $\text{ran}(\varphi) = \{y \mid \exists x, (x, y) \in \varphi\}$. If φ and ψ are relations, the *composition* of φ and ψ is the relation $\zeta = \{(x, z) \mid \exists y, (x, y) \in \varphi \text{ and } (y, z) \in \psi\}$, and ζ is denoted by $\psi \circ \varphi$. For relation φ on X to Y, the *upper section* of φ at $x \in X$ (*x-section of φ*) is the set $\{y \mid (x, y) \in \varphi\}$, which is denoted by $\varphi(x)$. Similarly, the *lower section* of φ at $y \in Y$ is the set $\{x \mid (x, y) \in \varphi\}$. We define φ^{-1} for relation φ as $\varphi^{-1} = \{(x, y) \mid (y, x) \in \varphi\}$. Then, the lower section of φ at $y \in \text{ran}(\varphi)$ is nothing but $\varphi^{-1}(y)$, which is the upper section of φ^{-1} at y.

---

[11] Let $T = \{x \mid x \notin x\}$. Consider whether T is an element of T. If $T \in T$, then by the definition of T, we have $T \notin T$, a contradiction. Hence, we have a proof for $T \notin T$. On the other hand, $T \notin T$ implies that T satisfies the sufficient condition for an element of T. Therefore, we have also a proof for $T \in T$. It follows that for the consistency of the theory, we cannot treat such T as a set (object) in the domain of discourse.

For two relations φ and ψ, φ is a *restriction* of ψ if $\mathrm{dom}\,(\varphi) \subset \mathrm{dom}\,(\psi)$ and $\varphi(x) = \psi(x)$ for all $x \in \mathrm{dom}\,(\varphi)$, and ψ is an *extension* of φ if $\varphi \subset \psi$.

Functions and correspondences

A *function* f on X to Y, denoted by $f : X \to Y$, is a relation on X to Y such that $\mathrm{dom}\,(f) = X$ and every upper section is a singleton. Function φ on X to 2^Y, where 2^Y denotes the family of all subsets of Y, is called a *correspondence* on X to Y and is also denoted by $\varphi : X \to Y$ or, more precisely, by $\varphi : X \ni x \mapsto \varphi(x) \subset Y$. For function f on X to Y, the unique element of the upper section (not the singleton itself) at x is traditionally denoted by $f(x)$, so we write $f : X \to Y$ and $f : X \ni x \mapsto f(x) \in Y$. Element $f(x)$ is also called the *image* of x under f. On the other hand, the lower section of f at $y \in Y$, $f^{-1}(y)$, itself is called the *inverse image* of y under f. Function $f : X \to Y$ is said to be *injective* (*one to one*) if for all x and x' in X, $x \neq x'$ means $f(x) \neq f(x')$ and is said to be *surjective* (*onto*) if for all $y \in Y$ there is element $x \in X$ such that $y = f(x)$. Two sets X and Y are said to have the same *cardinality* if there is a *bijective* (injective and surjective) function $f : X \to Y$. A set having the same cardinality with a subset of $N = \{0, 1, 2, \ldots\}$ is called a *countable set*.

Binary relations

A *binary relation* on X is a subset of $X \times X$. For binary relation $\mathscr{R} \subset X \times X$, we customarily write $x\mathscr{R}y$ instead of $(x, y) \in \mathscr{R}$. Binary relation $\mathscr{R}$ on X is said to be a *preordering* if it is *reflexive* ($\forall x \in X, x\mathscr{R}x$) and *transitive* ($\forall x, y, z \in X, (x\mathscr{R}y$ and $y\mathscr{R}z) \implies x\mathscr{R}z$). The pair of X and preordering $\mathscr{R}$ on X, $(X, \mathscr{R})$, is called a *preordered set*. A *directed* set is a non-empty preordered set such that for each of its elements i, j, element k satisfies $k\mathscr{R}i$ and $k\mathscr{R}j$. Preordering $\mathscr{R}$ on X is said to be an *ordering* if it is *antisymmetric* ($\forall x, y \in X, (x\mathscr{R}y$ and $y\mathscr{R}x) \implies x = y$). If preordering $\mathscr{R}$ on X is *symmetric* ($\forall x, y \in X, (x\mathscr{R}y) \implies y\mathscr{R}x$), it is called an *equivalence relation* on X. Given two preordered sets $(X, \mathscr{R})$ and $(Y, \mathscr{Q})$, mapping $f : X \to Y$ is said to be *monotone* (*isotone, order preserving*) if $x\mathscr{R}z$ implies $f(x)\mathscr{Q}f(z)$ for each $x, z \in X$.

Axiom of choice and products of a family of sets

Given family (set) of sets $\{X_i | i \in I\}$, the *Cartesian product* of the family of sets, $\prod_{i \in I} X_i$, is the set of functions on I to $\bigcup_{i \in I} X_i$ such that for each

$i \in I$, the image of i, x_i, belongs to X_i. Such a function, $f : I \to \bigcup_{i \in I} X_i$, is called a *choice function*. The existence of at least one choice function for each non-empty family of non-empty sets is assured in the theory of sets as an axiom called the *Axiom of Choice*. If there is binary relation $\mathcal{Q}_i$ on X_i for each $i \in I$, we may naturally define the *product relation* $\mathcal{Q}$ on $X = \prod_{i \in I} X_i$ as $f \mathcal{Q} g$ if and only if $f(i) \mathcal{Q}_i g(i)$ for all $i \in I$. Product relation $\mathcal{Q}$ is reflexive, transitive, and anti-symmetric if all $\mathcal{Q}_i$s are reflexive, transitive, and anti-symmetric respectively. Hence, $(X, \mathcal{Q})$ is a directed, preordered, and ordered set as long as all component spaces are directed, preordered, and ordered sets respectively, where for the non-emptiness of X and Q, the choice axiom is necessary.

Topology

A *topology* on space (set) X is a family of subsets of X, $\mathcal{T}$, satisfying the conditions that (1) $X \in \mathcal{T}$, (2) $\emptyset \in \mathcal{T}$, (3) for each non-empty finite subset $\mathcal{U} \subset \mathcal{T}$, the intersection $\bigcap \mathcal{U} = \bigcap_{U \in \mathcal{U}} U$ is an element of $\mathcal{T}$, and (4) for each subset $\mathcal{U} \subset \mathcal{T}$, the union $\bigcup \mathcal{U} = \bigcup_{U \in \mathcal{U}} U$ is an element of $\mathcal{T}$. Pair $(X, \mathcal{T})$ is called a *topological space*, and each element $U \in \mathcal{T}$ is said to be an *open* set in topological space $(X, \mathcal{T})$. The complement of an open set, $X \setminus U$, $U \in \mathcal{T}$, is called a *closed* set. For each point x in a topological space, set V including open set $U \ni x$ is called a *neighborhood* of x. For subset A of topological space $(X, \mathcal{T})$, we define the *relativization* $\mathcal{T}_A$ of $\mathcal{T}$ on A as $\mathcal{T}_A = \{U \cap A | U \in \mathcal{T}\}$. (Verify that $\mathcal{T}_A$ is a topology on A.)

Closure and interior

By the definition of topology, it is clear that (1) $\emptyset$ is closed, (2) total space X is closed, (3) the finite union of closed sets is closed, and (4) an arbitrary intersection of closed sets is closed. For subset A of topological space X, therefore, we may define the smallest closed set containing A, the *closure* of A, as $\text{cl}\, A = \bigcap \{B | A \subset B, B \text{ is closed in } X\}$. In the same way, we may define the largest open set contained in A, the *interior* of A, as $\text{int}\, A = \bigcup \{B | B \subset A, B \text{ is open in } X\}$.

Continuity

If $(X, \mathcal{T}_X)$ and $(Y, \mathcal{T}_Y)$ are topological spaces, function $f : X \to Y$ is *continuous* if, for each open set $U_Y \in \mathcal{T}_Y$, the *inverse image of set* U_Y, $f^{-1}(U_Y) = \{x \in X | f(x) \in U_Y\}$, is an element of $\mathcal{T}_X$. The condition is

1: Introduction

equivalent to saying that at each $x \in X$, for every open set $U \ni f(x)$, there is open set $V \ni x$ such that the *image of set V*, $f(V) = \{f(z)|z \in V\}$, is a subset of U. (Since a set is open iff for each of its elements there is an open neighborhood contained in the set, the latter condition is sufficient for the former. For the necessity, use the property that $f(f^{-1}(U)) \subset U$ for any $U \subset Y$.) It is easy to see that if $f : X \to Y$ and $g : Y \to Z$ are continuous, then their composition $g \circ f$ is also continuous.

Convergence

A *net* in topological space X is a function $S : D \to X$ whose domain $(D, \geq)$ is a directed set. If D is the set of all natural numbers with the ordinary $\geq$ relation, net is called a *sequence*. Net S in X *converges* to $x^* \in X$, if for each neighborhood U of x^* there exists $\bar{\nu} \in D$ such that $\forall \nu \geq \bar{\nu}, S(\nu) \in U$. (Net S is said to be *eventually in* U.) Net (also called a *generalized sequence*) is a useful concept to describe closedness, continuity of mappings, etc., for general topological spaces in exactly the same way as the notion of convergent sequence does in Euclidean spaces. One can verify that set $A \subset X$ is closed if and only if for every net in A converging to a certain point x in X, $x \in A$ necessarily follows. Furthermore, we may prove that function $f : X \to Y$ is continuous if and only if for every net $S : D \to X$ on X, net $f \circ S : D \to Y$ converges to $f(x^*) \in Y$ as long as S converges to $x^* \in X$. (Use the second condition, $\forall U \ni f(x), \exists V \ni x, f(V) \subset U$, for the continuity. The necessity of this third net-characterization condition is trivial. For the sufficiency, define net S on the directed set of neighborhoods of $x^* \in X$ at which the second condition for the continuity is not satisfied.)

Subnet and cluster point

A *subnet* of net $S : D \to X$ is net $T : E \to X$ such that mapping M exists on directed set E to D satisfying $T = S \circ M$ and the condition that for all $m \in D$ element $\bar{n} \in E$ exists such that $M(n) \geq m$ for all $n \geq \bar{n}$. The condition is typically satisfied when M is monotone and for all $m \in D$ element $n \in E$ exists such that $M(n) \geq m$. (More specifically, when E is a subset of D such that for all $m \in D$ there is an element $n \in E$, i.e., E is a *cofinal* subset of D. Although this may seem a standard way of constructing subnets, such a simple class of subnets is not sufficient for all purposes, unfortunately.) For net $S : D \to X$, point $x \in X$ is called a *cluster point* of S if for all neighborhoods U of x, for all $\bar{\nu}$ in D, there is $\nu \geq \bar{\nu}$ such that $S(\nu) \in U$. (Net S is said to be *frequently in U*.) One may prove that if x is a cluster point of net $S : D \to X$, then there is a subnet of S converging to x. (To see this, let $\mathcal{N}$ be the set of all open neighborhoods of x directed by the inclusion, and for each $N \in \mathcal{N}$ let D_N be the cofinal subset of D such that $S(\nu) \in N$ for all $\nu \in D_N$. Consider mapping $M : \mathcal{N} \times \prod_{N \in \mathcal{N}} D_N \ni (N, f) \mapsto f(N) \in D$ on the product directed set and subnet $T = S \circ M$. Or let $E \subset D \times \mathcal{N}$ be the set of all pairs (d, N) such that $S(d) \in N$ under product ordering, define M on E to D as $M(d, N) = d$, and consider subnet $T = S \circ M$.)

Base for a topology

Let $(X, \mathscr{T})$ be a topological space. A *base* for topology $\mathscr{T}$, $\mathscr{B}$, is a subset of $\mathscr{T}$ such that the set of arbitrary unions of elements of $\mathscr{B}$, $\{\bigcup \mathscr{C} | \mathscr{C} \subset \mathscr{B}\}$, equals $\mathscr{T}$. A *subbase* for topology $\mathscr{T}$, $\mathscr{S}$, is a subset of $\mathscr{T}$ such that the set of finite intersections of the members of $\mathscr{S}$, $\{\bigcap \mathscr{C} | \mathscr{C}$ is a finite subset of $\mathscr{S}\}$, is a base for topology $\mathscr{T}$. The concept of subbase (or base) for a topology is important because it characterizes such properties as minimal requirements in various topological arguments for a given topology. For example, we can see that net $S : D \to X$ in X converges to $x^* \in X$ if and only if for every neighborhood U of x^* belonging to a subbase for the topology, S is eventually in U.

Product topology

Consider family of sets $\{X_i | i \in I\}$. If each X_i is a topological space with topology $\mathscr{T}_i$, the *product topology* on $\prod_{i \in I} X_i$ is a topology whose subbase is the family that consists of set $\{f | f : I \to \bigcup_{i \in I} X_i, \forall i \in I \setminus \{j\}, f(i) \in X_i, f(j) \in U_j\}$ for some $j \in I$ and $U_j \in \mathscr{T}_j$. By considering the definition of subbase, product topology may be characterized as the weakest topology such that for every $j \in I$, the *projection* $\mathrm{pr}_j : \prod_{i \in I} X_i \ni (\cdots, x_j, \cdots) \mapsto x_j \in X_j$ is continuous. It can also be verified that net S in product space $\prod_{i \in I} X_i$ (the product set under the product topology) converges to x^* if and only if each net $\mathrm{pr}_j \circ S$ in j-th coordinate space, X_j, converges to the j-th coordinate $x_j^* = \mathrm{pr}_j(x^*)$ of x^*.

Quotient topology

Assume that $\mathscr{R}$ is an equivalence relation on topological space X. For each $x \in X$, denote by $[x]$ the equivalence class of x, i.e., $[x] = \{y \in X | y \mathscr{R} x\}$. The family of all such equivalence classes, $\{[x]|x \in X\}$, gives a *decomposition (partition)* of X, i.e., a disjoint family of subsets of X whose union is X. Decomposition $\{[x]|x \in X\}$, which is also denoted by $X/\mathscr{R}$, is called the *quotient set* of X with respect to $\mathscr{R}$. On X to $X/\mathscr{R}$, we may naturally define function $P : X \to X/\mathscr{R}$ to assign each $x \in X$ to its equivalence class $[x] \in X/\mathscr{R}$. P is called the *projection (quotient map)* of X onto quotient set $X/\mathscr{R}$. The *quotient topology* on quotient set $X/\mathscr{R}$ of topological space X is family $\{O|O \subset X/\mathscr{R}, P^{-1}(O) \text{ is open}\}$, which is the finest topology such that quotient map $P : X \to X/\mathscr{R}$ is continuous.

Other concepts

For finite set A, we denote by $\sharp A$ the number of elements of A. The set of real numbers is denoted by R. We assume that readers have basic knowledge of the topological and algebraic features of R as a *conditionally complete ordered field*.[12] Denote by R_+ (resp., by R_{++}) the set of all non-negative reals (resp., strictly positive reals) and by R^n the n-th product of the set of real numbers. If there are no additional explanations, R^n is supposed to have the product of the usual (order) topology of R with vector-space, inner-product, and Euclidean-metric structures (n-dimensional Euclidean space). For easily understanding this book, the reader needs the most basic knowledge of Euclidean spaces.

1.3.2 Compact sets, open coverings, and partition of unity

Since the open *covering* of a space is an extremely important concept throughout this book, it is appropriate to use one subsection here to state several inherent concepts and properties that are repeatedly used in later chapters.

Let X be a topological space. A family of open subsets of X, $\{M_i | i \in I\}$, is said to be a *covering* of X if $\bigcup_{i \in I} M_i = X$. For two coverings, $\mathscr{M} = \{M_i | i \in I\}$ and $\mathscr{N} = \{N_j | j \in J\}$ of X, $\mathscr{N}$ is a *subcovering* (resp., *refinement*) of $\mathscr{M}$, if and only if for all $N_j \in \mathscr{N}$, there exists $M_i \in \mathscr{M}$ such that $N_j = M_i$ (resp., $N_j \subset M_i$). Covering $\mathscr{M}$ is said to be *finite* if $\mathscr{M}$ is a finite set.

---

[12] If unsure, see Debreu (1959).

Topological space X is *compact* if each covering of X has a finite subcovering. Equivalently, it can be said that space X is compact if and only if arbitrary non-empty family $\{F_i | i \in I\}$ of closed sets in X having the *finite intersection property* (every finite intersection among sets in $\{F_i | i \in I\}$ has a non-empty intersection) has a non-empty intersection. The compactness can also be characterized through the convergence of nets in the space.

THEOREM 1.3.1: (Net Characterization of Compactness) *Topological space X is compact iff every net in X has a converging subnet.*

PROOF: To see that every net $S : (D, \geq) \to X$ in compact set X has a converging subnet, use the finite intersection property of the family of closures of sets $A_m = \{S(n) | n \geq m\}$, $m \in D$. To see the sufficiency, suppose that every net in X has a converging subnet. Then for arbitrary family $\{F_i | i \in I\}$ of closed sets in X having the *finite intersection property*, if we consider a net on the set of finite subsets of I directed by inclusion as $S : \mathscr{F}(I) \ni A \mapsto S(A) \in \bigcap_{i \in A} F_i$, the limit point of a converging subnet of S is easily seen to belong to all F_i, $i \in I$. ∎

In Euclidean n-space, a closed bounded set is compact. (The fact is known as the Heine–Borel covering theorem.)

In this book, we base many theorems on Brouwer's classical fixed-point theorem (Theorem 2.1.1) that may be applicable to all sets *homeomorphic* to a non-empty compact convex subset of Euclidean space R^n.[13] So it is useful to remember the next property on the homeomorphism between compact spaces. (Topological space X is said to be *Hausdorff* if for all $x, y \in X$, $x \neq y$, two open sets U_x and U_y exist such that $x \in U_x$, $y \in U_y$ and $U_x \cap U_y = \emptyset$.)

THEOREM 1.3.2: (Isomorphism Between Compact Sets) *A continuous bijection on compact space X to Hausdorff space Y is a homeomorphism.*

PROOF: Let $f : X \to Y$ be a continuous bijection. (Note that by the continuity of bijection f, Y is also compact and X is also a Hausdorff

---

[13] Topological spaces X and Y are said to be *homeomorphic* if continuous bijection $f : X \to Y$ exists such that f^{-1} is also continuous. (Function f is called a homeomorphism between X and Y.) One can prove that if X has the fixed-point property (i.e., every continuous mapping on the space to itself has a fixed point), space Y homeomorphic to X also has the fixed-point property.

1: Introduction

space.) We have to show that f^{-1} is continuous. Consider net $\{y^\nu\}$ in Y that converges to $y^* \in Y$. Since X is compact, net $x^\nu = \{f^{-1}(y^\nu)\}$ in X has a subnet $\{f^{-1}(y^{\nu(\mu)})\}$ in X converging to point $x^* \in X$. Since f is continuous, $f(x^{\nu(\mu)})$ must converge to $f(x^*)$, so $f(x^*)$ is a cluster point of converging net $\{y^\nu\}$; i.e., $f(x^*)$ must equal y^* since Y is a Hausdorff space. It remains to be shown that net $\{f^{-1}(y^\nu)\}$ converges to x^*. The above argument ensures that every converging subnet of $\{f^{-1}(y^\nu)\}$ must converge to the same point, $x^* = f^{-1}(y^*)$. If $\{f^{-1}(x^\nu)\}$ does not converge to x^*, again by the compactness of X, net $\{f^{-1}(y^\nu)\}$ has a subnet that converges to a point different from x^*: a contradiction. ∎

By definition, every closed subset of a compact space is obviously also compact under the relativized topology. One can also prove that compact subset X of topological space Y is closed if the topology of Y is Hausdorff.

Hausdorff space X is said to be *normal* if for any two closed subsets, A and B, such that $A \cap B = \emptyset$, there are two open sets, U_A and U_B, such that $U_A \supset A$, $U_B \supset B$ and $U_A \cap U_B = \emptyset$. From the definition, in normal space X, every open neighborhood U of $x \in X$ clearly includes closed neighborhood C of x. (Consider two closed sets, $X \setminus U$ and $\{x\}$.) It is also easy to prove that every compact Hausdorff space is normal.

THEOREM 1.3.3: (Partition of Unity) *Let X be a normal space, and let $\mathscr{U} = \{U_1, \ldots, U_n\}$ be a finite covering of X. It is known that a family of non-negative real valued continuous functions exists, $f_1 : X \to R_+, \ldots, f_n : X \to R_+$, such that $f_i(x) = 0$ for all $x \in X \setminus U_i$ for each i, and $\sum_{i=1}^n f_i(x) = 1$ for all $x \in X$.*

The family of functions stated in the above theorem is called *a partition of unity* on space X subordinate to covering $\mathscr{U}$. The theorem is an immediate consequence of the so-called Urysohn's Lemma on two closed subsets of a normal space.[14] A complete proof is given in Mathematical Appendix I.

1.3.3 Vector space duality and hyperplane

We denote by R^n_+ (resp., by R^n_{++}) the set $\{(x_1, \ldots, x_n) | x_1 \in R_+, \ldots, x_n \in R_+\}$ (resp., $\{(x_1, \ldots, x_n) | x_1 \in R_{++}, \ldots, x_n \in R_{++}\}$) in n-dimensional

---

[14] The proof of this theorem is easy when the topology of X is given through a metric as in the Euclidean spaces. Let $F_i(x)$ be the distance from x to $X \setminus U_i$ and define $f_i(x)$ as normalization $F_i(x)/\sum_{j=1}^n F_j(x)$ for each i and $x \in X$.

Euclidean space R^n. Readers are expected to have the most basic knowledge of vector-space structure in Euclidean spaces.

A *vector space* over real field R is a set L on which mapping $(x,y) \mapsto x+y$ on $L \times L$ to L, called *addition*, and mapping $(a,x) \mapsto ax$ on $R \times L$ to L, called *scalar multiplication*, are defined to satisfy the following axioms: (In the following, x, y, z and a, b are arbitrary elements of L and R, respectively.)

(1) $(x+y)+z = x+(y+z)$
(2) $x+y = y+x$
(3) $a(x+y) = ax + ay$
(4) $(a+b)x = ax + bx$
(5) $a(bx) = (ab)x$
(6) $\exists 0, 0 + x = x + 0 = x$
(7) $\forall x, \exists -x, x + (-x) = (-x) + x = 0$
(8) $1x = x$

Mapping f on vector space L over R to vector space M over R is *linear* if $f(ax + by) = af(x) + bf(y)$ for all $x \in L$, $y \in L$, and $a, b \in R$.

For m points $x^1, \ldots, x^m$ of vector space L over R and m scalars $a_1, \ldots, a_m$ in R, point $x = \sum_{i=1}^m a_i x^i$ in L is a *linear combination* (under coefficients $a_1, \ldots, a_m$) of points $x^1, \ldots, x^m$. Points $x^1, \ldots, x^m$ are *linearly independent* if $\sum_{i=1}^m a_i x^i = 0 \implies a_1 = 0, a_2 = 0, \ldots, a_m = 0$. In other words, points $x^1, \ldots, x^m$ are linearly independent if no x^i can be represented as a linear combination of other points. More generally, if subset A of L is such that no element x of A can be represented as a linear combination of other (finite) points in A, then set A of the points is *linearly independent*. If A is not linearly independent, it is *linearly dependent*.

Subset M of vector space L is a *linear subspace* of L if all additions between points in M and all scalar multiplications of points in M are also points in M. For subspace M of L, the subset of form $x + M = \{x + z \mid z \in M\}$ for some $x \in L$ is called an *affine subspace* of L. If A is a linearly independent subset of vector space L over R, the set of all linear combinations of points in A, $L(A)$, forms a subspace of L. Linearly independent subset A is called a *basis* (*Hamel basis*) of $L(A)$. Linear mapping on $L(A)$ is uniquely determined by the images of elements of the basis.

In vector space L over R, if m coefficients $a_1, \ldots, a_m$ for m points $x^1, \ldots, x^m$ belong to R_+ and satisfy $\sum_{i=1}^m a_i = 1$, linear combination

1: Introduction

$\sum_{i=1}^{m} a_i x^i$ is called a *convex combination* (under coefficients $a_1, \ldots, a_m$) of points $x^1, \ldots, x^m$. Subset X of vector space L over R is *convex* if all convex combinations of two points in X are also elements of X. Given set A of vector space L, co A denotes the set of all convex combinations among points in A. One may prove that co A is the smallest convex set that includes A, which is also equal to the intersection of all convex sets that include A. (Use the fact that an arbitrary intersection of convex sets is also convex.[15])

A *topological vector space* over R is a vector space having a topology on which the addition and scalar multiplication are continuous. (Since $(a^\nu x^\mu + b^\eta y^\varsigma) - (a^* x^* + b^* y^*) = (a^\nu x^\mu - a^* x^*) + (b^\eta y^\varsigma - b^* y^*)$, one can verify that they are indeed jointly continuous.) Therefore, if $A = \{x_1, \ldots, x_\ell\}$ is a linearly independent subset of topological vector space over R, bijective linear mapping $f : R^\ell \ni (a_1, \ldots, a_\ell) \mapsto a_1 x_1 + \cdots + a_\ell x_\ell \in L(A)$ is continuous. A family of neighborhoods of $x \in X$, $\mathscr{U}$, such that for each neighborhood U_x of x, member U of family $\mathscr{U}$ included in U_x exists, is called a *neighborhood base* at x. Neighborhood base at $0 \in X$ is called a 0-neighborhood base. This concept is important since the topological features of a topological vector space are completely determined by a 0-neighborhood base. A *locally convex space* is a Hausdorff topological vector space with a 0-neighborhood base consisting of convex sets.

For vector space E, real valued linear function f is called a *(real) linear form* (or a linear functional) on E. The set of all real linear forms, E^*, may also be considered a vector space by defining $(f+g)(x)$ as $f(x) + g(x)$ and $(\alpha f)(x)$ as $\alpha f(x)$. E^* is called the *algebraic dual space* (or *algebraic dual*) of E. On topological vector space E, the set of all continuous real linear forms, E', is also recognized as a vector space and is called the *topological dual space* (or *topological dual*) of E.[16] The *weak topology* on E, $\sigma(E, E')$, is a topology whose subbase is constructed by sets of form $\{y \in E | f(y) < \alpha\}$ for some $f \in E'$ and $\alpha \in R$. It is the weakest locally convex topology under

---

[15] As stated in Section 1.2, although the "convexity" concept in this book is often used in the generalized sense, it may not be so harmful to give priority to the vector-space interpretation over the generalized one when a vector space structure is explicitly given.
[16] For example, let R^∞ be the set of the countably infinite product of R and let R_∞ be the subspace of R^∞ that consists of points whose coordinates are all 0 except for finite components. By considering the duality operation, $\langle (1, 1, \cdots), (x_1, \cdots, x_n, 0, \cdots, 0) \rangle = 1 x_1 + \cdots + 1 x_n$ for $(x_1, \cdots, x_n, 0, \cdots) \in R_\infty$, we can recognize $(1, 1, \cdots) \in R^\infty$ as an algebraic linear form on R_∞. The element $(1, 1, \cdots) \in R^\infty$ is not continuous, however, if we relativize the product topology on R^∞ to R_∞.

which every $f \in E'$ is continuous. On the other hand, the topology on E', whose subbase is constructed by sets of form $\{f \in E'|f(y) < \alpha\}$ for some $y \in E$ and $\alpha \in R$, is called the *weak star topology* on E', $\sigma(E', E)$.

If f is a real linear form on vector space E, set H of form $\{y \in E|f(y) = \alpha\}$ for some $\alpha \in R$ is called a *hyperplane* in E. In topological vector space E, hyperplane H is closed if and only if it is associated with continuous linear form f. We say that two sets, A and B, in vector space E are *separated* (resp., *strictly separated*) by a hyperplane if hyperplane $H = \{y|f(y) = \alpha\}$ exists such that $\forall a \in A, \forall b \in B, f(a) \leq \alpha \leq f(b)$, (resp., $f(a) < \alpha < f(b)$). The next theorem is especially critical for economic arguments. (For the proof, see Mathematical Appendix II. See also Schaefer (1971, p. 64, 9.1).)

THEOREM 1.3.4: (First Separation Theorem) *In topological vector space E, if A is a convex set whose interior $\operatorname{int} A$ is non-empty and B is a non-empty convex set such that $\operatorname{int} A \cap B = \emptyset$, then closed hyperplane H exists that separates A and B. If both A and B are open, we may choose H so that A and B are strictly separated.*

THEOREM 1.3.5: (Second Separation Theorem) *In locally convex space E, if A is a non-empty closed convex set and B is a non-empty compact convex set such that $A \cap B = \emptyset$, then a closed hyperplane exists that strictly separates A and B.*

PROOF: Under the basic property of vector space topology, set $-A = \{-a | a \in A\}$ is closed. Since B is compact, we can also verify that $B + (-A) = \{b + (-a) | b \in B, a \in A\}$ is closed. (Use a net and a converging subnet in compact set B.) Then, there is convex 0-neighborhood U that does not intersect with $B+(-A)$. (In the following, for subsets in a vector space, such notations as $A + B$, $-A$ and $B + (-A) = B - A$ will be used without any explanations. If one such set is a singleton, we often write $x + A$ instead of $\{x\}+A$.) Without loss of generality, we may assume U to be open. (Note that the interior of a convex set is always open under vector space topology.) Let $W = U \cap -U$ and define V as $V = (1/3)W = \{(1/3)w| w \in W\}$. Then, $A+V$ and $B+V$ are two disjoint convex open sets satisfying all conditions in Theorem 1.3.4, and thus the result is an immediate consequence of the First Separation Theorem. ∎

Notes on References

Since this is a research monograph, many theorems and arguments must be supplemented with sources to establish priority or confidence. At the same time, I want this book to be readable as a text for graduate students in economics who are concerned with rigorous mathematical arguments. Therefore, in the main sections of this book, references to the literature for every important (especially mathematical) theorem and concept have been minimized as suggestions for further reading from an educational viewpoint. References necessary for research-level arguments are given in the last section of each chapter as Bibliographic Notes.

Chapter 2
Fixed-Point Theorems

2.1 Classical Results and Basic Extensions

2.1.1 Classical fixed-point arguments

Let X be a topological space. Function $f : X \to X$ is said to have a *fixed point* if the element x^* of X exists such that $f(x^*) = x^*$. For set valued mapping (or a correspondence) $\varphi : X \to 2^X$, x^* is a *fixed point* of φ if $x^* \in \varphi(x^*)$. The set of all fixed points (*fixed-point set*) of f (resp., of φ) is denoted by $\boldsymbol{Fix}(f)$ (resp., by $\boldsymbol{Fix}(\varphi)$). The next theorem from L. Brouwer is one of the most fundamental results in the theory of fixed points.

THEOREM 2.1.1: (Brouwer's Fixed-Point Theorem) *If X is a non-empty compact convex subset of an n-dimensional Euclidean space R^n, then for every continuous function $f : X \to X$, $\boldsymbol{Fix}(f) \neq \emptyset$.*

A proof of Brouwer's fixed-point theorem is given in Chapter 6. Until then, we use it as a given result to show other important theorems on multivalued mappings and their extensions.

Most of the results in this book are related to fixed-point arguments on correspondences. Therefore, it is appropriate here to start by defining several basic notions and properties that characterize the topological property (the continuity) of correspondences.

Let X and Y be topological spaces. Correspondence $\varphi : X \to Y$ has the *open-lowersection property* if for each $y \in Y$, lower section $\varphi^{-1}(y)$ is open. We also say that correspondence $\varphi : X \to Y$ has the *local-intersection property* at $x \in X$ on $K \subset X$ if there is at least one $y^x \in Y$ and an open neighborhood $U(x)$ of x such that for all $z \in U(x) \cap K$, $y^x \in \varphi(z)$. Correspondence φ is said to have the *local-intersection property* on $K \subset X$ if it simply has the local-intersection property at each $x \in K$ on K. It is clear that if correspondence $\varphi : X \to X$ has the open-lowersection property, then it has the local-intersection property on $\{x \in X | \varphi(x) \neq \emptyset\}$. If non-empty valued $\varphi : X \to X$ has an open graph in $X \times X$, i.e., the

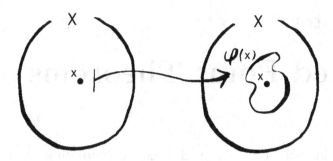

Figure 7: Fixed points

graph, $G_\varphi = \{(x,y) \in X \times X | y \in \varphi(x)\}$, is open under the product topology on $X \times X$, then φ has the open-lowersection property on X.[1] Fixed-point theorems on such mappings (open graph) constitute one of the most important arguments in the fixed-point theory of correspondences. The basic theorem given by Browder (1968) is that on a non-empty compact convex subset of Hausdorff topological vector space, every non-empty convex valued correspondence having an open graph (in $X \times X$ under the product topology) has a fixed point, or more generally:

THEOREM 2.1.2: (Browder) *Let X be a non-empty compact convex subset of Hausdorff topological vector space E. If $\varphi : X \to X$ is a non-empty convex valued correspondence having the open-lowersection property, then φ has a fixed point.*

The proof of this theorem will be given in the next subsection. (Local continuous selections and the partition of unity theorem play essential roles.) Such mappings (with more generality including mappings with local-intersection properties) are later called mappings of the Browder type.

Another important kind of correspondence in fixed-point theory is the class of closed correspondences. Correspondence $\varphi : X \to Y$, where X and Y are topological spaces, is said to be *closed* if the graph of φ, $G_\varphi = \{(x,y) | y \in \varphi(x)\}$, is closed in $X \times Y$ under the product topology. The class of closed correspondences has great significance since it includes all continuous functions (as single valued correspondences), so a fixed-point argument may be regarded as an extention of Brouwer's theorem. A representative theorem

---

[1] Rigorously speaking, since a correspondence is defined in this book as a relation defined as a set of ordered pairs (Subsection 1.3.1), no difference exists between a correspondence φ and its graph G_φ (or function f and its graph) as a mathematical entity.

of this type was developed by Kakutani (1941), which was then generalized by Fan (1952) and Glicksberg (1952) in locally convex topological vector spaces.

THEOREM 2.1.3: **(Kakutani–Fan–Glicksberg)** *Let X be a non-empty compact convex subset of Hausdorff locally convex topological vector space E and φ be a non-empty convex valued correspondence on X to X having a closed graph. Then φ has a fixed point.*

The proof of Theorem 2.1.3 is also given in the later part of this section with more general versions of it. As with the previous Browder's case, the generalized class of correspondences, including the above mappings with closed graphs, is called the Kakutani type. The generalization and proofs in this book are given through the following concept of *upper semicontinuity* that abstracts a certain property from mappings with closed graph and relates it to the same arguments in cases with mappings having an open graph (on local continuous selections and partition of unity method). Let X and Y be topological spaces. Correspondence $\varphi : X \to Y$ is *upper semicontinuous* if for each $x \in X$ and for each open set $U \supset \varphi(x)$, open set $V \ni x$ exists such that $\forall z \in V, U \supset \varphi(z)$. It is said to be *lower semicontinuous* if for each $x \in X$ and for each open set U such that $U \cap \varphi(x) \neq \emptyset$, open set $V \ni x$ exists such that $\forall z \in V, U \cap \varphi(z) \neq \emptyset$, and it is said to be *continuous* if it is both upper and lower semicontinuous. If range of correspondence $\varphi : X \to Y$ is contained in a compact set (especially if Y is compact), we may restate the property having a closed graph through the upper semicontinuity as follows.

THEOREM 2.1.4: (Closed Graph and Upper Semicontinuity) *Let X and Y be Hausdorff spaces. If Y is compact, then $\varphi : X \to Y$ is closed if and only if it is closed valued and upper semicontinuous.*

PROOF: Suppose that φ is closed valued and upper semicontinuous and that S is a converging net in graph G_φ of φ to point $(x,y) \in X \times Y$ under the product topology. Assume that $y \notin \varphi(x)$. Since Y is normal, two disjointed open sets, $A \ni y$ and $B \supset \varphi(x)$, exist in Y. By the upper semicontinuity of φ, however, net S is eventually in $X \times B$, so projection $\text{pr}_Y \circ S$ is eventually in B, a contradiction. Consequently, we have $y \in \varphi(x)$, and G_φ is closed. If graph G_φ of $\varphi : X \to Y$ is closed in $X \times Y$, every $x \in X$ section of G_φ is obviously closed under the product topology, and thus φ is closed valued. Arbitrarily take $x \in X$ and open set U including $\varphi(x)$. Let y be a point in $Y \setminus U$. Since $(x,y) \in X \times Y \setminus G_\varphi$, neighborhoods $V_{x,y}$ of x and U_y of y exist such that $V_{x,y} \times U_y \cap G_\varphi = \emptyset$. Since $\{U\} \cup \{U_y | y \in Y \setminus U\}$ is a covering of Y, there is a finite subcovering, $\{U\} \cup \{U_{y_1}, \ldots, U_{y_n}\}$. Let $V = \bigcap_{i=1}^n V_{x,y_i}$. Then, from $\varphi(z) \subset U$ for all $z \in V$, the upper semicontinuity of φ follows. ∎

The notion of closed correspondence, therefore, may be generalized through an extension of the closed valued upper semicontinuity. The details are given in the next subsection.

Brouwer's Theorem 2.1.1 is so fundamental that all the fixed-point results in this book, including both Browder and Kakutani types, may essentially be considered merely corollaries to it. Under the simplest case of the argument, let us see this point as a summary to this subsection. If X is the unit interval, $I = \{x \in R \mid 0 \leq x \leq 1\}$, and f is a continuous function on I into I. On product space $I \times I$, the graph of f is a continuous curve that begins at a point in $\{0\} \times I$ and ends at a point in $\{1\} \times I$ (Figure 8a). Brouwer's fixed-point theorem, in this case, maintains the obvious fact that every graph of continuous function necessarily intersects with diagonal $\Delta = \{(x,x) | x \in I\}$. Of course if f is not continuous, the intersection may not necessarily exist (Figure 8b). It is easy to check that for a function, the condition of continuity is equivalent to the closedness of its graph. In general, if we consider a correspondence, however, the closedness of its graph is not sufficient to assure the existence of fixed points (Figure 9a). Even in this case, if the correspondence is convex valued, we expect the existence of fixed points as before (Figure 9b). This is the case with Kakutani's fixed-point theorem.

Rigorously speaking, to obtain Kakutani-type results, we must use some limit arguments that relate the conditions of convexity and closedness to known fixed-point results like Brouwer's theorem. The difference between

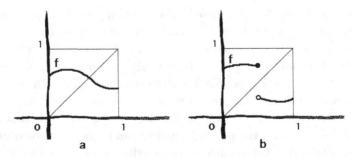

Figure 8: Graph of continuous function in $I \times I$

2: Fixed-Point Theorems

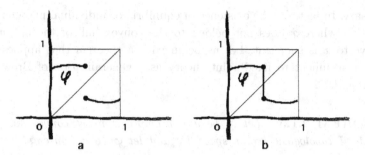

Figure 9: Mapping with closed graph in $I \times I$

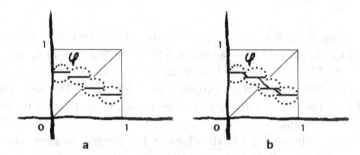

Figure 10: Mapping with closed graph in $I \times I$

Figures 8a and 9b is not as simple as we can intuitively see in the figures. In this book, we base our approach on an argument of continuous selections that are more directly related to the following Browder-type theorems. The case of correspondence φ on I to I having an open graph is illustrated in Figure 10a, where the open graph property (or the local intersection property) ensures the existence of local constant mapping selections of φ. Local selections may be amalgamated (through the "partition of unity") into a continuous selection of φ as long as φ is convex valued. Hence, by Brouwer's fixed-point theorem, we have a fixed-point of $\varphi : I \to I$.

2.1.2 *Extensions of classic results*

In economics, one of the most general settings assumes the open-lowersection property for mappings like preference correspondences or better set correspondences. (Precise definitions for such mappings are given in later chapters.) It is also known that in such equilibrium arguments, the full convexity for the value of such mappings, that is, better sets, is not necessary. In fact, for the existence of equilibria or individual maxima, the condition where x does not belong to the convex hull of the better set relative to x is sufficient. Let us begin with a theorem that applies this type of argument to fixed-point theory as a generalization of Browder's Theorem 2.1.2.

THEOREM 2.1.5: *Let X be a non-empty compact convex subset of Hausdorff topological vector space E, and let φ be a non-empty valued mapping having the open-lowersection property. Assume that for each $x \notin \mathcal{F}ix(\varphi)$, convex set $\Psi(x)$ exists such that $\varphi(x) \subset \Psi(x)$ and $x \notin \Psi(x)$. Then φ has a fixed point.*

In Theorem 2.1.5, convexity for the value of φ in Browder's theorem is replaced by the existence of convex set $\Psi(x)$ including $\varphi(x)$ at each x such that $x \in X \setminus \mathcal{F}ix(\varphi)$. Note also that Ψ may be considered a correspondence satisfying a local intersection property on $X \setminus \mathcal{F}ix(\varphi)$. The key concept behind the theorem is a characterization of mappings relative to the set of their fixed points.

To prove Theorem 2.1.5, therefore, it is sufficient to show the next theorem, which is written through further general concepts and properties for correspondence φ relative to the set of its fixed points. As we see in later chapters (e.g., discussions for Lefschetz's number in Chapter 6 and fixed-point degree in Chapter 7), this type of argument is not only quite useful but also important for relating our fixed-point results in earlier chapters to algebraic ones. We say that correspondence $\Psi : X \to X$ is a *fixed-point-free extension* of $\varphi : X \to X$ on $A \subset X$ if Ψ is an extension of φ having no fixed point on A (i.e., if $\varphi(x) \subset \Psi(x)$ and $x \notin \Psi(x)$ for all $x \in A$).

Figure 11: Convex set $\Psi(x)$ such that $\varphi(x) \subset \Psi(x)$ and $x \notin \Psi(x)$

THEOREM 2.1.6: *Let X be a non-empty compact convex subset of Hausdorff topological vector space E, and let φ be a non-empty valued correspondence on X to X satisfying the following condition:*

> (K*) φ has a convex valued fixed-point-free extension having a local intersection property on $X\backslash\boldsymbol{Fix}(\varphi)$.

Then φ has a fixed point.

PROOF: Suppose that φ does not have a fixed point. Let Ψ be a fixed-point-free extension of φ satisfying condition (K^*) on $X = X\backslash\boldsymbol{Fix}(\varphi)$. For each $x \in X$, let $U(x)$ be an open neighborhood of x on which the local intersection property of Ψ holds. Since $X\backslash\boldsymbol{Fix}(\varphi) = X$ is compact, by taking a finite subcovering of covering $\{U(x)|\, x \in X\}$ of X, we have finite points $x^1, \ldots, x^n \in X$ and their open neighborhoods $U(x^1), \ldots, U(x^n)$ that cover X, and for each $t = 1, \ldots, n$, point y^{x^t} satisfies $\forall z \in U(x^t)$, $z \notin \Psi(z)$ and $y^{x^t} \in \Psi(z)$ by the local intersection property of correspondence Ψ. Let $\beta_t : X \to [0,1]$, $t = 1, \ldots, n$ be a partition of unity subordinate to $U(x^1), \ldots, U(x^n)$ (see Theorem 1.3.3). Denote by A set $\{y^{x^1}, \ldots, y^{x^n}\}$ and by D set $\Delta^A = \{e|e : A \to R, \forall a \in A, e(a) \geq 0, \sum_{a \in A} e(a) = 1\}$ identified with the $(n-1)$-dimensional unit simplex in Euclidean space R^n under isomorphism that maps e^i, the mapping satisfying $e^i(y^{x^i}) = 1$, to the i-th unit vector in R^n for each $i = 1, \ldots, n$. Let $g : D \to X$ be the continuous mapping $e \mapsto \sum_{t=1}^n e(y^{x^t})y^{x^t}$ and let us consider function f on D to itself such that $f(e) = \sum_{i=1}^n \beta_i(g(e))e^i$. Then, since f is a continuous function on finite dimensional compact set D to itself, f has fixed-point e^* by Brouwer's fixed-point theorem. On the other hand, for all i such that $g(e^*) \in U(x^i)$, $y^{x^i} \in \Psi(g(e^*))$. Moreover, since Ψ is convex valued, we have $g(e^*) = g(f(e^*)) = g(\sum_{i=1}^n \beta_i(g(e^*))e^i) = \sum_{t=1}^n[\sum_{i=1}^n \beta_i(g(e^*))e^i](y^{x^t})y^{x^t} = \sum_{t=1}^n[\beta_t(g(e^*))e^t](y^{x^t})y^{x^t} = \sum_{t=1}^n \beta_t(g(e^*))y^{x^t} \in \Psi(g(e^*))$, which contradicts condition $g(e^*) \notin \Psi(g(e^*))$. ∎

The theorem immediately yields the following special cases, Theorems 2.1.7 and 2.1.8, which may also be considered direct extensions of Theorems 2.1.5 and 2.1.2, respectively.

THEOREM 2.1.7: *Let X be a non-empty compact convex subset of Hausdorff topological vector space E, and let φ be a mapping having a local intersection property on $X\backslash\boldsymbol{Fix}(\varphi)$. Assume that for each $x \in X\backslash\boldsymbol{Fix}(\varphi)$, x does not belong to $C(\varphi(x))$, the set of all convex combinations among points in $\varphi(x)$. Then φ has a fixed point.*

PROOF: For each $x \notin \mathcal{F}ix(\varphi)$, define $\Psi(x)$ as $\Psi(x) = C(\varphi(x))$ and for each $x \in \mathcal{F}ix(\varphi)$, define $\Psi(x)$ as $\Psi(x) = \varphi(x)$. Then, correspondence $\Psi : X \to X$ is a fixed-point-free extension of φ satisfying condition (K*) in Theorem 2.1.6. ∎

THEOREM 2.1.8: *Let X be a non-empty compact convex subset of Hausdorff topological vector space E, and let φ be a convex valued correspondence on X to X. If φ has a local intersection property on $X \backslash \mathcal{F}ix(\varphi)$, then φ has a fixed point.*

PROOF: In Theorem 2.1.6, define Ψ as $\Psi = \varphi$. ∎

In the proof of Theorem 2.1.6, it is an essential feature that the compactness of X enables us to choose finite covering $\{U(x^1), \ldots, U(x^n)\}$ such that on each member, the key property of local intersection of the mapping is ensured. Accordingly, let us define *mappings of the Browder type* for correspondences in Hausdorff topological vector spaces as follows. Let X be a normal convex subset of Hausdorff topological vector space. Correspondence $\varphi : X \to X$ is said to be of the Browder type if the next condition is satisfied:

> (Browder-Type Mapping: Hausdorff t.v.s.) For each open set $U \supset \mathcal{F}ix(\varphi)$ of X, finite open sets $U^1, \ldots, U^n$ of X exist such that (1) $\{U, U^1, \ldots, U^n\}$ covers X, and (2) convex valued fixed-point-free extension Ψ of φ on $\bigcup_{t=1}^n U^t$ to X satisfies that $\bigcap_{x \in U^t} \Psi(x) \neq \emptyset$ for each $t = 1, \ldots, n$.

Note that condition (2) ensures the existence of a fixed-point-free convex valued extension Ψ of φ having the local-intersection property on $\bigcup_{t=1}^n U^t$. One can easily verify that the proof of Theorem 2.1.6 actually shows the

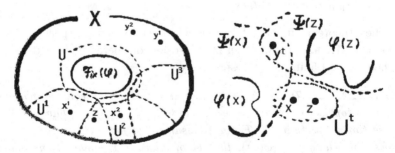

Figure 12:   Mappings of Browder type — $y^t \in \bigcap_{x \in U^t} \Psi(x)$

existence of fixed points for all non-empty valued Browder-type mappings on convex subset X of the Hausdorff topological vector space to itself.[2]

THEOREM 2.1.9: *Every non-empty valued Browder type mapping on a normal convex subset of Hausdorff topological vector space to itself has a fixed point.*

We next consider the case with mappings with closed graphs and an extension of Kakutani–Fan–Glicksberg's theorem. As shown in the previous subsection (Theorem 2.1.4), the correspondence on the Hausdorff compact set to itself having a closed graph is closed valued and upper semicontinuous. Moreover, if φ is convex valued and if the topology is locally convex, we can strictly separate x and $\varphi(x)$ by closed hyperplane H_x (continuous linear functional) as long as $x \in X \backslash \boldsymbol{\mathcal{F}ix}(\varphi)$ (Second Separation Theorem 1.3.5). The upper semicontinuity of φ means that such a hyperplane may be chosen in common at least for all z near x (left illustration in Figure 13). Hence, in the same way as with Browder's case, we may define a generalized class of mappings including those treated in the Kakutani–Fan–Glicksberg Theorem as follows. Let X be a normal subset of Hausdorff topological vector space. Correspondence $\varphi : X \twoheadrightarrow X$ is of the Kakutani type if the next condition is satisfied:

(Kakutani-Type Mapping: Hausdorff t.v.s.) For each open set $U \supset \boldsymbol{\mathcal{F}ix}(\varphi)$ of X, finite open sets $U^1, \ldots, U^n$ of X with continuous

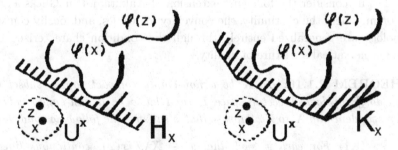

Figure 13: Closed hyperplane H_x strictly separates z and $\varphi(z)$ for all z near x

---

[2] In this theorem, even closedness for domain X is not necessary. Characterization based on coverings (instead of continuity arguments like (K*)) has advantages in its generality for the domain of mappings and its development of discussion for Čech and Vietoris homological treatment of this problem in later chapters. The condition is not necessarily weaker than (K*) since each U^t must be open in X (not in $X \backslash \boldsymbol{\mathcal{F}ix}(\varphi)$).

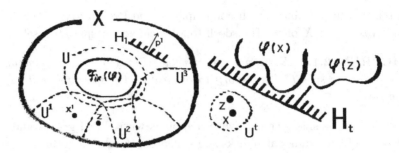

Figure 14:   Mappings of Kakutani type

linear functionals $p^1, \ldots, p^n$ exist such that (1) $\{U, U^1, \ldots, U^n\}$ covers X, and (2) for each t, a hyperplane defined by p^t strictly separates U^t and $\bigcup_{z \in U^t} \varphi(z)$.

Figure 14 shows mapping φ of the Kakutani type, where H_t denotes the hyperplane defined by p^t. The concept of Kakutani-type mapping is extended in the last section of this chapter based on the condition shown by the illustration on the right in Figure 13, where K_t denotes a general closed convex set. This concept will be further generalized through the notion of acyclicity in Chapter 6 to spaces without convex structures. In this section, based on the vector space structure and the system of duality, we show a fixed-point theorem for a wide class of mappings, including the Kakutani type.

Let us consider the following extension of Kakutani–Fan–Glicksberg's theorem, where the continuity, the convexity for values, and locally convex topology are generalized entirely through the condition characterized by the mathematical structure of duality.

THEOREM 2.1.10: *Let X be a non-empty compact convex subset of Hausdorff topological vector space E, and let φ be a non-empty valued correspondence on X to X. Suppose that φ satisfies the following condition:*

> (K1) *For each x such that $x \in X \backslash \mathcal{F}ix(\varphi)$, continuous linear functional $p \in E'$ and open neighborhood U of x exist such that for all $z \in U \cap (X \backslash \mathcal{F}ix(\varphi))$, we have $p(v) > 0$ for all $v \in \varphi(z) - z$.*

Then φ has a fixed point.

PROOF: Assume that φ has no fixed point. Since $X = X \backslash \mathcal{F}ix(\varphi)$ is compact, there are finite continuous linear functionals, $p^1, \ldots, p^n \in E'$, and open covering $U^1, \ldots, U^n$ of X such that for all $t = 1, \ldots, n$ and $z \in U^t$, $p^t(v) > 0$ for all $v \in \varphi(z) - z$. Let $\beta_t : X \to [0, 1]$, $t = 1, \ldots, n$ be a partition of unity subordinate to $\{U^1, \ldots, U^n\}$. For each $z \in X$, define $\Psi(z)$ as

$$\Psi(z) = \left\{ y \in X \,\middle|\, \left[\sum_{t=1}^n \beta_t(z)p^t\right](y - z) > 0 \right\}.$$

For all $z \in X$, for all $y \in \varphi(z)$, and for all t such that $z \in U_t$, we have $p^t(y - z) > 0$, so that $\sum_{t=1}^n \beta_t(z)(p^t(y - z)) = [\sum_{t=1}^n \beta_t(z)p^t](y - z) > 0$. That is, $\Psi(z) \supset \varphi(z)$ for all $z \in X$. Therefore, $\Psi : X \to X$ is a fixed-point-free convex extension of φ on X. Moreover, Ψ has a local intersection property on $X \backslash \mathbf{\mathcal{F}ix}(\varphi) = X$, since if $x \in X$ and $y^x \in \varphi(x)$, open neighborhood $V(x)$ of x in X exists such that $\forall z \in V(x)$, $y^x \in \Psi(z)$. (Indeed, since $y^x \in \Psi(x)$, we have $[\sum_{t=1}^n \beta_t(z)p^t](y^x - z) > 0$ for all z near x by the continuity of value $[\sum_{t=1}^n \beta_t(z)p^t](y^x - z) = \sum_{t=1}^n \beta_t(z)[p^t(y^x - z)] = \sum_{t=1}^n p^t(\beta_t(z)(y^x - z))$ with respect to variable z.) Therefore, φ has a fixed point by Theorem 2.1.6. ∎

In the above proof, the essential feature is that the compactness of X ensures the existence of finite linear functionals, $p^1, \ldots, p^n$, and open covering $\{U^1, \ldots, U^n\}$. Consequently, as before, we have the following result for Kakutani-type mappings on a normal subset of Hausdorff topological vector space to itself.

THEOREM 2.1.11: *Every non-empty valued Kakutani-type mapping on a normal convex subset of Hausdorff topological vector space to itself has a fixed point.*

We have now completed all of the proofs for Theorems 2.1.2–2.1.11. Discussions about Browder- and Kakutani-type mappings comprise two important streams in fixed-point theory, so we mainly treat such mappings throughout this book. As obviously seen in the definitions of Browder- and Kakutani-type mappings, the arguments in the main theorems, 2.1.9, 2.1.10, and 2.1.11, merely depend on two concepts, "convexity" and "dual space," except for the general topological frameworks. Hence, if we abstract these concepts as structures independent of topological vector space structure, we may argue and extend these results freely in general topological spaces where we can base our economic equilibrium arguments on minimal mathematical requirements (Chapters 3–5) and also relate our fixed-point

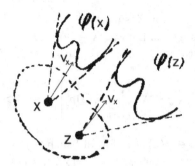

Figure 15: Condition $K\sharp$

methods to purely algebraic ones (Chapters 6 and 7). The following two Sections, 2.2 and 2.3, are devoted to this task.

We complete this section with an additional fixed-point theorem that can be treated under the same conditions as (K*) and (K1), but depends directly on the linear structure of vector spaces. Different from (K*) and (K1), condition (K#) in the next theorem asserts the existence of vectors that must be added at each point of the space, so the concept of vectorial addition is essential for characterizing this class of mappings.

THEOREM 2.1.12: *Let X be a non-empty compact convex subset of Hausdorff topological vector space E over R, and let φ be a correspondence on X to X. Suppose that φ satisfies the following condition:*

(K#) *For each $x \in X \backslash \boldsymbol{Fix}(\varphi)$ there is vector v_x with open neighborhood V_x of x such that for each $z \in V_x$, we may choose $\lambda_z > 0$ to satisfy $z + \lambda_z v_x \in \varphi(z)$.*

Then φ has a fixed point.

PROOF: Assume that $\boldsymbol{Fix}(\varphi) = \emptyset$. Since $X = X \backslash \boldsymbol{Fix}(\varphi)$ is compact, we obtain finite points $x^1, \ldots, x^n \in X$, their open neighbourhoods $U^1, \ldots, U^n$ which cover X, and vectors $v^1, \ldots, v^n \in E$ such that for all $t = 1, \ldots, n$ and for all $z \in U^t$, positive real number $\lambda^t(z) > 0$ satisfies that $\lambda^t(z)v^t \in \varphi(z) - z$. Let $\beta^t : U^t \to [0,1]$, $t = 1, \ldots, n$ be the partition of unity subordinate to U^t, $t = 1, \ldots, n$ such that $\beta^t(x) > 0$ iff $x \in U^t$ for all t. Denote by $\hat{X}$ the intersection of X and the finite dimensional topological vector space $\hat{E}$ spanned by $x^1, \ldots, x^n$ and $v^1, \ldots, v^n$. Clearly $\hat{X}$ is a compact convex subset of finite dimensional topological vector space $\hat{E}$. Let $f : \hat{X} \to \hat{X}$ be mapping defined by

$$f(z) = \sum_{t=1}^{n} \beta^t(z)(z + \hat{\lambda}^t(z)v^t),$$

where $\hat{\lambda}^t(z) = \max\{\lambda \in R_+ | z + \lambda v^t \in \hat{X}\} \geq \lambda^t(z) > 0$. It is easy to verify that $\hat{\lambda}^t$ is continuous for each t so that f is a continuous function on the finite dimensional compact convex set $\hat{X}$ to itself. Hence, f has fixed-point z^* by Brouwer's fixed point theorem. On the other hand, for all t such that $z^* \in U^t$, $z^* + \lambda^t(z^*)v^t$ is an element of $\varphi(z^*)$, so that $z^* + \hat{\lambda}^t(z^*)v^t$ is an element of $\hat{\varphi}(z^*) = \{y \in \hat{X} | y = z^* + \lambda w, \lambda \in R_{++}, w \in \varphi(z^*) - z^*\}$. Therefore, since $\hat{\varphi}(z^*)$ is convex, we have $\sum_{t=1}^{n} \beta^t(z^*)(z^* + \hat{\lambda}^t(z^*)v^t) \in \hat{\varphi}(z^*)$. That is, we have $f(z^*) = z^* \in \hat{\varphi}(z^*)$, which is impossible by the definition of $\hat{\varphi}(z^*)$ since $z^* \notin \varphi(z^*)$. ∎

2.2 Convexity and Duality for General Spaces

As seen in the classical fixed-point theorem of Eilenberg and Montgomery (1946) and its further generalization by Begle (1950b), the vector-space structure for topological space is not a necessary setting for fixed-point arguments. Even for cases with the classical Gale–Nikaido–Debreu lemma for which a certain kind of duality structure seems necessary, the vector structure is superfluous and the necessary setting is "a compact ... space in which the convex linear combination of finitely many points depends continuously on its coefficients" (Nikaido 1959, p. 362, Theorem 4). Since we finally relate our fixed-point arguments to such homological settings in Chapters 6 and 7, convexity and duality must be defined independently of the linear (vector-space) structure.

Let A be a non-empty finite subset of space X. Recall that for finite set A, $\sharp A$ denotes the number of elements of A. Denote by Δ^A the set of all functions $e : A \to R_+$ such that $\sum_{a \in A} e(a) = 1$. For each $a \in A$, denote by e^a the element of Δ^A such that $e^a(a) = 1$ and $e^a(a') = 0$ for each $a' \in A \setminus \{a\}$. Identify Δ^A with the $(\sharp A - 1)$-dimensional standard simplex in $R^{\sharp A}$, the $\sharp A$-dimensional Euclidean space, by identifying e^a with an appropriately chosen element of the standard basis, $\{(1,0,\ldots,0),(0,1,0,\ldots,0),\ldots,(0,\ldots,0,1)\}$, of $R^{\sharp A}$. Moreover, for each finite set $A' \supset A$, regard Δ^A as a subset of $\Delta^{A'}$ by identifying $e \in \Delta^A$

with element $e' \in \Delta^{A'}$ such that $e'(a) = e(a)$ for each $a \in A$ and $e'(a') = 0$ for each $a' \in A' \setminus A$.

2.2.1 Convex structure and convex combination

The abstract setting for convexity in this book is simply an extraction of the concept of "convex combination" from vector-space structure. Let X be a topological space. Denote by $\mathscr{F}(X)$ the set of all non-empty finite subsets of X. We say that on space X a *convex structure* is defined if for each $A \in \mathscr{F}(X)$, there is function f_A on Δ^A to X (intuitively, defining set-A-dependent weighted sums among points in A) and set $C(A) \subset X$ (representing the set of all convex combinations among points in A), satisfying the following conditions:

(C1) For each $A \in \mathscr{F}(X)$, $f_A : \Delta^A \to X$ is continuous.
(C2) For each $A \in \mathscr{F}(X)$, $C(A) \subset X$ is non-empty.
(C3) For each $A \in \mathscr{F}(X)$, there is at least one $\bar{A} \supset A$ such that $f_{\bar{A}}(\Delta^B) \subset C(B)$ for all non-empty $B \subset A$.

We call f_A a *set-dependent weighted sum function* for each A, and C a *convex combination operator*. Condition (C3) is the minimum requirement for set-dependent weighted sums to be consistent with the global convex combination concept.

A convex structure on X, therefore, is a list $(\{f_A | A \in \mathscr{F}(X)\}, C)$ of set-dependent weighted sum functions giving local candidates for weighted sums among finite points that continuously depend on their coefficients, and total operator $C : \mathscr{F}(X) \to 2^X$, directly defining the set of all convex combinations among points in A, $C(A)$, for each $A \in \mathscr{F}(X)$.

Given convex structure $(\{f_A | A \in \mathscr{F}(X)\}, C)$, subset Z of X is said to be *convex* if for any non-empty finite subset A of Z, we have $C(A) \subset Z$.[3] It is easy to check that X is a convex set and that an arbitrary intersection of convex sets is also a convex set. Therefore, the smallest convex set containing Z, the *convex hull*, $\operatorname{co} Z$, of $Z \subset X$ always exists. It is natural to

---

[3] The convexity concept defined here includes the concept of *L-convexity*, which is one of the most general settings in such treatments (Ben-El-Mechaiekh et al., 1998). General convexity settings will be utilized in later chapters. For example, when X has a special point such that the implications of "convex combination" change crucially depending on whether the point is included.

2: Fixed-Point Theorems

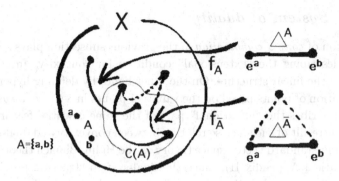

Figure 16: Convex combination $C(A)$

extend the convex combination operator on infinite subsets, $Z \subset X$, $Z \neq \emptyset$, $Z \notin \mathscr{F}(X)$ as

$$C(Z) = \bigcup_{A \subset Z, A \in \mathscr{F}(X)} C(A).$$

Then, by definition, we have $C(Z) \subset \operatorname{co} Z$ for all $Z \subset X$. Note, however, that conditions (C1)–(C3) are not sufficient to ensure that $C(Z)$ is convex. In this sense, $C(Z) = \operatorname{co} Z$ is not necessarily the case, although it is always true for ordinary vector-space convexity.

REMARK 2.2.1: (**Standard Vector-Space Convexity**) If X is a subset of topological vector space, standard vector-space convexity (defined in Subsection 1.3.3) can be obtained by identifying for each $A \in \mathscr{F}(X)$, convex combination function $f_A : \Delta^A \mapsto X$ with the ordinary vector-space convex combination and the value of convex combination operator $C(A)$ with the set of all convex combinations of points in A. Even in such a standard convexity case, the fixed-point results in this book have sufficient generality. Hence, although "convexity" will be used in an abstract sense for the remainder of this book, it is unnecessary for readers (especially those who are unconcerned with the "relation" between our general fixed-point arguments and algebraic homological arguments) to be fully aware of the significance of abstract convexity. For such cases, remember the following: (1) a "convexity" or "convex combination" may be defined even with no linear (vector) space structure, (2) $C(A)$ denotes the set of all convex combinations among points in finite set A, and (3) in proofs, read f_A as the ordinary operator to define the convex combination among finite points in A as $e \mapsto \sum_{a \in A} e(a) a$ for each $e \in \Delta$.

2.2.2 System of duality

The notion of *convex combination* in the previous subsection plays a central role in describing the "extensional" condition for convexity, an essential feature of the linear structure. On the other hand, by defining hyperplanes or separation of points based on the values of points in a space, a *system of duality* describes the "intentional" part of the same essential feature. One important result of such a close relation between convexity and dual-system is known in standard vector-space theory as the Hahn–Banach theorem (see Mathematical Appendix II). Since the dual-system structure provides an appropriate way to describe the world based on the commodity/price, the action/message, or the fact/value classifications, generalizing the notion of duality in a suitable way is necessary to incorporate changes in the notion of convexity.

We generally defined in Chapter 1 the duality system of two topological spaces, X and W, as function g on $X \times X \times W \times W$ to real field R so that each point $w \in W$ can be identified with a function $g(x_0, \cdot, w_0, w) : X \to R$ for each $(x_0, w_0) \in X \times W$. In Chapters 2–5, we discuss a special case where correspondence $V : X \times W \to X$ is defined for each $x_0 \in X$ and $w \in W$ as subset $V(x_0, w) \subset X$, which may be interpreted as the set of points positively evaluated by w at x_0 (Figure 17a). It can also be said that $V(x_0, w)$ is the set of points for improvement at x_0 under value w. $V(x_0, w)$ can be obtained as $V(x_0, w) = \{z \in X \mid g(x_0, z, w, w) > 0\}$ for a certain $g : X \times X \times W \times W \to R$, if necessary.

If X and Y are linear (vector) spaces, under the standard concept of dual-system (X, Y, f), where f is a bilinear form, we may define V as $V(x_0, y) = \{x \in X \mid f(x - x_0, y) > 0\}$ (Figure 17b). In view of economics, such a standard (linear) case corresponds to ordinary linear commodity/price duality. We do not depend on such linearity, though in Chapters 2–5, we suppose the value of V to be convex. In Chapters 6

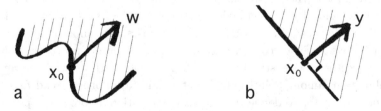

Figure 17: Set of positions in direction w or y from x_0

and 7, we discuss more general cases.[4] Until then, we regard the concept of generalized duality structure through correspondence $V : X \times W \to X$.

Assume that X and W are sets having convex structures. Correspondence $V : X \times W \to X$ is called the (*generalized*) *dual system structure* on X and W if V satisfies the following three conditions:

(V1) $V(x,w)$ is a convex subset of X.
(V2) $x \notin V(x,w)$.
(V3) $y \in V(x,w^0) \cap V(x,w^1) \cap \cdots \cap V(x,w^n) \Rightarrow y \in V(x,w)$ for all $w \in C(\{w^0, w^1, \ldots, w^n\})$.

In condition (V3), $C(\{w^0, w^1, \ldots, w^n\})$ denotes the set of all convex combinations among points in $\{w^0, w^1, \ldots, w^n\}$. Triplet (X, W, V) is called the *generalized dual system*.

In economics, structure (X, W, V) may be used to abstract relations among choice sets, values, and constraints, as we shall see in Chapters 4 and 5. Mathematically, we may use the structure to prepare a notion that gives an intentional (supporting) condition for convex sets, roughly speaking, an *incremental direction* associated with an element of dual space. By applying it to characterizing values $\varphi(x)$ at each x for mapping $\varphi : X \to 2^X$, we can obtain several weak conditions for the existence of fixed points.

2.2.3 Topological condition for dual-system structure

Let E be a topological vector space. *Algebraic dual* E^* of E is the set of all linear functionals on E and *topological dual* E' of E is the set of all continuous linear functionals on E. Element $p \in E'$ defines open half space $\{x \in E \mid p(x) > a\}$ (equivalently, closed hyperplane $\{x \in E \mid p(x) = a\}$) for each $a \in R$. Given finite functionals $p^1, \ldots, p^m$ in E' and m coefficients $a_1, \ldots, a_m$ in R, value $\sum_{i=1}^m a_i p^i(x)$ is jointly continuous with respect to $a_1, \ldots, a_m$ and x. (For every finite dimensional subspace L of E', canonical bilinear form $(x, p) \mapsto p(x)$ is jointly continuous on $E \times L$.)

---

[4]In Chapter 6, we discuss a more general case in which the value of V is acyclic, and in Chapter 7 we return to the original general concept of duality structure based on $g : X \times X \times W \times W \to R$, in relation to the concept of differentiability and cohomological arguments in general spaces.

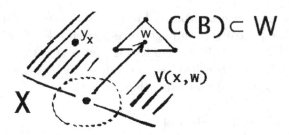

Figure 18: Condition (V4)

We say that the generalized dual system is *topological* if, after added to (V1), (V2), and (V3), the following condition is satisfied:

(V4) $V : X \times W \to X$ has the local-intersection property on $X \times C(B)$ for all $B \in \mathscr{F}(W)$.

This condition drastically weakens the above joint-continuity property with respect to x and finite coefficients in the canonical bilinear form through the local intersection property, so under (V1)–(V3), it generalizes the notion of *topological dual* space (Figure 18).

2.2.4 Differentiability and dual-space representation

Recognizing an element of dual space as a representation of a certain kind of direction at each point in the original space is sometimes useful (see again Figures 17 and 18). If we return to the concept that X represents "facts" (e.g., a commodity space) and W represents "values" (e.g., a price space), then area $V(x, w)$, evaluated positively (for example) by dual space element w relative to x, has an important meaning based on which every social equilibrium actions and states should be decided and measured. An element of dual space, in this sense, may be recognized as a definition of a *positively evaluated area* for each x.

Given duality system (X, W, V), therefore, it is possible for set A and point $x \in X$ to ask whether an element of dual space W defines a positively evaluated area for x to which set A belongs (Figure 19). If such an element exists, that is, there is dual space element $w \in W$ such that $A \subset V(x, w)$, we say that *directions* for points in A at x are *supported by w*, or w is a *supporting element* (or *supporting direction*) for set A at x. We denote by $\mathscr{D}(A, x) \subset W$ the set of all supporting directions for set A at x.

Figure 19:   Each w defines positively evaluated area to which set A belongs at x

For topological vector space X, function f (resp., a correspondence φ) on X to itself may be identified with a definition of vector $f(x) - x$ (resp., a set of vectors $\varphi(x) - x$) at each $x \in X$, a *vector field* (resp., a *multi-valued vector field*) on X, and fixed points are characterized as 0-points of such vector fields. When dual-space elements may be thought to stand for a supporting direction in X and may be used for describing such 0-points, we have various possibilities to deal with fixed-point arguments through general and abstract settings.

An important example of this kind is the notion of gradient vectors and (multi-valued) vector fields containing them. Consider duality system (X, W, V) and real valued function $u : X \to R$. When the duality system represents the standard finite dimensional cases with $X = R^n$ and $W = R^n$, the differentiability of u at each $x \in X$ defines gradient vector $\operatorname{grad} u(x) = (\frac{\partial u}{\partial x_1}(x), \ldots, \frac{\partial u}{\partial x_n}(x))$ at each $x \in X$, i.e., the *gradient vector field* on X. If we consider correspondence $\varphi : X \ni x \mapsto \{y \in X \,|\, u(y) > u(x)\} \subset X$, the better set correspondence defined by u, and if $\varphi(x)$ has a supporting element at each $x \in X$ as long as $\varphi(x) - x \neq \emptyset$, then the greatest elements under u are characterized by the emptiness of φ, and a non-greatest element is always characterized by the existence of dual-space supporting elements. A non-zero gradient vector for u is necessarily one such supporting element.

We can repeat the same argument without using the basic real valued function u on X. Generally, given correspondence $\varphi : X \to X$ and point $x \in X$, consider the set of all supporting elements in W for $\varphi(x)$ at x, $\mathscr{D}(\varphi(x), x)$ (Figure 20a). If φ has an economic or game-theoretic interpretation as a certain kind of better-set correspondences, and if a supporting element of

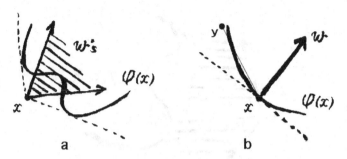

Figure 20: Supporting elements for $\varphi(x)$ at x

$\varphi(x)$ exists at each x such that $x \notin \varphi(x)$, then the existence of a maximal element with respect to the preference defined by φ (i.e., $x \in X$ such that $\varphi(x) \neq \emptyset$) may appropriately be characterized by the field of dual-space supporting elements on X. In such cases, it is sometimes helpful to say that the *preference defined by correspondence* $\varphi : X \to X$ *is differentiable* at x if neighborhood U of x exists such that $\mathscr{D}(U \cap \varphi(x), x) \neq \emptyset$ and $\forall w \in \mathscr{D}(U \cap \varphi(x), x)$, $\forall y \in U \cap V(x, w)$, $C(\{y, x\}) \cap \varphi(x) \neq \emptyset$ (Figure 20b).

2.3 Extension of Classical Results to General Spaces

In this section, we reformulate the results in Section 2.1 under the abstract settings of convexity and duality in general spaces (without linear structures) introduced in the previous section. The domain of mappings is denoted by X, a subset of Hausdorff space E. The abstract convex structure on E (or on X) is denoted by $(\{f_A | A \in \mathscr{F}(E)\}, C)$ (or $A \in \mathscr{F}(X)$ for the case with X). Generalized duality (if necessary) is represented as $(E, W, V : W \to E)$ or $(X, W, V : W \to X)$ with convex structure $(\{g_B | B \in \mathscr{F}(W)\}, C)$ on W. (For convex combination operators, common notation C on W and E or X are used as long as there is no risk of confusion.)

The definitions of the mappings of Browder and Kakutani types are also extended. Although the discussion for Browder-type mappings is straightforward, generalizing the concepts in Kakutani-type arguments is not a routine task, since it depends both on convex and dual-system structures. If we decide to base our studies on topological dual settings, we will obtain a generalized theorem with respect to condition (K1). On the other hand, if we merely consider the convexity setting, we may obtain another natural extension of Kakutani–Fan–Glicksberg's theorem.

First, we generalize Theorem 2.1.6, the basic fixed-point theorem for Browder type mappings.

THEOREM 2.3.1: *Let X be a non-empty compact convex subset of Hausdorff space E having a convex structure, and let φ be a non-empty valued correspondence on X to X satisfying the following condition:*

(K*) φ *has a convex valued fixed-point-free extension having the local intersection property on $X \backslash \mathcal{F}ix(\varphi)$.*

Then φ has a fixed point.

PROOF: Suppose that φ does not have a fixed point and let Ψ be the convex valued fixed-point-free extension of φ in condition (K*). Since $X \backslash \mathcal{F}ix(\varphi) = X$ is compact, by the local intersection property of correspondence Ψ on X, we have points $x^1, \ldots, x^n \in X$ and their open neighborhoods $U(x^1), \ldots, U(x^n)$ that cover X, and for each $t = 1, \ldots, n$, point y^{x^t} exists such that $\forall z \in U(x)$, $z \notin \Psi(z)$ and $y^{x^t} \in \Psi(z)$. Let A be the finite set $\{y^{x^1}, \ldots, y^{x^n}\}$ and $\alpha_t : X \to [0,1]$, $t = 1, \ldots, n$, be the partition of unity subordinate to $U(x^1), \ldots, U(x^n)$. Take $\bar{A} \in \mathcal{F}(X)$ for A in condition (C3) of the convex structure and define continuous mapping F on Δ^A to itself as

$$F(e) = \sum_{t=1}^{n} \alpha^t(f_{\bar{A}}(e))e^{y^t},$$

where e^{y^t} denotes the vertex of Δ^A corresponding to y^{x^t} for each $t = 1, \ldots, n$. Then F has fixed-point $e^* = F(e^*)$ by Brouwer's fixed-point theorem. Let $x^* = f_{\bar{A}}(e^*)$ and $B = \{y^{x^t} | \alpha^t(e^*) > 0\}$. Since $\alpha^t(x^*) > 0$ implies that $y^{x^t} \in \Psi(x^*)$, we have $f_{\bar{A}}(\Delta^B) \subset \Psi(x^*)$. It follows that $x^* = f_{\bar{A}}(e^*) = f_{\bar{A}}(\sum_{t=1}^n \alpha^t(f_{\bar{A}}(e^*))e^{y^t}) \in f_{\bar{A}}(\Delta^B) \subset \Psi(x^*)$. Hence, we have $x^* \in \mathcal{F}ix(\Psi) \subset \mathcal{F}ix(\varphi)$ so that φ has a fixed point contrary to the assumption. ∎

Theorem 2.3.1 easily gives generalizations of Theorems 2.1.5, 2.1.2, 2.1.7, and 2.1.8 in Hausdorff space E having convex structure in the same way as Theorem 2.1.6 does in Hausdorff topological vector spaces. We omit the details here except for the next extension of Theorem 2.1.7:

THEOREM 2.3.2: *Let X be a non-empty compact convex subset of Hausdorff space E having a convex structure, and let $\varphi : X \to X$*

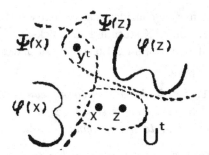

Figure 21:   Mappings of Browder type — $y^t \in \bigcap_{x \in U^t} \Psi(x)$

be a correspondence having the local-intersection property on $X \backslash \mathcal{F}ix(\varphi)$. Assume that for each $x \in X \backslash \mathcal{F}ix(\varphi)$, x does not belong to $C(\varphi(x))$, the set of all convex combinations among points in $\varphi(x)$. Then φ has a fixed point.

PROOF: For each $x \notin \mathcal{F}ix(\varphi)$, define $\Psi(x)$ as $\Psi(x) = C(\varphi(x))$ and for each $x \in \mathcal{F}ix(\varphi)$, define $\Psi(x)$ as $\Psi(x) = \varphi(x)$. Note that under the abstract-convexity setting, set $C(\varphi(x))$ is not necessarily a convex set. Based on the supposition that φ does not have a fixed point and $\Psi(x) = C(\varphi(x))$, we can repeat exactly the same argument in the previous proof until we obtain fixed-point $e^* = F(e^*)$, point $x^* = f_{\bar{A}}(e^*)$ and set $B = \{y^{x^t} | \alpha^t(e^*) > 0\}$. Since $\alpha^t(x^*) > 0$ implies that $y^{x^t} \in \Psi(x^*)$, and since y^{x^t} is an element of $\varphi(x^*)$ now, we have $f_{\bar{A}}(\Delta^B) \subset \Psi(x^*) = C(\varphi(x^*))$. It follows that $x^* = f_{\bar{A}}(e^*) = f_{\bar{A}}(\sum_{t=1}^n \alpha^t(f_{\bar{A}}(e^*))e^{y^t}) \in f_{\bar{A}}(\Delta^B) \subset \Psi(x^*)$. Hence, we have $x^* \in \mathcal{F}ix(\Psi) \subset \mathcal{F}ix(\varphi)$ so that φ has a fixed point, which is contrary to the assumption. ■

The characterization for Browder-type mappings based on coverings is also given in exactly the same way as before. Let X be a normal subset of Hausdorff space having a convex structure. Correspondence $\varphi : X \to X$ is of the *Browder type* if the next condition is satisfied:

(Browder-Type Mapping: Hausdorff Space with Convex Structure) For each open set $U \supset \mathcal{F}ix(\varphi)$ of X, there are finite open sets $U^1, \ldots, U^n$ of X such that (1) $\{U, U^1, \ldots, U^n\}$ covers X, and (2) there is a convex valued fixed-point-free extension Ψ of φ on $\bigcup_{t=1}^n U^t$ to X satisfying that $\bigcap_{x \in U^t} \Psi(x) \neq \emptyset$ for each $t = 1, \ldots, n$.

When space X is compact, a typical sufficient condition for correspondence φ to be of the Browder type is that φ is convex valued and has a local intersection property at each $X \backslash \mathcal{F}ix(\varphi)$ on X (or condition (K*) with closed $\mathcal{F}ix(\varphi)$, etc.). One can verify that the proof of Theorem 2.3.1 actually shows the existence of fixed points for all non-empty valued Browder-type mappings on normal convex subset X of Hausdorff space with convex structure to itself.[5]

THEOREM 2.3.3: *Every non-empty valued Browder-type mapping on a normal convex subset of Hausdorff space with convex structure to itself has a fixed point.*

Next, we consider cases with Kakutani-type mappings. The arguments for the Kakutani-type conditions in Section 2.1 were uniformly characterized as corollaries to Theorem 2.1.10. Note, however, that Condition (K1) in Theorem 2.1.10 depends on the concept of a dual system, which by no means is restrictive as long as our considerations are based on the vector-space structure; furthermore, we do not necessarily expect the dual system in this section to be a topological one satisfying condition (V4). On the other hand, as we see later, the concept of Kakutani-type mappings can naturally be extended without using the dual-system structure. Therefore, it is desirable to treat the problem of generalizing Theorem 2.1.10 (which is based on the topological dual system and convex structures) and Kakutani–Fan–Glicksberg's Theorem (which is based merely on topological and convex structures) as different issues.[6] Theorem 2.1.10 can be generalized as follows.

THEOREM 2.3.4: *Let X and W be non-empty compact Hausdorff spaces having convex structures, and let φ be a non-empty valued correspondence on X to X such that the following condition, $(K1)$, is satisfied under dual-system structure $(X, W, V : X \times W \to 2^X)$ satisfying the topological condition $(V4)$.*

---

[5] As before (Theorem 2.1.9), closedness for domain X is not necessary in the next theorem. Such characterization (based on coverings instead of global continuity arguments) has an advantage in its development of discussion for Čech and Vietoris homological treatments of this problem in Chapters 6 and 7.
[6] More precisely, it is not the settings (V1)–(V3) of the dual-system structure but the topological condition (V4) that makes (K1) fail to be a unified condition including all generalized Kakutani-type mappings. (See discussions after Theorem 2.3.5 below.) Condition (V4) also makes (K1) fail to be a unified condition including all generalized Browder-type mappings. (On this point, see Notes for Section 2.2 (p. 55) and arguments after Theorem 3.1.4 (p. 62) in the next chapter.)

(K1) *For each x such that $x \in X \backslash \mathcal{F}ix(\varphi)$, element $w \in W$ and open neighborhood U of x exist such that for all $z \in U \cap (X \backslash \mathcal{F}ix(\varphi))$, we have $\varphi(z) \subset V(z, w)$.*

Then φ has a fixed point.

PROOF: Suppose that φ has no fixed point. Since $X \backslash \mathcal{F}ix(\varphi) = X$ is compact, we may obtain (through points and neighborhoods in condition (K1)) finite open covering $U^1, \ldots, U^m$ of X and points $w^1, \ldots, w^m \in W$ such that for each $i = 1, \ldots, m$ and for each $z \in U^i$, we have $\varphi(z) \subset V(z, w^i)$. Let $B = \{w^1, \ldots, w^m\}$ and $\beta^1, \ldots, \beta^m$ be the partition of unity subordinate to $U^1, \ldots, U^m$. Define mapping $\hat{\varphi} : X \to 2^X$ as

$$\hat{\varphi}(x) = V\left(x, f_{\bar{B}}\left(\sum_{i=1}^m \beta^i(x) e^i\right)\right) \supset \varphi(x),$$

where e^i is the vertex of $\Delta^B \subset \Delta^{\bar{B}}$ corresponding to w^i for each $i = 1, \ldots, m$, and the inclusion follows from (V3). Hence, by (V1) and (V2), $\hat{\varphi}$ is a convex valued fixed-point-free extension of φ. Moreover, under (V4), $\hat{\varphi}$ has a local intersection property on X. It follows that $\hat{\varphi}$ has a fixed point by Theorem 2.3.1, which contradicts condition (V2). ∎

In view of mathematical economics, the dual-space setting in Theorem 2.3.4 is quite natural as a generalized setting of commodity/price duality. The theorem suggests in such cases that a certain kind of joint continuity among systems related to facts and values (like condition (V4)) will ensure the existence of equilibrium.

Note also that Condition (K1) includes the most representative cases of Browder type mappings (having open graphs) as well as (K1) in the Theorem 2.1.10 type; accordingly, we have obtained a unified viewpoint on both the classical Browder and Kakutani fixed-point theorems. Indeed, a convex valued correspondence having an open graph, $\varphi : X \to X$, induces a dual-space structure satisfying (V4), V_φ, such that $V_\varphi(x, y) = \varphi(x)$ for $y \in \varphi(x)$ and $V_\varphi(x, y) = \emptyset$ for $y \notin \varphi(x)$. Hence, the theorem includes typical cases of both Kakutani–Fan–Glicksberg's theorem and Browder's classical theorem.

We now define Kakutani-type mappings for general spaces without vector-space structures and consider a natural extension of Kakutani–Fan–Glicksberg's theorem. Mapping φ on subset X of Hausdorff topological space having a convex structure to itself is of the *Kakutani type* if the

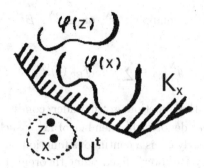

Figure 22: Kakutani-type mappings

following condition is satisfied:

(Kakutani-Type Mapping: Hausdorff Space with Convex Structure) For each open set $U \supset \mathcal{F}ix(\varphi)$ of X, there are finite open sets $U^1, \ldots, U^n$ of X with closed convex sets $K_1, \ldots, K_n$ such that (1) $\{U, U^1, \ldots, U^n\}$ covers X, and (2) for each t, $U^t \cap K_t = \emptyset$ and for all $z \in U^t$, $\varphi(z) \subset K_t$.

If we restrict our attention to compact domain X, a typical sufficient condition for correspondence $\varphi : X \to X$ to be of the Kakutani type is that for each $x \in X \backslash \mathcal{F}ix(\varphi)$, closed convex set K_x exists such that $x \notin K_x$ and for all z near x, we have $\varphi(z) \subset K_x$. It is possible to show that every Kakutani-type mapping in compact Hausdorff spaces has a fixed point.

THEOREM 2.3.5: (**Kakutani's Theorem in Compact Hausdorff Space with Convex Structure**) *Let X be a non-empty compact Hausdorff space having convex structure. If $\varphi : X \to 2^X \backslash \{\emptyset\}$ is of the Kakutani type, then φ has a fixed point.*

PROOF: If φ has no fixed point, then for each $x \in X$, we have $x \notin \varphi(x)$ so that closed convex set K_x and open neighborhood $U(x)$ of x satisfy the condition that for all $z \in U(x)$, $\varphi(z) \subset K_x$. Since K_x is closed, we may assume that $U(x) \cap K_x = \emptyset$ without loss of generality. Since X is compact, covering $\{U(x) | x \in X\}$ has finite subcovering $\{U(x_1), \ldots, U(x_m)\}$. Let $N = \{N_1, \ldots, N_n\}$ be an arbitrary refinement of $\{U(x_1), \ldots, U(x_m)\}$ and $a_1^N, \ldots, a_n^N$ be arbitrary points of $N_1, \ldots, N_n$, respectively. Denote by $\beta_1^N : X \to [0,1], \ldots, \beta_n^N : X \to [0,1]$ the partition of unity subordinate to $\{N_1, \ldots, N_n\}$. Take $b_1^N, \ldots, b_n^N$ as arbitrary points of $\varphi(a_1^N), \ldots, \varphi(a_n^N)$, respectively, and define mapping g^N on Δ^{B^N}, $B^N = \{b_1^N, \ldots, b_n^N\}$ to itself as

$$g^N(e) = \sum_{i=1}^n \beta_i^N(f_{\bar{B}^N}(e))e^i,$$

where $f_{\bar{B}^N}$ denotes the function given under condition (C3) of convex structure on X and e^i denotes the member of Δ^{B^N} such that the value of b_i^N is 1 for each i. Clearly g^N is a continuous function on Δ^{B^N} to itself, so it has fixed-point e^N by Brouwer's fixed-point theorem. Let $x^N = f_{\bar{B}^N}(e^N)$. Since X is compact, by considering the set of all finite coverings of X with the refinement relation as a directed set, we may consider that there is converging subnet $x^N \to x^*$.

Since $\{U(x_1), \ldots, U(x_m)\}$ is a cover of X, we may assume that $x^* \in U(x_1)$ without loss of generality. By taking refinement $\mathfrak{N} = \{N_1, \ldots, N_n\}$ so fine that $x^N \in U(x_1)$, we may suppose that

$$\bigcup\{N_i | x^N \in N_i, N_i \in N\} \subset U(x_1).$$

However, point $x^N = f_{\bar{B}}(e^N)$ satisfies

$$x^N = f_{\bar{B}}(g^N(e^N)) = f_{\bar{B}}\left(\sum_{i=1}^n \beta_i^N(x^N)e^i\right),$$

where e^i denotes the members of Δ^{B^N} such that the value of b_i^N is 1. Note that $\beta_i^N(x^N) > 0$ means that $x^N \in N_i \subset U(x_1)$. That is, $\beta_i^N(x^N) > 0$ means that $b_i^N \in K_{x_1}$. This means, however, that $x^N = f_{\bar{B}}(\sum_{i=1}^n \beta_i^N(x^N)e^i)$ is an element of K_{x_1}, since K_{x_1} is convex. Since $x^N \in U(x_1)$, we have $x^N \in U(x_1) \cap K_{x_1}$, a contradiction. ■

The above result for Kakutani-type mappings can be recaptured through a dual-system structure or a Browder-type condition. In the proof of the previous theorem, we define correspondence Ψ on space X to X as follows:

$$\Psi(x) = \bigcap_{i \in \{j | x \in U(x_j)\}} K_{x_i}.$$

Then $\Psi : X \to 2^X$ is a convex-valued fixed-point-free extension of φ. If we define induced dual-system structure V_Ψ as $V_\Psi(x, y) = \Psi(x)$ for $y \in \Psi(x)$ and $V_\Psi(x, y) = \emptyset$ for $y \notin \Psi(x)$, then V_Ψ satisfies (V1), (V2), and (V3). Unfortunately, we cannot expect Ψ to have a local-intersection property

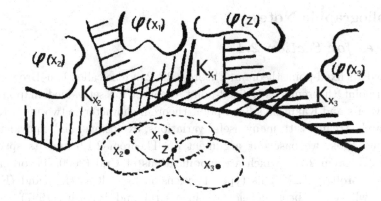

Figure 23: Browder-type characterization for Kakutani-type mapping φ

or V_Ψ to satisfy condition (V4) (see point z in Figure 23 at which Ψ fails to have a local intersection) and thus a Browder- or (K1)-type Theorem (e.g., Theorems 2.3.1 or 2.3.4) is not directly applicable. As we see in the next chapter (Theorem 3.1.3), however, this problem is not serious since in normal spaces we can take a certain kind of refinement that excludes the situation of point z described in Figure 23.[7] The above Browder-type or (K1)-type dual-system characterization through K_x's for Kakutani-type mapping φ plays a significant role in Chapters 6 and 7 in relating ordinary convexity methods to algebraic (homological) fixed-point arguments.

Mathematically, fixed-point theorems for Kakutani-type mappings are more important than those of the Browder type, since they are applicable for all continuous single-valued mappings. In cases with locally convex topological vector spaces, every continuous function $f : X \to X$ automatically satisfies the conditions for Kakutani-type mapping, so it has a fixed point. Without locally convex topology, however, the fixed-point property is a mathematical open question. Theorem 2.3.5 argues that even in general spaces (not necessarily locally convex), the existence of convex area K_x at each $x \in X \backslash \mathcal{F}ix(\varphi)$ locally separating points $z \in U(x)$ and their images $\varphi(z)$ is sufficient to assure the existence of fixed points for mapping φ.

---

[7] Theorem 3.1.3 in the next chapter assures the existence of closed refinement $\{C(x_1), \ldots, C(x_m)\}$ of $\{U(x_1), \ldots, U(x_m)\}$ such that $\bigcup_{i=1}^{m} \operatorname{int} C(x_i) \supset X$. Let us redefine Ψ as $\Psi(x) = \bigcap_{i \in \{j | x \in C(x_j)\}} K_{x_i}$. Then, Ψ satisfies the open lower-section property and the corresponding V_Ψ satisfies topological condition (V4*) in the next chapter. Hence, Theorem 2.3.1 or a fixed-point theorem for generalized dual-system structure under (V1), (V2), (V3), and (V4*) in the next chapter (Theorem 3.1.4) is applicable.

Bibliographic Notes

Notes for Section 2.1

Browder's fixed-point theorem, Theorem 2.1.2, is also called Fan–Browder's fixed-point theorem. As a fixed-point theorem for set-valued mappings, Browder's and Kakutani's fixed-point theorem characterize two important research streams with many useful variations, extensions, and equivalents. In this book, we base our argument on Theorem 2.1.6 and its special case, Theorem 2.1.5, which are modifications of Urai (2000, Theorem 1, (K*), Corollary 1.2). This type of extension as well as (K1) and (K#) type, which has been developed since Urai and Hayashi (1997), aims to acquire a minimal condition on the convexity and continuity of individual preference and constraint correspondences for the existence of economic and game-theoretic equilibria. Further generalization has also been given by Park (2004) (arguments without compact domains like a type of Theorem 2.1.9), Urai and Yoshimachi (2004) (extension of (K1) condition for upper semicontinuous local selection of dual elements — an upper semicontinuously differentiable correspondence treated in the next chapter), etc.

Theorem 2.1.8 has an elegant form as a direct extension of Theorem 2.1.2. As a modification of the Browder fixed-point theorem, such a theorem has many equivalents. One of the oldest analogues of such results is the fixed-point theorem of Tarafdar (1977). In mathematical economics, the well-known result on the equilibrium existence in Shafer and Sonnenschein (1975) uses essentially the same argument on treatments of convexity. See also maximal element existence theorems in Yannelis and Prabhakar (1983), theorems in Mehta and Tarafdar (1987), etc.

Kakutani–Fan–Glicksberg's fixed-point theorem (Theorem 2.1.3) was first developed by Kakutani (1941) as a generalization of Brouwer's fixed-point theorem to closed-graph mappings, which was then generalized independently by Fan (1952) and Glicksberg (1952) in locally convex spaces. The closed-graph property is usually related to the concept of upper semicontinuity through such arguments as Theorem 2.1.4. (This theorem was from Nikaido (1959, Section 4.1).) The generalizations of Kakutani–Fan–Glicksberg's theorem in this section are based on condition (K1) in Theorem 2.1.10 (Urai 2000, Theorem 1, (K1)). Condition (K1) may also be considered an extension of the concept of *upper demicontinuity* in Fan (1969).

Notes for Section 2.2

Convexity in this book is treated in an abstract sense. This type of argument has been developed by Takahashi (1970), Komiya (1981), Horvath (1991), Park and Kim (1996), Ben-El-Mechaiekh et al. (1998), etc. In this book, the concept of "convex hull" (the smallest convex set including itself) is not used in a generalized sense, since the concept is not necessary for this book's fixed-point arguments. With respect to "convex combination" (the set of all convex combinations among points in it), however, the definition presented here gives one of the most general frameworks in this field. (In our treatment of convexity, if we consider the special case that $\bar{A}$ in (C3) may be chosen as any set including A, the concept is called *L-convexity* defined by Ben-El-Mechaiekh et al. (1998).) The author thanks Professor Hidetoshi Komiya (Keio University) for an excellent brief introduction, Komiya (1999), to the theory of abstract convexity.

As well as the present usage of convexity, the concept of a dual-system structure is also original to this book. A special case of a generalized dual system, however, adopting (X, X, V) with a standard linear structure on X, is given in Urai (2000, Section 6, Theorem 21) and is extended in Urai and Yoshimachi (2004) as a "directional structure" and in Urai and Yokota (2005) as a dual-system structure. Under the structure of a generalized dual system, the condition (K1) in the previous section may also be captured as the existence of a *locally constant selection* for the correspondence defined by supporting elements of $\varphi(x)$ at x for each $x \in X \backslash \boldsymbol{Fix}(\varphi)$. The notion could be easily extended to cases where the correspondence defined by supporting elements has a *locally continuous* (or *locally upper semicontinuous*) selection. The idea will be treated in the next chapter as an upper semicontinuously differentiable correspondence. (In defining differentiability for general spaces, concepts in Rubinstein (2006, p. 48) were utilized.) Note also that under the dual-system structure, the (K1) condition may be utilized to integrate Kakutani- and Browder-type (having open lower sections) fixed-point theorems, as discussed in Section 2.3.

Notes for Section 2.3

Extensions of Fan–Browder's theorem to abstract convexity settings are treated in Park (2001) (see Corollary 4.5 proved under G-convex space),

Ben-El-Mechaiekh et al. (1998) (Proposition 3.8 proved under L-convexity), Ding (2000) (under a special contractible condition), Luo (2001) (see Theorem 3.2 proved under δ-convexity with semilattice structure), etc. Our treatment of abstract convexity in this book is more general than all of the corresponding concepts in those papers, so our results, Theorems 2.3.1 and 2.3.3, add new aspects to this problem. I also note that by using open coverings, Browder and Kakutani mappings may be characterized without compactness of the domain. For this point, I am indebted to Park for his idea of theorems in Park (2004).

As with Browder-type theorems, our results for Kakutani-type theorems also depend on one of the most general convexity concepts introduced in this book. As shown in Section 2.1, Theorem 2.3.4 implies one of the most general forms of Kakutani's fixed-point theorem. In this theorem, the necessary condition is given through the distinctive notion of a generalized dual system defined in this book without using concepts under a vector-space structure. Although Ben-El-Mechaiekh et al. (1998) (Theorem 4.2 and Corollary 4.7) treat similar problems (including Kakutani-type fixed-point results under L-convexity), the approach in this book has the advantage of not using uniform and locally convex structures in the space. Theorem 2.3.5, another very general type of Kakutani's theorem, may also be compared with the above results in Ben-El-Mechaiekh et al. (1998). Again, in our theorem, the convexity is weaker than their L-convexity, and moreover, uniform and locally convex structures are not necessary.

Chapter 3
Nash Equilibrium and Abstract Economy

3.1 Multi-Agent Product Settings for Games

In this chapter, the Nash equilibrium and the generalized Nash equilibrium (social equilibrium) existence problems are reexamined. This section treats the basic concepts and the mathematical tools for non-cooperative games with theorems on the maximal-element existence for multi-agent product settings relating our fixed-point arguments to game-theoretic equilibrium existence problems.

The simplest case of a Nash equilibrium is nothing but a fixed point as a list of players' best responses based on a list of all players' strategies. In the argument, we have to use the list or the sequence of players' *strategies* (*strategy profile*) in the product space of all players' *strategy sets*. Let us begin with a precise definition of these game-theoretic concepts and prepare important mathematical tools for such product spaces.

A *non-cooperative game* in *strategic form* is specified as a list of the following terms:

(G1) Names of players
(G2) Sets of strategies for all players
(G3) Payoff structure among players based on their choices of strategies

Let I be the set of players.[1] For each i in I, the *strategy set* of i, X_i, is assumed to be a non-empty compact convex subset of a Hausdorff topological space having convex structure (e.g., a topological vector space). Denote by X product space $\prod_{i \in I} X_i$ with the product topology. By definition, the product of the Hausdorff spaces is obviously also a Hausdorff space. The product of the compact spaces is also compact (Tychonoff's Theorem), so we have X as a compact Hausdorff space.

---
[1] In this chapter, we do not restrict I to be a finite set.

THEOREM 3.1.1: (Tychonoff's Product Theorem) *The product of a family of compact spaces is compact under the product topology.*

PROOF: Let $\{X_i | i \in I\}$ be a family of compact sets and $S : D \to X = \prod_{i \in I} X_i$ be a net on directed set $(D, \leqq)$ into X. S has a converging subnet. Indeed, since each X_i is compact, there is a subnet $S \circ T_i : D_i \to X$ such that $\mathrm{pr}_i \circ S \circ T_i$ converges to point x_i^* in X_i for each $i \in I$, where D_i is a set directed by $\precsim_i$, T_i is a map on D_i to D, and $\mathrm{pr}_i : X \to X_i$ denotes the i-th projection. Consider directed set $E = (I \times \prod_{i \in I} D_i, \precsim)$, where $(j, (d_i)_{i \in I}) \precsim (k, (e_i)_{i \in I})$ if and only if $d_i \precsim_i e_i$ for all $i \in I$. Define map T on E to D as $T(j, (d_i)_{i \in I}) = T_j(d_j) \in D$ and consider subnet $S \circ T : E \to X$ of S. Since for each $i \in I$, $\mathrm{pr}_i \circ S \circ T$ converges to x_i^* for each $i \in I$, $S \circ T$ converges to $(x_i^*)_{i \in I} \in X$ under the product topology, so $S \circ T$ is a converging subnet of S. ∎

In this book, the *payoff structure* for each i is represented by a *better-set correspondence* on X to X_i (i.e., a preference on his/her strategy set, X_i, given other players' strategies) as

$$P_i : X \ni x \mapsto P_i(x) \subset X_i.$$

Suppose that for each i, the payoff structure for i, P_i, satisfies the following three conditions:

(P1) Irreflexivity: For each $x \in X$, the i-th coordinate of x, $x_i = \mathrm{pr}_i(x)$, does not belong to $P_i(x)$.
(P2) Local Intersection Property: $P_i : X \to X_i$ has the local intersection property on $\{x \in X | P_i(x) \neq \emptyset\}$.
(P3) Convex Combination Property: For each $x \in X$, the i-th coordinate of x, $x_i = \mathrm{pr}_i(x)$, does not belong to $C(P_i(x))$, the set of all convex combinations among points in $P_i(x)$.

Figure 24 shows a typical better-set correspondence, $P_i : X \ni x = (\ldots, x_i, \ldots) \mapsto P_i(x) \subset \Psi_i(x) \subset X_i$, where $\Psi_i(x)$ denotes a set that includes all convex combinations among points in $P_i(x)$ (a convex set). Except for the situation where the domain is not X_i but product $\prod_{j \in I} X_j$, one can see that the condition we suppose is essentially the same as those in Theorem 2.1.5 (Figure 11) in Chapter 2. It is also possible to consider better-set correspondences that satisfy more general conditions like (K*), (K1), (Browder-Type mapping), and (Kakutani-Type mapping); but, for now, we concentrate on this special setting.

3: Nash Equilibrium and Abstract Economy

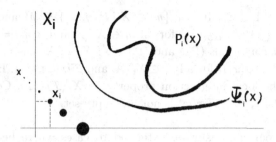

Figure 24: Better-set correspondence $P_i : \prod_{j \in J} X_j \to X_i$

When continuous functions $f_i : X \to R$, $i \in I$ (called *payoff functions*) define for each $i \in I$, player i's payoff $f_i(x)$ under strategy profile $x = (x_j)_{j \in I} \in \prod_{j \in I} X_i$, we may obtain better-set correspondences that satisfy conditions (P1)–(P3) as

$$P_i(x) = \{y_i \in X_i | f_i(y_i; x_{\hat{i}}) > f_i(x)\},$$

where $f_i(y_i; x_{\hat{i}})$ denotes i's payoff under the (strategy profile) vector obtained by replacing i-th coordinate x_i of x with y_i, $i \in I$.[2] Each P_i has the open lower-section property by the continuity of f_i. Since the convexity of $P_i(x)$ is usually assured by the quasi-concavity of f_i, settings (P1)–(P3) for better-set correspondences include ordinary settings in non-cooperative game theory with payoff functions and much more.

If $P_i(x) = \emptyset$, there is no better strategy $y_i \in X_i$ for i than $x_i = \mathrm{pr}_i(x) \in X_i$ as long as the strategies of other players are $x_j = \mathrm{pr}_j(x)$ for all $j \neq i$, $j \in I$. In this sense, given other players' strategies $x_{\hat{i}}$, player i's better-set correspondence $P_i(\cdot; x_{\hat{i}})$ may be identified with binary relation $\succ_i$ on X_i as $z_i \succ_i y_i$ if and only if $z_i \in P_i(y_i; x_{\hat{i}})$. If $P_i(x) = \emptyset$, x_i is a $\succ_i$-*maximal element* of X_i (no y_i in X_i such that $y_i \succ_i x_i$). Our setting (P1)–(P3) for better-set correspondences is sufficient to guarantee the existence of such maximal elements for each i.

THEOREM 3.1.2: (Maximal Element Existence) *If P_i satisfies (P1), (P2), and (P3), then for any $x \in \prod_{j \in I} X_j$ and $i \in I$, element $x_i^* \in X_i$ exists such that $P_i(x_i^*; x_{\hat{i}}) = \emptyset$.*

---

[2] Under the multi-agent product setting, $X = \prod_{i \in I} X_i$, as long as there is no risk of misunderstanding or confusion, notation x_i will represent the i-th coordinate of $x \in X$ and $x_{\hat{i}}$ will denote the vector obtained by eliminating i-th coordinate x_i from x.

PROOF: Let $A \subset X_i$ be set $\{y_i \in X_i | P_i(y_i; x_{\hat{i}}) = \emptyset\}$ and define $\varphi : X_i \to X_i$ as $\varphi(y_i) = P_i(y_i; x_{\hat{i}})$ for all $y_i \in X_i \backslash A$ and $\varphi(y_i) = \{y_i\}$ for all $y_i \in A$. Under conditions (P1) and (P3), $\varphi : X_i \to X_i$ has no fixed point on $X_i \backslash A$, $x_i \notin C(\varphi(x_i))$ for all $x_i \in X_i \backslash A$, and $\boldsymbol{\mathcal{F}ix}(\varphi) = A$. By condition (P2), φ has the local intersection property on $X_i \backslash \boldsymbol{\mathcal{F}ix}(\varphi)$. Consequently, by Theorem 2.3.2, $A = \boldsymbol{\mathcal{F}ix}(\varphi)$ is not an empty set. ∎

The theorem can easily be extended to cases where better-set correspondence P_i is characterized by more general conditions such as (K*), (K1), (Browder-Type mapping), and (Kakutani-Type mapping); the general cases are discussed in the last part of this section.

Given a non-cooperative game in strategic form, $((X_i, P_i)_{i \in I})$, a strategic profile $x^* = (x_i^*)_{i \in I} \in X = \prod_{i \in I} X_i$ is called a *Nash equilibrium* if for each $i \in I$, $P_i(x^*) = \emptyset$, or, equivalently, $x_i^* \in X_i$ is a maximal element for preference relation $\succ_i$ defined by $P_i(\cdot, x_{\hat{i}}^*)$. It is sometimes more desirable to describe the possibility that the strategy of one person affects another person's available set of strategies (like prices in markets or externality among agents with respect to possible actions). We can obtain the structure by defining *constraint correspondence* $K_i : \prod_{j \in I} X_j \ni x \mapsto K_i(x) \subset X_i$ for each $i \in I$. Non-cooperative game $((X_i, P_i)_{i \in I})$ with constraint correspondences $(K_i)_{i \in I}$ is called an *abstract economy*. An *equilibrium* (sometimes called a *generalized Nash equilibrium*) for an abstract economy is strategy profile $x^* = (x_i^*)_{i \in I}$ such that for each $i \in I$, $x_i^* \in \varphi_i(x^*)$ and $P_i(x^*) \cap K_i(x^*) = \emptyset$.

The existence of the Nash and the generalized Nash equilibria are the main topics in the subsequent two sections. As preparation, we discuss several important theorems and concepts in the rest of this section. The first theorem, which addresses the property of open coverings in normal spaces, will be used repeatedly throughout this book. The second is a fixed-point theorem with an alternative setting for the topological features of the dual-system structure. The third is a definition of general individual preferences and special kinds of products of correspondences based on conditions used for fixed-point theorems in the previous and current chapters. The rest are the theorems and notes on the existence of maximal elements or equilibrium actions for such general settings of correspondences.

Recall that in normal space X, every open neighborhood U of $x \in X$ includes closed neighborhood C of x (see Figure 25a). It follows that in normal spaces, the class of closed neighborhoods is sufficient to characterize topological features like convergence. The situation indeed is a common

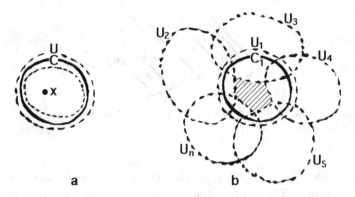

Figure 25:   Closed refinement in normal space

property for *regular* spaces,[3] however, and for normal spaces, we can go further. That is, if U is a neighborhood of a closed set, then closed neighborhood $C \subset U$ of the closed set exists. (Consider two disjoint closed sets, $X \setminus U$ and the closed subset of U, and use the definition of normal space. See, for example, the relation among U_1, the closed shaded area, and C_1 in Figure 25b.) By repeatedly using this property, we can see the following result on the closed refinement for coverings in normal spaces.

THEOREM 3.1.3: (Closed Refinement in Normal Spaces) *If $\{U_1, \ldots, U_n\}$ is an open covering of normal space X, then there are n closed sets $C_1, \ldots, C_n$ such that $C_1 \subset U_1, \ldots, C_n \subset U_n$, and $\{\mathrm{int}\, C_1, \ldots, \mathrm{int}\, C_n\}$ covers X.*

PROOF: Since $\{U_1, \ldots, U_n\}$ is a covering of X, closed set $X \setminus \bigcup_{i=2}^{n} U_i$ is a subset of U_1 (see shaded area in Figure 25b). Since X is normal, two disjoint closed sets, $X \setminus U_1$ and $X \setminus \bigcup_{i=2}^{n} U_i$, can be separated by two disjoint open sets, $V_1 \supset X \setminus U_1$ and $W_1 \supset X \setminus \bigcup_{i=2}^{n} U_i$. Define C_1 as $C_1 = X \setminus V_1$. Obviously, C_1 is closed, $C_1 \subset U_1$, and $\mathrm{int}\, C_1 \supset W_1 \supset X \setminus \bigcup_{i=2}^{n} U_i$, so family $\{\mathrm{int}\, C_1, U_2, \ldots, U_n\}$ covers X. Given covering $\{\mathrm{int}\, C_1, U_2, \ldots, U_N\}$, we can also define $C_2 \subset U_2$ in exactly the same way as $C_1 \subset U_1$ so that family $\{\mathrm{int}\, C_1, \mathrm{int}\, C_2, U_3, \ldots, U_n\}$ covers X. In general, if for $k \leq n - 1$, we can define $C_1, \ldots, C_k$ so that $\{\mathrm{int}\, C_1, \ldots, \mathrm{int}\, C_k, U_{k+1}, \ldots, U_n\}$ covers X, we can

---

[3] Hausdorff space X is called *regular* if for every point $x \in X$ and closed set $A \subset X$ such that $x \notin A$, two open sets U and V exist such that $x \in U$, $A \subset V$, and $U \cap V = \emptyset$.

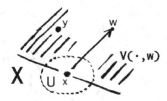

Figure 26: Condition (V4*): For all $y \in V(x,w)$, there is open set U such that for all $z \in U$, $y \in V(x,w)$

also define closed $C_{k+1} \subset U_{k+1}$ so that $\{\text{int}\, C_1, \ldots, \text{int}\, C_k, \text{int}\, C_{k+1}, U_{k+2}, \ldots, U_n\}$ covers X. Hence, the theorem follows by mathematical induction. ∎

At each point of the space, the existence of such closed neighborhoods is useful for constructing global mapping in a continuous way based on local mappings. The theorem will be used repeatedly in the next section to prove the existence of Nash equilibria under several weaker conditions on preferences or profit functions. In this section, note that Theorem 3.1.3 can be utilized for fixed-point results where the dual-system structure is given an alternative topological condition.

Let X and W be non-empty compact Hausdorff spaces having convex structures. Suppose dual-system structure $(X, W, V : X \times W \to 2^X)$ satisfies the following condition (V4*) with (V1), (V2), and (V3):

(V4*) If $V(x,w) \neq \emptyset$, for all $y \in V(x,w)$ there is open neighborhood U_x of $x \in X$ such that $y \in V(z,w)$ for all $z \in U_x$.

That is, for each $w \in W$, correspondence $V(\cdot, w) : X \to X$ has the open lowersection property. Since (V4) is a condition based on the local-intersection property, (V4*) is not a generalization of (V4). Condition (V4), however, assumes a sort of joint continuity for two variables of V, which is replaced by a partial continuity in (V4*). (Compare the next theorem with Theorem 2.3.4.)

THEOREM 3.1.4: (Fixed-Point Theorem: Dual-System (V4*) (K1)) *Let X and W be non-empty compact Hausdorff spaces having convex structures, and let φ be a non-empty valued correspondence on X to X such that the following condition, $(K1)$, is satisfied under a dual-system structure satisfying $(V1)$, $(V2)$, $(V3)$, and $(V4*)$.*

(K1) *For each x such that $x \in X\backslash\boldsymbol{\mathcal{F}ix}(\varphi)$, element $w \in W$ and open neighborhood U of x exist such that for all $z \in U \cap (X\backslash\boldsymbol{\mathcal{F}ix}(\varphi))$, we have $\varphi(z) \subset V(z,w)$.*

Then φ has a fixed point.

PROOF: Suppose that φ has no fixed point. Since $X\backslash\boldsymbol{\mathcal{F}ix}(\varphi) = X$, there is a finite open covering $U^1, \ldots, U^m$ of X, points $w^1, \ldots, w^m \in W$ such that for each $t = 1, \ldots, m$ and for each $z \in U^t$, we have $\varphi(z) \subset V(z, w^t)$. By Theorem 3.1.3, we also have refinement $\{V^1, \ldots, V^m\}$ of $\{U^1, \ldots, U^m\}$ such that closure $\bar{V}^t$ is included in U^t for each $t = 1, \ldots, m$. Let $\bar{N}(x) = \{t \mid x \in \bar{V}^t\} \subset \{1, \ldots, m\}$ and $\hat{N}(x) = \{t \mid x \in U^t\} \subset \{1, \ldots, m\}$. For each $x \in X$, define $\Psi(x) \subset X$ as $\Psi(x) = \bigcap_{t \in \bar{N}(x)} V(x, w^t) \supset \varphi(x)$. Then $\Psi : X \to X$ is a non-empty convex valued correspondence having the local-intersection property. Indeed, if $x \in X$ is not a boundary point of any V^t, $t = 1, \ldots, m$, the assertion is obvious since $\bar{N}(\cdot)$ is locally fixed near x and V satisfies (V4*), the open-lowersection property. If $x \in X$ is a boundary point of some V^t's, we can choose neighborhood U of x as a subset of $\bigcap_{t \in \hat{N}(x)} U^t$ since $\bar{V}^t \subset U^t$ for all $t = 1, \ldots, m$. Then for each $z \in U$ and $t \in \hat{N}(x) \supset \bar{N}(x)$ we have $z \in U^t$, so $\varphi(z) \subset \bigcap_{t \in \hat{N}(x)} V(z, w^t) \subset \bigcap_{t \in \bar{N}(x)} V(z, w^t)$. It follows that with respect to an element $y \in \varphi(x) \subset \bigcap_{t \in \hat{N}(x)} V(z, w^t)$, open set U with the open lowersection property of V ensures the local-intersection property of $\Psi(z) = \bigcap_{t \in \bar{N}(x)} V(z, w^t)$. Hence, Ψ has a fixed point by Theorem 2.3.1, which contradicts condition (V2). ∎

This theorem gives a unified viewpoint on Browder- and Kakutani-type fixed-point arguments in a more general way than similar discussions after Theorem 2.3.4 in the previous chapter. Needless to say, condition (V4*) is automatically satisfied in every topological dual-system for a topological vector space. By appealing to the second separation theorem, the theorem generalizes Kakutani–Fan–Glicksberg's fixed-point theorem. On the other hand, a convex valued correspondence having an open-lowersection property, $\varphi : X \to X$, induces a dual-space structure satisfying (V4*), V_φ, such that $V_\varphi(x,y) = \varphi(x)$ for $y \in \varphi(x)$ and $V_\varphi(x,y) = \emptyset$ for $y \notin \varphi(x)$. Hence, the theorem includes typical cases of both Kakutani–Fan–Glicksberg's and Browder's theorems. Note also that Theorem 3.1.4 gives an alternative proof for Theorem 2.3.5 (p. 53, Footnote 7).

We now treat general individual preferences (better-set correspondences) and some important kinds of product correspondences that are directly related to the conditions prepared for general fixed-point theorems seen in Chapter 2 and in this section. In the proof of Theorem 3.1.2, the maximal-element existence problem for preferences (or the empty-value existence problem for better-set correspondences) is converted into a fixed-point problem. Roughly speaking, since a preference relation for i is naturally assumed to be irreflexive with respect to the i-th coordinate (in other words, better-set correspondence never has a fixed point with respect to the i-th coordinate), we may interpret a fixed-point theorem for non-empty valued correspondences as a theorem maintaining the existence of an empty value.

Consider a class of mappings $\{\varphi_i | i \in I\}$, where I is an arbitrary index set, and for each $i \in I$, φ_i is a correspondence on $X = \prod_{j \in I} X_j$ to a non-empty compact convex subset X_i of a Hausdorff topological space having convex structure. Define correspondence $\varphi : X \to X$ as $\varphi(x) = \prod_{i \in I} \varphi_i(x)$. Call $\varphi : X \ni x \mapsto \prod_{i \in I} \varphi_i(x) \subset X$ the *product correspondence* of family $\{\varphi_i | i \in I\}$ and each φ_i the *i-th coordinate correspondence* of φ. In the following, suppose that setting X_i, $i \in I$ constructs part of a non-cooperative game structure. Each coordinate correspondence φ_i will also be recognized as a necessary structure to specify the conditions for individual choices like the better-set correspondence of i or the constraint correspondence for i in non-cooperative games or abstract economy settings. For each strategy profile $x = (x_j)_{j \in I} \in X$, $\varphi_i(x) \subset X_i$ may be interpreted as the set of actions for i that are more attractive than x_i or merely available under $(x_j)_{j \in I}$. For the former interpretation, it is natural to assume for each i and $x = (x_j) = j \in I \in X$, that $x_i \notin \varphi_i(x)$, and for the latter, it is appropriate to assume that φ_i is non-empty valued for each i.

For the maximal-element existence, we have already seen the case in which each φ_i satisfies conditions (P1), (P2), and (P3). The discussion is easily extended or modified into the following cases of general preferences. (In the following list of conditions, for a point $x \in X = \prod_{i \in I} X_i$, notation x_i means the i-th coordinate of x. Given product correspondence $\varphi = \prod_{i \in I} \varphi_i$, denote by $\mathcal{E}q_i(\varphi)$ the set $\{x \in X | \varphi_i(x) = \emptyset \text{ or } x_i \in \varphi_i(x)\}$ for each $i \in I$.)

(Product K*) Convex valued extension Ψ_i of φ_i satisfies the following: (1) for each $x \in X \backslash \mathcal{E}q_i(\varphi)$, $x_i \notin \Psi_i(x)$, and (2) Ψ_i has the local-intersection property on $X \backslash \mathcal{E}q_i(\varphi)$. (This is an extension of (K*) in Chapter 2, including "(P1)–(P3)," "convex valued mappings having the local intersection property," and "convex valued mappings having open graphs.")

(Product B) For each open set $U \supset \mathcal{E}q_i(\varphi)$, there are finite open sets $U^1, \ldots, U^n$ of X such that (1) $\{U, U^1, \ldots, U^n\}$ covers X, (2) on $\bigcup_{t=1}^n U^t$, there is a convex valued extension Ψ_i of φ_i satisfying $x_i \notin \Psi_i(x)$ for each $x \in \bigcup_{t=1}^n U^t$, and (3) $\bigcap_{x \in U^t} \Psi_i(x) \neq \emptyset$ for each $t = 1, \ldots, n$. (This includes the condition (Browder-Type mappings) in Chapter 2 as a special case $I = \{i\}$.)

(Product K) For each open set $U \supset \mathcal{E}q_i(\varphi)$, there are finite open sets $U^1, \ldots, U^n$ of X with closed convex sets $K_1, \ldots, K_n$ such that (1) $\{U, U^1, \ldots, U^n\}$ covers X, (2) $U^t \cap K_t = \emptyset$ for each $t = 1, \ldots, n$, and (3) for all $z \in U^t$, $\varphi_i(z) \subset K_t$ for each $t = 1, \ldots, n$. (This is a direct extension of condition (Kakutani-Type mappings) in Chapter 2, including "closed convex valued upper semicontinuous correspondences under locally convex topological vector space structures.")

(Product K1–V4) There dual-system structure $(X_i, W_i, V_i : X_i \times W_i \to X_i)$ satisfies (V4). For each $x \in X \setminus \mathcal{E}q_i(\varphi)$, element $w_i \in W_i$ and open neighborhood U of x exist such that for all $z = (z_j)_{j \in I} \in U$, $\varphi_i(z) \subset V_i(z_i, w_i)$. (This is an extension of condition (K1) in Chapter 2, including "closed convex valued upper semicontinuous correspondences under locally convex topological vector space structures.")

(Product K1–V4*) There dual-system structure $(X_i, W_i, V_i : X_i \times W_i \to X_i)$ satisfies (V4*). For each $x \in X \setminus \mathcal{E}q_i(\varphi)$, element $w_i \in W_i$ and open neighborhood U of x exist such that for all $z = (z_j)_{j \in I} \in U$, $\varphi_i(z) \subset V_i(z_i, w_i)$. (This is an alternative extension of condition (K1) in Chapter 2, including "closed convex valued upper semicontinuous correspondences under locally convex topological vector space structures.")

Notations K*, B, K, K1–V4, and K1–V4* represent that they are closely related to those conditions in fixed-point theorems, Theorems 2.3.1 (K*), 2.3.3 (Browder-Type), 2.3.5 (Kakutani-Type), 2.3.4 (K1)–(V4), and 3.1.4 (K1)–(V4*), respectively. We may classify these conditions into the following three (wide-sense) categories:

(1) **Browder-type:** Conditions (P1)–(P3), (Product K*), and (Product B) for product mappings, and (K*) and (Browder-Type) with or without vector space structures in Chapter 2.

(2) **Kakutani-type:** Conditions (Product K) for product mappings and (Kakutani-Type) in Chapter 2.
(3) Others: **K1-type** conditions like (Product K1–V4) and (Product K1-V4*) for product mappings and (K1) with standard or generalized dual-system structures in Chapter 2, and other conditions like (USC differentiable) in Section 3.3 and (K#) in Chapter 2.

Now we prove the theorem of maximal-element existence based on these conditions.

THEOREM 3.1.5: (Maximal-Element Existence) *For each $i \in I$, let X_i be a compact Hausdorff space and $\varphi_i : X = \prod_{i \in I} \to X_i$ be a correspondence satisfying one of the above conditions for product correspondence, K*, B, K, K1–V4, and K1–V4*. Then for any $x \in X = \prod_{j \in I} X_j$ and $i \in I$, element $x_i^* \in X_i$ of X_i exists such that $(x_i^*; x_{\hat{i}}) \in \mathcal{E}q_i(\varphi)$.*

PROOF: If no $x_i^* \in X_i$ exists such that $(x_i^*; x_{\hat{i}}) \in \mathcal{E}q_i(\varphi)$, mapping $\varphi_i(\cdot, x_{\hat{i}}) : X_i \to X_i$ is a non-empty valued correspondence having no fixed point. By identifying X_i with compact subset $\{y = (y_j)_{j \in I} \in X \mid y_j = x_j$ for all $j \neq i \}$ of X, however, conditions (Product K*), (Product B), (Product K), (Product K1–V4), and (Product K1–V4*) imply conditions (K*), (Browder-Type), (Kakutani-Type), (K1) under (V4), and (K1) under (V4*) in Theorems 2.3.1, 2.3.3, 2.3.5, 2.3.4, and 3.1.4, respectively. Hence, $\varphi_i(\cdot, x_{\hat{i}}) : X_i \to X_i$ has a fixed point, and we have a contradiction. ∎

If φ_i is interpreted as a better-set correspondence (i.e., it is natural to assume that $x_i^* \in \varphi_i(x_i^*; x_{\hat{i}})$ never occurs), the above theorem warrants the existence of maximal elements for i under x. Condition (Product K*) generalizes almost all kinds of ordinary assumptions for preferences, e.g., "(P1)–(P3)," "to be convex valued and to have the local-intersection (or open-lower-section) property" and "to be convex valued and to have an open graph." Condition (Product B) is written in this general form as an extension of the (Browder-type mapping) condition discussed in Chapter 2, Section 2.3. Hence, with respect to condition (Product B), the compactness for X_i is superfluous. When each X_i is compact and $\mathcal{E}q_i(\varphi)$ is closed, (Product K*) implies (Product B). Condition (Product K) is also intended to be an extension of the (Kakutani-type mapping) condition in Chapter 2, Section 2.3, so it includes all closed convex valued upper semicontinuous correspondences when each X_i is compact and has a locally convex topological vector space structure. Both conditions (Product K) and (Product K1–V4) are more suitable for conditions like constraint correspondences than better-set correspondences. They as well as (Product K1–V4*) are generalized concepts of closed convex valued upper semicontinuity. Any better-set correspondences obtained from differentiable utility functions fail to satisfy conditions (Product K) and (Product K1–V4) unless their gradient vectors are locally constant. To utilize the generality of (Product K) and (Product K1–V4) for better-set correspondences, therefore, we must consider non-ordered preferences such as non-transitive preferences with thick indifferent classes at every point. Condition (Product K1–V4*) can also be used as a general condition for better-set correspondence if we consider the generalized dual-system structure induced by $\varphi_i(\cdot\,;x_{\hat{\imath}})$ itself (recall the unified viewpoint discussed immediately after Theorem 3.1.4).

3.2 Nash Equilibrium

Once a non-cooperative game $(X_i, P_i)_{i\in I}$ is given with correspondence $P\colon \prod_{i\in I} X_i \ni x \mapsto \prod_{i\in I} P_i(x) \subset \prod_{i\in I} X_i$ on product set $X = \prod_{i\in I} X_i$ to itself, we can characterize the Nash-equilibrium of the game with the fixed point of product correspondence P. As in the previous section, let us denote by $\mathcal{E}q_i(P)$ the set $\{x \in X\,|\, x_i \in P_i(x) \text{ or } P_i(x) = \emptyset\}$ for each $i \in I$, where by x_i we ordinarily represent the i-th coordinate of $x \in \prod_{i\in I} X_i$. (This notational convention is used only when X has product setting $X = \prod_{i\in I} X_i$ and there is no fear of confusion.) Let $\mathcal{E}q(P)$ be set $\bigcap_{i\in I} \mathcal{E}q_i(P)$. If for each P_i and $x \in X$, $x_i \in P_i(x)$ is impossible, then $x \in \mathcal{E}q(P)$ satisfies $P_i(x) = \emptyset$ for all $i \in I$, so $\mathcal{E}q(P)$ is the set of Nash equilibrium points. On the other hand, if for each P_i and $x \in X$, $P_i(x) \neq \emptyset$, then $x \in \mathcal{E}q(P)$ is the point such that $x_i \in P_i(x)$ for all $i \in I$, i.e., $x \in P(x)$, a fixed point of correspondence P.

In the previous section, we extended the notion of Browder-, Kakutani-, and K1-type conditions to cases with the correspondences defined on product space $X = \prod_{i\in I} X_i$. Now we utilize them as conditions for ensuring the existence of the fixed points of P and Nash equilibrium points for game $(X_i, P_i)_{i\in I}$. Since the results in this section can also be extended or interpreted as necessary conditions on constraint correspondences as well as better-set correspondences for the generalized Nash equilibrium existence treated in the next section, we use notations φ and φ_i instead of P and P_i for the correspondences in the theorems.

Throughout this section, we assume that each strategy set X_i is a compact Hausdorff space. Adding to the basic (Browder-, Kakutani-, and K1-types) conditions for correspondences, as a special assumption for such product-space (multi-agent) settings, we consider the next *closedness condition* on coordinate correspondence $\varphi_i : X \to X_i$ for each $i \in I$.

(Closedness) For each $x \in X$ such that $\varphi_i(x) \neq \emptyset$ and $x_i \notin \varphi_i(x)$, open neighborhood $U(x)$ of x exists such that for all $z \in U(x)$, $\varphi_i(z) \neq \emptyset$ and $z_i \notin \varphi_i(z)$, i.e., $\mathcal{E}q_i(\varphi)$ is closed in X.

The condition assures that if $x \in X$ is not a fixed point of φ (Nash-equilibrium for $(X_i, \varphi_i)_{i \in I}$) at least one coordinate $i \in I$ assures that all z near x are not fixed points (equilibrium points). We use the condition (Closedness) mainly for the technical reason that the non-equilibrium situation for multi-agent settings can be reduced to a problem of one agent at least locally, so it enables us to relate assumptions on coordinate correspondences to those on $\varphi : X \to X$ and to utilize the fixed-point theorems on X seen in the previous chapter and the previous section in this chapter. Note, however, that the condition also has a natural economic interpretation as a *minimal requirement for the continuity* of preferences and/or constraint correspondences. This condition is automatically satisfied when the open-lowersection property or the closed graph condition for each coordinate correspondence is assumed. Moreover, if non-empty valuedness for each φ_i can naturally be supposed (e.g., when each φ_i is interpreted as constraint correspondences under abstract economy settings), conditions (Product K), (Product K1–V4*), and (Product K1–V4) imply condition (Closedness). Condition (Closedness) is, therefore, quite a weak condition for constructing economic arguments. We will use it (in a minimum level) to obtain a unified perspective on several conditions.

With (Closedness), the conditions we mainly use for product correspondences are (Product B), (Product K), (Product K1–V4), and (Product K1–V4*). Conditions (P1)–(P3) and (Product K*) can be treated as special cases of (Product B) under (Closedness) for each i and the compactness of X. Conditions (Product B) and (Product K) are generalizations of the products of convex valued mappings with open and closed graphs, respectively. (Product K1–V4) is a dual-space characterization of mappings having fixed points including both open- and closed-graph mappings, and (Product K1–V4*) is an alternative dual-space characterization based on the open-lowersection property (stronger than the local intersection property) with a partial continuity condition (weaker than the joint continuity condition) on the duality structure. Now, let us prove the four

Nash-equilibrium existence theorems (fixed-point theorems for products of correspondences).

THEOREM 3.2.1: (**Nash-Equilibrium or Fixed-Point Existence for Products of Browder-Type Mappings**) *Let I be a set. For each $i \in I$, let X_i be a non-empty compact convex subset of Hausdorff space E having convex structure and let φ_i be a correspondence on $X = \prod_{i \in I} X_i$ to X_i. Define correspondence $\varphi : X \to X$ as $\varphi(x) = \prod_{i \in I} \varphi_i(x)$ for each $x \in X$. Then, $\mathcal{E}q(\varphi) \neq \emptyset$ if each φ_i satisfies (Product B) and (Closedness). (Especially as stated above, $\mathcal{E}q(\varphi) \neq \emptyset$ if each φ_i satisfies "(Product K*) and (Closedness)" or each φ_i satisfies "(P1)–(P3)."*)

PROOF: Suppose that $\mathcal{E}q(\varphi) = \emptyset$. Then for each $x = (x_i)_{i \in I} \in X$, at least one $i \in I$ exists such that $x \notin \mathcal{E}q_i(\varphi)$. Since every $\mathcal{E}q_i(\varphi)$ is closed and since X is compact and Hausdorff (i.e., X is normal), two open sets $U(x) \ni x$ and $V(x) \supset \mathcal{E}q_i(\varphi)$ separate x and $\mathcal{E}q_i(\varphi)$. Then, by the compactness of X, the covering, $\{U(x) | x \in X\}$ of X has a finite subcovering $\{U(x^1), \ldots, U(x^m)\}$ with agents $j^1, \ldots, j^m$ such that $V(x^t) \supset \mathcal{E}q_{j^t}(\varphi)$ for each $t = 1, \ldots, m$. For each $i \in I$ such that $j^t = i$ for some $t = 1, \ldots, m$, let U_i^0 be set $\bigcap_{i=j^t} V(x^t) \supset \mathcal{E}q_i(\varphi)$. Then under (Product B) for each $i \in I$, we can obtain a finite refinement of covering $\{U(x^1), \ldots, U(x^m)\}$, $\{U^1, \ldots, U^k\}$, and a finite sequence of indices $i^1, \ldots, i^k$ with points $y_{i^1}^1 \in X_{i^1}, \ldots, y_{i^k}^k \in X_{i^k}$ such that for each $t = 1, \ldots, k$ and for each $z = (z_i)_{i \in I} \in U^t$, $z_{i^t} \notin \Psi_{i^t}(z)$, $\varphi_{i^t}(z) \subset \Psi_{i^t}(z)$, and $y_{i^t}^t \in \Psi_{i^t}(z)$, where $\Psi_1, \ldots, \Psi_k$ are correspondences given in Condition (Product B). By Theorem 3.1.3, open covering $\{V^1, \ldots, V^k\}$ of X exists such that for each $t = 1, \ldots, k$ closure $\bar{V}^t$ is included in U^t.

Let J be set $\{i^1, \ldots, i^k\}$. For each $x \in X$, let $J(x)$ be set $\{i^t | x \in V^t\} \subset J$ and let $N(x) = \{t | x \in V^t\} \subset \{1, \ldots, k\}$. Denote by Ψ the convex valued correspondence defined as $\Psi(x) = \prod_{i \in J(x)} \Psi_i(x) \times \prod_{i \in I, i \notin J(x)} X_i$, where convex structure $(\{f_A | A \in \mathscr{F}(X)\}, C)$ on X is defined by using the convex structures on coordinate spaces, $(\{f_A^i | A \in \mathscr{F}(X_i)\}, C^i)$, $i \in I$, as $f_A(x) = (f_{\text{pr}_i(A)}^i(x_i))_{i \in I}$ and $C(A) = \prod_{i \in J} C_i(A) \times \prod_{i \in I, i \notin J} X_i$. (One can verify that C satisfies condition (C3) by defining $\bar{A}$ for each $A \in \mathscr{F}(X)$ as $\bar{A} = \{x \in X | \exists a \in A, \forall i \in I \setminus J, x_i = a_i, \forall j \in J, \exists b_j \in \overline{\text{pr}_j A}, x_j = b_j\}$.[4]) Then, at each $x \in X$, Ψ has a local intersection property on X. Indeed, if x is not a boundary of any V^t, $t = 1, \ldots, k$, set $J(z)$ is locally fixed

---

[4] It is appropriate to call convex structure $(\{f_A | A \in \mathscr{F}(X)\}, C)$ on $X = \prod_{i \in I} X_i$ the *product convex structure* given by convex structures of its component spaces, $X_j, j \in J$.

near x, so the local intersection property of Ψ_i for each $i \in J(x)$ is clearly sufficient to ensure the same property of Ψ by the definition of Ψ. If x is a boundary of some V^t's, let $\bar{N}(x) \subset \{1, \ldots, k\}$ be set $\{t|\, x \in \bar{V}^t\} \supset N(x)$ and $\bar{J}(x) \subset J$ be set $\{i^t|\, x \in \bar{V}^t\} \supset J(x)$. Since $\bar{V}^t \subset U^t$ for all $t = 1, \ldots, k$, we may choose open neighborhood $U(x)$ of x such that $U(x) \subset U^t$ for all $t \in \bar{N}(x)$. Define $y(x) = (y_j(x))_{j \in I} \in X$ by letting $y_j(x)$ be $y_{i^m}^m$ for a certain $m \in \bar{N}(x)$ such that $i^m = j$ for $j \in \bar{J}(x)$, and $y_j(x)$ be an arbitrary element of $X_j(x)$ for $j \notin \bar{J}(x)$. Then, for all $z \in U(x)$, we have $y_i(z) \in \Psi_i(z)$ for all $i \in I$, so Ψ has a local intersection property at x on X. It follows that $\Psi : X \to X$ is a non-empty convex valued correspondence having a local intersection property at each $x \in X \backslash \mathcal{F}ix(\Psi)$, where $\mathcal{F}ix(\Psi) = \emptyset$. This contradicts Theorem 2.3.1. ∎

In the theorem, if we add an assumption that $x_i \notin \varphi_i(x)$ for each $x \in X$ and $i \in I$ (a natural assumption for better-set correspondences), then $x \in \mathcal{E}q(\varphi)$ may be considered a Nash equilibrium for game $(X_i, \varphi_i)_{i \in I}$. On the other hand, if we add an assumption that $\varphi_i(x) \neq \emptyset$ for each $x \in X$ and $i \in I$ (this is also a standard situation for constraint correspondences), then $x \in \mathcal{E}q(\varphi)$ is a fixed point of product mapping $\varphi : X \ni x \mapsto \prod_{i \in I} \varphi_i(x) \subset X$. The fixed points of the product of constraint correspondences, $\varphi : X \ni x \mapsto \prod_{i \in I} \varphi_i(x) \subset X$, under the game-theoretic (multi-agent abstract economy) setting, $(X_i, P_i, \varphi_i)_{i \in I}$, may also be interpreted as a feasible strategy profile allowing for *externality* among agent's actions (strategies). Such interpretations for coordinate correspondences (as better-set or constraint correspondences) and elements of $\mathcal{E}q(\varphi)$ are also valid for all of the other three cases of conditions for correspondences (Product K), (Product K1–V4), and (Product K1–V4*).

THEOREM 3.2.2: (Nash-Equilibrium or Fixed-Point Existence for Product of Kakutani-Type Mappings) *Let I be a set. For each $i \in I$, let X_i be a non-empty compact convex subset of Hausdorff space E having convex structure and let φ_i be a correspondence on $X = \prod_{i \in I} X_i$ to X_i. Define correspondence $\varphi : X \to X$ as $\varphi(x) = \prod_{i \in I} \varphi_i(x)$ for each $x \in X$. Then, $\mathcal{E}q(\varphi) \neq \emptyset$ if each φ_i satisfies (Product K) and (Closedness). (Especially when E is an ordinary locally convex space, $\mathcal{E}q(\varphi) \neq \emptyset$ if each φ_i is a closed convex valued upper semicontinuous correspondence or, equivalently, a convex valued correspondence having a closed graph. If each φ_i is non-empty valued, condition (Closedness) is automatically satisfied, so it may be reduced.)*

3: Nash Equilibrium and Abstract Economy

PROOF: Assume that $\mathcal{E}q(\varphi) = \emptyset$. Then for each $x = (x_i)_{i \in I} \in X$, at least one $i \in I$ exists such that $x \notin \mathcal{E}q_i(\varphi)$. Since every $\mathcal{E}q_i(\varphi)$ is closed and since X is normal, two disjoint open sets, $U(x) \ni x$ and $V(x) \supset \mathcal{E}q_i(\varphi)$, separate x and $\mathcal{E}q_i(\varphi)$. By the compactness of X, covering $\{U(x) | x \in X\}$ has finite subcovering $\{U(x^1), \ldots, U(x^m)\}$ with agents $j^1, \ldots, j^m$ such that $V(x^t) \supset \mathcal{E}q_{j^t}(\varphi)$ for each $t = 1, \ldots, m$. For each $i \in I$ such that $j^t = i$ for some $t = 1, \ldots, m$, let U_i^0 be set $\bigcap_{i=j^t} V(x^t) \supset \mathcal{E}q_i(\varphi)$. Then under (Product K) for each $i \in I$, we obtain a finite refinement of the above covering $\{U_i^0, U^1, \ldots, U^k\}$ with a finite sequence of indices $i^1, \ldots, i^k \in I$ and closed sets $K_{i^1}^1, \ldots, K_{i^k}^k$ in $X_{i^1}, \ldots, X_{i^k}$, respectively, satisfying the properties stated in condition (Product K). Moreover, we have by Theorem 3.1.3 open covering $\{V^1, \ldots, V^k\}$ of X such that for each $t = 1, \ldots, k$, closure $\bar{V}^t$ is included in U^t.

Let J be set $\{i^1, \ldots, i^k\}$ and, for each $x \in X$, $J(x) = \{i^n | x \in V^n\} \subset I$, $N(x) = \{n | x \in V^n\} \subset \{1, \ldots, k\}$, and $\bar{N}(x) = \{n | x \in \bar{V}^n\} \subset \{1, \ldots, k\}$. Define for each $x \in X$, closed set K_x in X as $K_x = \bigcap_{t \in \bar{N}(x)} \left(K_{i^t}^t \times \prod_{i \in I, i \neq i^t} X_i \right)$, which may also be considered a convex subset of X under the same product convex structure on X given by coordinates in J defined in the proof of the previous theorem (p. 69, Footnote 4). Then for each $x \in X$, we can define open neighborhood U_x of x such that $U_x \cap K_x = \emptyset$ and for each $z \in U_x$, $\varphi(z) \subset K_x$. Indeed, if x is not a boundary point of any V^t, $t = 1, \ldots, k$, $\bar{N}(z) = N(z)$ is locally fixed near x, so condition (Product K) for every coordinate correspondence is sufficient to ensure that $\varphi(z) \subset K_x$ for all z near x. If x is a boundary point of some V^t's, take open neighborhood U_x of x as a subset of $\bigcap_{t \in \bar{N}(x)} U^t$. This is possible since $\bar{V}^t \subset U^t$ for each $t \in \{1, \ldots, k\}$. The condition (Product K) for each coordinate correspondence at x implies that $\varphi_i(z) \subset K_{i^t}$ for each $t \in \bar{N}(x)$ such that $i^t = i$ and $z \in U_x$, so $\varphi(z) \subset K_x$ for all $z \in U_x$. Consequently, by the compactness of X, $\varphi : X \to X$ is shown to be a non-empty valued correspondence of the Kakutani type. Theorem 2.3.5 shows that φ has a fixed point, so we have $\mathcal{E}q(\varphi) \neq \emptyset$: a contradiction. ∎

Let us consider the theorems based on condition (K1). We consider first a theorem based on dual-system condition (V4*) and then a theorem based on (V4). Even for these two cases, condition (Closedness) is necessary to assure the existence of Nash equilibria. With respect to the fixed-point interpretation, however, (Closedness) can be dropped if each φ_i is non-empty valued since in such a case both (Product K1–V4*) and

(Product K1–V4) automatically assure the closedness of each $\mathcal{E}q_i(\varphi)$ under condition (V2) of the dual-system structure.

THEOREM 3.2.3: (Nash-Equilibrium or Fixed-Point Existence: Product K1–V4*) *Let I be a set. For each $i \in I$, let X_i and W_i be non-empty compact Hausdorff spaces having convex structures and dual-system structure $V_i : X_i \times W_i \to X_i$ on them, and let φ_i be a correspondence on $X = \prod_{i \in I} X_i$ to X_i. Define correspondence $\varphi : X \to X$ as $\varphi(x) = \prod_{i \in I} \varphi_i(x)$ for each $x \in X$. Then, $\mathcal{E}q(\varphi) \neq \emptyset$ if each φ_i satisfies (Product K1-V4*) and (Closedness). (Especially when E is an ordinary locally convex space with duality (E, E'), $\mathcal{E}q(\varphi) \neq \emptyset$ if each φ_i is a closed convex valued upper semicontinuous correspondence or, equivalently, a convex valued correspondence with closed graph. If each φ_i is non-empty valued, (Closedness) is automatically satisfied, so it is not necessary.)*

PROOF: Assume that $\mathcal{E}q(\varphi) = \emptyset$. Then for each $x = (x_i)_{i \in I} \in X$, at least one $i \in I$ exists such that $x \notin \mathcal{E}q_i(\varphi)$. Since every $\mathcal{E}q_i(\varphi)$ is closed and since X is compact, under (Product K1–V4*), we have finite points $x^1, \ldots, x^k$ of X, open covering $\{U(x^1), \ldots, U(x^k)\}$ of X, finite sequence of indices $i^1, \ldots, i^k \in I$, and points in the dual space, $w_{i^1}^1, \ldots, w_{i^k}^k$ in $W_{i^1}, \ldots, W_{i^k}$, respectively, satisfying for each coordinate $i = i^1, \ldots, i^k$, the properties stated in (Product K1–V4*). Moreover, we have by Theorem 3.1.3 open covering $\{V(x^1), \ldots, V(x^k)\}$ of X such that for each $t = 1, \ldots, k$, $x^t \in V(x^t)$ and closure $\bar{V}(x^t)$ is included in $U(x^t)$. Let J be set $\{i^1, \ldots, i^k\}$ and, for each $x \in X$, $J(x) = \{i^m \mid x \in V(x^m)\} \subset I$, $N(x) = \{n \mid x \in V(x^n)\} \subset \{1, \ldots, k\}$, $\bar{N}(x) = \{n \mid x \in \bar{V}(x^n)\} \subset \{1, \ldots, k\}$, and $\hat{N}(x) = \{n \mid x \in U(x^n)\} \subset \{1, \ldots, k\}$.

For each $x \in X$, let $\Psi(x)$ be set $\bigcap_{t \in \bar{N}(x)} (V_{i^t}(x_{i^t}, w_{i^t}^t) \times \prod_{i \in I, i \neq i^t} X_i)$. $\Psi(x)$ may be considered a convex subset of X under the same product convex structure on X given by the coordinates in J defined in the proof of the Browder-type Theorem 3.2.1. By considering condition $\bar{V}(x^t) \subset U(x^t)$ for each $t = 1, \ldots, k$, we can verify that $\Psi(x) \supset \varphi(x) \neq \emptyset$ for each $x \in X$ and Ψ has the local intersection property. (Indeed, at each $x \in X$, every point $y \in \bigcap_{t \in \hat{N}(x)} (V_{i^t}(x_{i^t}, w_{i^t}^t) \times \prod_{i \in I, i \neq i^t} X_i) \subset \bigcap_{t \in \bar{N}(x)} (V_{i^t}(x_{i^t}, w_{i^t}^t) \times \prod_{i \in I, i \neq i^t} X_i) = \Psi(x)$ belongs to $\Psi(z)$ for all z near x, i.e., for all $z \in \bigcap_{t \in \hat{N}(x)} U(x^t)$.) Then, by Theorem 2.3.1, Ψ has a fixed point, which contradicts condition (V2). ∎

3: Nash Equilibrium and Abstract Economy

THEOREM 3.2.4: (Nash-Equilibrium or Fixed-Point Existence: Product K1–V4) *Let I be a set. For each $i \in I$, let X_i and W_i be non-empty compact Hausdorff spaces having convex structures and dual-system structure $V_i : X_i \times W_i \to X_i$ on them, and let φ_i be a correspondence on $X = \prod_{i \in I} X_i$ to X_i. Define correspondence $\varphi : X \to X$ as $\varphi(x) = \prod_{i \in I} \varphi_i(x)$ for each $x \in X$. Then $\mathcal{E}q(\varphi) \neq \emptyset$ if each φ_i satisfies (Product K1-V4) and (Closedness). (Especially when E is an ordinary locally convex space with duality (E, E'), $\mathcal{E}q(\varphi) \neq \emptyset$ if each φ_i is a closed convex valued upper semicontinuous correspondence or, equivalently, a convex valued correspondence having closed graph. If each φ_i is non-empty valued, condition (Closedness) is automatically satisfied under (Product K1-V4), hence we may reduce it.)*

PROOF: Assume that $\mathcal{E}q(\varphi) = \emptyset$. Then by (Closedness) for each $x \in X$ open set $U(x) \ni x$ exists such that for all $z \in U(x)$, $z \notin \mathcal{E}q_i(\varphi)$. Since X is compact, there are finite points $x^1, \ldots, x^k$, open covering $\{U(x^1), \ldots, U(x^k)\}$ of X, finite sequence of indices $i^1, \ldots, i^k \in I$, and points in the dual space, $w_{i^1}^1, \ldots, w_{i^k}^k$ in $W_{i^1}, \ldots, W_{i^k}$, respectively, satisfying the condition stated in the (Product K1-V4). Again, by Theorem 3.1.3, we have open covering $\{V(x^1), \ldots, V(x^k)\}$ of X such that for each $t = 1, \ldots, k$, $x^t \in V(x^t)$ and closure $\bar{V}(x^t)$ is included in $U(x^t)$. Let $\beta_1 : X \to [0,1], \ldots, \beta_k : X \to [0,1]$ be the partition of unity subordinate to $V(x^1), \ldots, V(x^k)$. Without loss of generality (if necessary by redefining all $V(x^t)$s) we may assume that $\beta^t(x) > 0$ iff $x \in V(x^t)$ for every $t = 1, \ldots, k$. Let J be finite set $\{i^1, \ldots, i^k\} \subset I$ and, for each $x \in X$, $J(x) = \{i^t | x \in V(x^t)\} \subset J$, $\bar{J}(x) = \{i^t | x \in \bar{V}(x^t)\} \subset J$, $\hat{J}(x) = \{i^t | x \in U(x^t)\} \subset J$, $N(x) = \{t | x \in V(x^t)\} \subset \{1, \ldots, k\}$, $\bar{N}(x) = \{t | x \in \bar{V}(x^t)\} \subset \{1, \ldots, k\}$, and $\hat{N}(x) = \{t | x \in U(x^t)\} \subset \{1, \ldots, k\}$.

Define for each $x \in X$ and $i \in \bar{J}(x)$, point $w_i(x) \in X_i$ as

$$w_i(x) = f_{\bar{B}_i}\left(\sum_{t \in \bar{N}(x), i^t = i} \frac{\beta^t(x)}{\sum_{s \in \bar{N}(x), i^s = i} \beta^s(x)} e_{i^t}^t \right),$$

where $B_i = \{w_{i^t}^t | i^t = i\}$, $\bar{B}_i$ and $f_{\bar{B}_i}$ are determined under the convex structure on W_i, and $e_{i^t}^t$ is the vertex of $\Delta^{B_i} \subset \Delta^{\bar{B}_i}$ corresponding to $w_{i^t}^t$ for each t. For each $x \in X$ and $i \in I$, define set $\Psi_i(x) \subset X_i$ as

$$\Psi_i(x) = \begin{cases} V_i(x_i, w_i(x)) & \text{if } i \in \bar{J}(x), \\ X_i & \text{otherwise.} \end{cases}$$

Moreover, for each $x \in X$, define $\Psi(x) \subset X$ as

$$\Psi(x) = \prod_{i \in I} \Psi_i(x).$$

For each $x \in X$ and $i \in I$, $\Psi_i(x)$ is a convex subset of X_i, so $\Psi(x)$ may be considered a convex subset of X under the same product convex structure on X given by coordinates in J defined in the proof of Theorem 3.2.1. By considering that $\bar{V}(x^t) \subset U(x^t)$ for each $t = 1, \ldots, k$, we can verify under condition (V3) that $\Psi_i(x) \supset \varphi_i(x) \neq \emptyset$ for each $x \in X$ and $i \in I$. Furthermore, we can see that Ψ has the local-intersection property. Indeed, for each $x \in X$, if $i \notin \bar{J}(x)$, then $\Psi_i(z) = X_i$ for all z near x. If $i \in \bar{J}(x)$, then since $\hat{N}(z)$ is locally fixed for all z near x (and since $\beta^t(z) = 0$ as long as $z \notin \bar{V}(x^t)$), we have for all z near x that

$$\sum_{t \in \bar{N}(z), i^t = i} \frac{\beta^t(z)}{\sum_{s \in \bar{N}(z), i^s = i} \beta^s(z)} e_{i^t}^t = \sum_{t \in \hat{N}(x), i^t = i} \frac{\beta^t(z)}{\sum_{s \in \hat{N}(x), i^s = i} \beta^s(z)} e_{i^t}^t.$$

If we take z near to x, the right hand side may be arbitrarily taken near

$$\sum_{t \in \hat{N}(x), i^t = i} \frac{\beta^t(x)}{\sum_{s \in \hat{N}(x), i^s = i} \beta^s(x)} e_{i^t}^t = \sum_{t \in \bar{N}(x), i^t = i} \frac{\beta^t(x)}{\sum_{s \in \bar{N}(x), i^s = i} \beta^s(x)} e_{i^t}^t.$$

Hence, $w_i(z)$ can be arbitrarily taken near $w_i(x)$ in $C_{\bar{B}_i}$, where $B_i = \{w_{i^t}^t \mid i^t = i\}$, and by (V4), $\Psi_i(z) = V_i(z_i, w_i(z))$ has the local intersection property near x. It follows that Ψ has the local intersection property. Consequently, by Theorem 2.3.1, Ψ has a fixed point, which contradicts condition (V2). ∎

In view of economic equilibrium theory, Browder- and Kakutani-type conditions generalize natural properties for preference (or better-set) correspondences and constraint (or budget) correspondences, respectively. Fixed-point results under the unified viewpoint for Browder- and Kakutani-type mappings under K1-type conditions, therefore, have important meanings for the economic equilibrium existence problem that is characterized as a fixed-point argument for the product of mappings including both Browder and Kakutani types.[5]

---

[5] From a purely mathematical viewpoint, the difference between Browder- and Kakutani-type mappings is transformed into a topological necessary condition for dual-space structure under which a fixed-point result under the K1-type condition is guaranteed.

As we shall discuss in the concluding Chapters 9 and 10 of this book, an equilibrium concept in the social sciences is not only a tool for describing society but also a base for our *views of the world*, i.e., a base for our rationality, justice, morals, and so on. From this context, the existence of equilibrium is important as a minimal requirement for the mathematical (model theoretic) *consistency* of such a base or a view of the world. Conditions about such individual preferences as (Product B), (Product K), (Product K1–V4*), and (Product K1–V4), therefore, may be interpreted on the one hand as a restriction for the range to which our models and theorems can be applicable, and on the other hand, as a restriction to our preferences (our rationality or morals) when our models and theorems describe mechanisms that our society really or desirably has.

Particularly for the latter interpretation for irreflexive φ_is, note that for Nash equilibrium existence theorems (3.2.1, 3.2.2, 3.2.3, 3.2.4) and the maximal element existence theorem (3.1.5), the conditions are sufficient to be checked for a certain irreflexive correspondence $\psi_i : X \to X_i$ (e.g., a certain selection of $\varphi_i : X \to X_i$) such that $\psi_i(x) \neq \emptyset$ whenever $\varphi_i(x) \neq \emptyset$. (Indeed, for the emptiness of $\varphi_i(x)$, the emptiness of $\psi_i(x)$ is sufficient.) These theorems indicate that any types of convexity with continuity conditions for φ_i may be replaced by an argument for the possibility obtaining ψ_i (e.g., a certain selection of φ_i) with a sort of fixed-direction conditions stated in those theorems. For example, when E is a finite dimensional topological vector space, gradient vector at x for pseudo-utility function u^x (as defined in Shafer and Sonnenschein (1975)) relative to φ_i might be used as a dual-space element for K1-type conditions to represent the direction of (a selection of) $\varphi_i(x)$ at x. Lexicographic preferences can also be treated as a special case that satisfies the Browder-type condition. Even the Kakutani-type condition, which is rather suitable for correspondences having a closed graph, may be utilized as a natural condition for non-ordered preferences with thick indifference curves. We discuss and treat the details of these examples in Chapter 5.

3.3 Abstract Economy

This section is devoted to the equilibrium existence problem for abstract economies (a generalized strategic form game: c.f. Debreu (1952), Shafer and Sonnenschein (1975)). For a non-cooperative strategic form game, we add a structure of constraint correspondences describing the situation where the feasible action of each player is restricted by other agent strategy choices on a certain subset of his/her own strategy set. Let us consider correspondence $K_i : \prod_{j \in I, j \neq i} X_j \to X_i$ for each $i \in I$, and given other players' strategies, $(x_j)_{j \in I, j \neq i}$, restrict the choice of i's strategy on subset $K_i((x_j)_{j \in I, j \neq i})$ of X_i. We call strategy profile $x_* = (x_i^*)_{i \in I}$ a *social equilibrium* (a generalized Nash equilibrium) if (1) $x_i^* \in K_i((x_j^*)_{j \in I, j \neq i})$ for each i, and (2) $P_i(x^*) \cap K_i(x^*) = \emptyset$ for all $i \in I$.

For notational convenience, the domain of each K_i should be recognized as $X = \prod_{j \in I} X_j$ instead of $\prod_{j \in I, j \neq i} X_j$ and characterized as coordinate correspondences of class (Product B), (Product K), (Product K1-V4), and (Kakutani Type) in exactly the same way as P_i's. In the following, a generalized non-cooperative strategic form game (abstract economy) will be denoted by $(X_i, P_i, K_i)_{i \in I}$, where for each $i \in I$, X_i is a non-empty compact subset of Hausdorff space having a convex structure, P_i (*better-set correspondence*) is a correspondence on X to X_i, and K_i (*constraint correspondence*) is a correspondence on X to X_i.

Theorems 3.2.1, 3.2.2, 3.2.3, and 3.2.4 in the previous section sufficiently illustrate the general situations for the Nash equilibrium existence problem for non-cooperative game $(X_i, P_i)_{i \in I}$. For the abstract economy setting, $(X_i, P_i, K_i)_{i \in I}$, however, it would be more desirable to consider situations where conditions for P_is (typically of the Browder type) and for K_is (typically of the Kakutani type) are given independently. Let us begin this section by extending the previous section's Nash-equilibrium existence results to the case with a mixture of Browder- and Kakutani-type conditions. (As before, we use notation φ and φ_i instead of P and P_i and denote by $\mathcal{E}q_i(\varphi)$ set $\{x \in X \mid x_i \in \varphi_i(x) \text{ or } \varphi_i(x) = \emptyset\}$. Set $\mathcal{E}q(\varphi)$ is the intersection of $\mathcal{E}q_i(\varphi)$, $i \in I$, and, when every $\varphi_i : X \to X_i$ satisfies the irreflexivity, $x_i \notin \varphi(x)$ for all $x \in X$ denotes the set of Nash equilibrium points. A dual-system structure for each X_i, $i \in I$, if necessary, is assumed to exist and is denoted by (X_i, W_i, V_i).)

(Product BK1) There is a fixed-point-free convex valued extension Ψ_i of φ_i on $X \setminus \mathcal{E}q_i(\varphi)$ such that for each $x \notin \mathcal{E}q_i(\varphi)$, open neighborhood $U(x)$ of x in X exists such that at least one of the following two conditions is satisfied: (1) $\bigcap_{z \in U(x)} \Psi_i(z) \neq \emptyset$, or (2) for one $w \in W = \prod_{i \in I} W_i$, we have $\Psi_i(z) \subset V_i(z, w)$ and $\Psi_i(z) \neq \emptyset$ for all $z \in U(x)$.

3: Nash Equilibrium and Abstract Economy

LEMMA 3.3.1: (Nash Equilibrium or Fixed-Point Existence: Product BK1 under V4*)[6] *Let I be a set. For each $i \in I$, let X^i and W^i be non-empty compact Hausdorff spaces having convex structures and dual-system structure $V_i : X_i \times W_i \to X_i$ on them. Let φ^i be a correspondence on $X = \prod_{i \in I} X^i$ to X^i and define correspondence $\varphi : X \to X$ as $\varphi(x) = \prod_{i \in I} \varphi_i(x)$ for each $x \in X$. Then $\mathcal{E}q(\varphi) \neq \emptyset$ if each φ_i satisfies (Product BK1) under (V4*).*

PROOF: Assume that $\mathcal{E}q(\varphi) = \emptyset$. Then for each $x \in X = \prod_{i \in I} X_i$ at least one $i \in I$ and open set $U(x)$ exist such that common element $y_i(x) \in X_i$ or dual-space element $w^i(x) \in W$ satisfies, respectively, (1) or (2) of condition (Product BK1). Since X is compact, we have finite set $\{x^1, \ldots, x^k, x^{k+1}, \ldots, x^\ell\} \subset X$, finite open covering $\{U(x^1), \ldots, U(x^k), U(x^{k+1}), \ldots, U(x^\ell)\}$ of X, and finite sequence of points $y_{i^1}(x^1), \ldots, y_{i^k}(x^k)$ in X_i and points $w^{i^{k+1}}(x^{k+1}), \ldots, w^{i^\ell}(x^\ell)$ in W with the sequence of player indices, $i^1, \ldots, i^\ell$, such that each of $(U(x^m), y_{i^m}(x^m))$, $m = 1, \ldots, k$, satisfies condition (1) of (Product B or K1), and each of $(U(x^m), w^{i^m}(x^m))$, $m = k+1, \ldots, \ell$, satisfies condition (2) of (Product B or K1). By Theorem 3.1.3, we have open covering $\{V(x^1), \ldots, V(x^\ell)\}$ of X such that for each $t = 1, \ldots, \ell$, $x^t \in V(x^t)$ and closure $\bar{V}(x^t)$ is included in $U(x^t)$. Let J be set $\{i^1, \ldots, i^\ell\}$ and, for each $x \in X$, $J(x) = \{i^m | x \in V(x^m)\} \subset I$, $N(x) = \{n | x \in V(x^n)\} \subset \{1, \ldots, \ell\}$, and $\bar{N}(x) = \{n | x \in \bar{V}(x^n)\} \subset \{1, \ldots, \ell\}$. Define for each $x \in X$, $\Psi(x) \subset X$ as $\Psi(x) = (\bigcap_{t \in \bar{N}(x), 1 \leq t \leq k}(\Psi_{i^t}(x) \times \prod_{i \in I, i \neq i^t} X_i)) \cap (\bigcap_{t \in \bar{N}(x), k+1 \leq t \leq \ell}(V_{i^t}(x_{i^t}, w^{i^t}(x_{i^t})) \times \prod_{i \in I, i \neq i^t} X_i))$. $\Psi(x)$ may be considered a convex subset of X under the product convex structure on X given by coordinates in J. (For a precise definition, see the proof of Theorem 3.2.1.) Let us define $\varphi(x)$ as $\bar{\varphi}(x) = \bigcap_{t \in \bar{N}(x)}(\varphi_{i^t}(x) \times \prod_{i \in I, i \neq i^t} X_i)$. By considering condition $\bar{V}(x^t) \subset U(x^t)$ for each $t = 1, \ldots, \ell$, we can also verify that $\Psi(x) \supset \bar{\varphi}(x) \neq \emptyset$ for each $x \in X$ and Ψ has the local-intersection property. Then by Theorem 2.3.1, φ has a fixed point, that contradicts assumption $\mathcal{E}q(\varphi) \neq \emptyset$. ∎

The next theorem on social equilibrium existence is an immediate consequence of Lemma 3.3.1.

---

[6] Of course, this result (under condition (Product BK1)) is weaker than the theorem under (Product B) with (Closedness) (Theorem 3.2.1) and (Product K1–V4*) with (Closedness) (Theorem 3.2.3). In condition (Product BK1), the open neighborhood is taken in X (not in $X \setminus \mathcal{E}q_i(\varphi)$), so (Closedness) condition is unnecessary.

THEOREM 3.3.2: (Social Equilibrium Existence) *Abstract economy $(X^i, P^i, K^i)_{i \in I}$ has a social equilibrium if the following conditions are satisfied:*

(1) *For each $i \in I$, X_i is a compact set having convex and dual-system structures $(X_i, W_i, V_i : X_i \times W_i \to X_i)$ satisfying (V1)–(V3) and (V4*).*
(2) *For each $i \in I$, P_i is a (possibly empty valued) correspondence on $X = \prod_{j \in I} X_j$ to X_i having the open-lowersection property.*
(3) *For each $i \in I$, K_i is a non-empty valued correspondence on X to X_i of the (Product K1-V4*) type. (For example, each K_i is non-empty closed convex valued upper semicontinuous correspondence on X to X_i when each X_i is a subset of a locally convex space.)*
(4) *For each $i \in I$, correspondence $X \ni x \mapsto \operatorname{int} K_i(x) \subset X_i$ has the open-lowersection property.*
(5) *For all $x \in X$, $\operatorname{int} K_i(x) \cap P_i(x) = \emptyset$ means that $K_i(x) \cap P_i(x) = \emptyset$.*

PROOF: For each $i \in I$ and $x = (x_j)_{j \in I} \in X$, let $Q_i(x) = K_i(x)$ if $x_i \notin K_i(x)$, let $Q_i(x) = P_i(x) \cap \operatorname{int} K_i(x)$ if $x_i \in K_i(x)$. Then, by (5), $x^* \in X$ is a social equilibrium point for $(X_i, P_i, K_i)_{i \in I}$ iff $x^* \in X$ is a Nash equilibrium point of non-cooperative strategic-form game $(X_i, Q_i)_{i \in I}$. By (2) and (4), correspondence $x \mapsto P_i(x) \cap \operatorname{int} K_i(x)$ has the open lowersection property. So under (2)–(4), Q_i satisfies the condition (Product BK1), and the result directly follows from Lemma 3.3.1. ∎

Theorem 3.3.2 is a direct extension of classical standard cases. Condition (4) with (2) enables us to treat mapping $x \mapsto \operatorname{int} K_i(x) \cap P_i(x)$ (defining Q_i in the proof) having the open lowersection property. Here, we use $\operatorname{int} K_i(x)$ instead of $K_i(x)$ since constraint correspondences, e.g., budget constraints in the general equilibrium theory, usually fail to satisfy this condition while their interior points will satisfy it as long as x does not represent the minimum asset situation.[7] It is also true that such a minimum asset condition guarantees conditions like (5).

Here, by defining $Q_i(x)$ as $K_i(x)$ (if $x_i \notin K_i(x)$) or $P_i(x) \cap K_i(x)$ (if $x_i \in K_i(x)$), we directly reduced the equilibrium existence problem of generalized game $(X_i, P_i, K_i)_{i \in I}$ into the Nash equilibrium or fixed-point existence problem with respect to game $(X_i, Q_i)_{i \in I}$. This type of argument is quite standard (see, e.g., Gale and Mas-Colell (1975), Toussaint (1984)),

---

[7] See the Minimum Wealth Condition in Chapter 5, Section 1.

and the above theorem shows that essentially the same arguments and methods are applicable to more general topological spaces under abstract convexity without vector space structures. Note, however, that we cannot directly follow the discussion based on lower semicontinuity like Gale and Mas-Colell (1975; Corrections (1979)) since the continuous selection theorem for lower semicontinuous carriers (c.f. Michael (1956, Theorem 3.2″)) depends on vector-space (normed linear space) structure.[8]

We can further generalize the above type of theorems by directly weakening some conditions. Bagh (1998, The last theorem of Section 3) shows that the conditions of the lower-semicontinuity type for K_i can be removed. (The essential role of the lower semicontinuous condition for K_i can be replaced by the openness condition for the domain of mappings on which $K_i(x) \cap P_i(x) \neq \emptyset$.) Bagh's idea "fattens" the graph of K_i so that it is open (hence, to have open lowersection properties, etc.), using a local base at 0 and limit arguments.[9]

Since Browder-, K1-, and Kakutani-type conditions for coordinate correspondences are more general than merely assuming open- or closed-graph properties, it is not so restrictive to suppose that both P_i's and K_i's satisfy one of the Browder-, K1-, and Kakutani-type conditions. The following theorems give various possibilities for such abstract-economy models even in finite dimensional cases.

THEOREM 3.3.3: (Social Equilibrium Existence: Browder Type)
Abstract economy $(X^i, P^i, K^i)_{i \in I}$ has a social equilibrium if the following conditions are satisfied:

(1) *For each $i \in I$, X_i is a compact set having convex structure.*
(2) *For each $i \in I$, K_i is a non-empty valued correspondence on X to X_i of the (Product BK1)-type.*
(3) *For each $i \in I$, correspondence $x \mapsto P_i(x) \cap K_i(x)$ on X to X_i is of the (Product BK1)-type.*

---

[8]This is one important direction that generalizes the arguments in this book.
[9]One can extend Theorem 3.3.2 through Bagh's method by taking the net of open coverings (directed by refinements) instead of the 0-neighborhood base (e.g., see the limit argument in the proof of Theorem 2.3.5). In such cases, however, to assure the convexity of the "fattened" value of K_i, one must assume locally convex topology (under the abstract convexity) on X as in the proof of Kakutani's Theorem in Ben-El-Mechaiekh et al. (1998).

PROOF: For each $i \in I$ and $x = (x_j)_{j\in I} \in X$, let $Q_i(x) = K_i(x)$ if $x_i \notin K_i(x)$, let $Q_i(x) = P_i(x) \cap K_i(x)$ if $x_i \in K_i(x)$ and $P_i(x) \cap K_i(x) \neq \emptyset$, and let $Q_i(x) = \emptyset$ if otherwise. Then, $x^* \in X$ is a social equilibrium point for $(X_i, P_i, K_i)_{i\in I}$ iff $x^* \in X$ is a Nash equilibrium point of non-cooperative strategic-form game $(X_i, Q_i)_{i\in I}$. Under (2) and (3), Q_i satisfies condition (Product BK1), so the result follows from Lemma 3.3.1. ∎

Even when we use (Product B) instead of (Product BK1) in the above theorem, Condition (3) is not so restrictive as long as we are allowed to assume that $\mathrm{cl}(\mathrm{int}(K_i(x))) \supset K_i(x)$. Since conditions of the Browder-type in this book do not necessarily depend on the closedness and/or openness of the value of the correspondences, they can be utilized for general and uniform characterizations of both preference and constraint correspondences.

If preference correspondences can be described as the K1 type, we obtain the next theorem with a simple arrangement for the continuity of correspondences.

THEOREM 3.3.4: (Social Equilibrium Existence: K1 Type)
Abstract economy $(X^i, P^i, K^i)_{i\in I}$ has a social equilibrium if the following conditions are satisfied:

(1) *For each $i \in I$, X_i is a compact set having convex and dual-system structures $(X_i, W_i, V_i : X_i \times W_i \to X_i)$ satisfying $(V1)$, $(V2)$, $(V3)$, and $(V4)$ or $(V4^*)$.*
(2) *For each $i \in I$, P_i is a (possibly empty valued) correspondence on X to X_i of the (Product K1)-type with (Closedness).*
(3) *For each $i \in I$, K_i is a non-empty valued correspondence on X to X_i of the (Product K1)-type with (Closedness).*
(4) *For each $i \in I$, set $\{x \in X \mid K_i(x) \cap P_i(x) \neq \emptyset\}$ is open.*

PROOF: For each $x = (x_j)_{j\in I} \in X$ and $i \in I$, define $Q_i(x)$ as $Q_i(x) = K_i(x)$ if $x_i \notin K_i(x)$, $Q_i(x) = P_i(x)$ if $x_i \in K_i(x)$ and $P_i(x) \cap K_i(x) \neq \emptyset$, and $Q_i(x) = \emptyset$ if $x_i \in K_i(x)$ and $P_i(x) \cap K_i(x) = \emptyset$. Then, abstract economy $(X^i, P^i, K^i)_{i\in I}$ has a social equilibrium if and only if strategic-form game $(X_i, Q_i)_{i\in I}$ has a Nash equilibrium. By condition (4), Q_i satisfies (Product K1)-Type condition under (V4) or (V4*), so the result follows from Theorems 3.2.3 and 3.2.4. ∎

The K1-type condition may be further utilized for fixed-point, Nash-equilibrium, and abstract-economy arguments if the dual-system structure

3: Nash Equilibrium and Abstract Economy

has more desirable features like the following (V5). Let us consider dual-system structure $(X, W, V : X \times W \to X)$, where X and W are Hausdorff spaces having convex structures.

(V5) If $V(x,w) \neq \emptyset$, for all $y \in V(x,w)$ there is an open neighborhood U of $(x,w) \in X \times W$ such that $y \in V(z,v)$ for all $(z,v) \in U$.

That is, correspondence $V : X \times W \to X$ has the open lowersection property. Condition (V5) is automatically satisfied when spaces X and W are subsets of finite dimensional topological vector spaces, or for compact subset X of Hausdorff topological vector space E, a uniform convergence topology on compact sets is supposed on subset W of algebraic dual E^*, etc. Of course, condition (V5) is stronger than conditions (V4) and (V4*). The condition, however, is the simplest of the three and is appropriate for most kinds of economic (commodity-price) duality arguments. We obtain under (V5) the following powerful extension of K1-type condition ensuring the fixed-point, Nash-equilibrium, and social-equilibrium existence results. Consider a class of dual-system structures, $(X_i, W_i, V_i : X_i \times W_i \to X_i)$, $i \in I$, satisfying (V1)–(V3) and (V5), and non-empty valued (coordinate) mappings, $\varphi_i : X = \prod_{i \in I} \to X_i$, $i \in I$.

(Upper Semicontinuously Differentiable Coordinate Correspondences) For each $x = (x_j)_{j \in I} \in X$ such that $x_i \notin X_i \backslash \boldsymbol{Fix}_i(\varphi_i)$, open neighborhood U of x and upper semicontinuous correspondence $\eta_i : U \to W_i$ exist such that for all $z = (z_j)_{j \in I} \in U$ and $w_i \in \eta_i(z)$, we have $\varphi_i(z) \subset V_i(z_i, w_i)$.

In other words, at x such that $x_i \notin \varphi_i(x)$, set $\varphi_i(x)$ is supported by all elements in $\eta_i(x)$, locally and upper semicontinuously. Under (V5), condition *upper semicontinuously (USC) differentiable coordinate correspondences* may be utilized to assure the existence of fixed points and Nash equilibria. The next result summarizes them as a social equilibrium existence theorem.

THEOREM 3.3.5: (Social Equilibrium Existence: USC Differentiable Case) *Abstract economy $(X^i, P^i, K^i)_{i \in I}$ has a social equilibrium if the following conditions are satisfied:*

(1) *For each $i \in I$, X_i is a compact set having convex and dual-system structures $(X_i, W_i, V_i : X_i \times W_i \to X_i)$ satisfying $(V1)$, $(V2)$, $(V3)$, and $(V5)$.*

(2) For each $i \in I$, P_i is an upper semicontinuously differentiable (possibly empty valued) coordinate correspondence on X to X_i.
(3) For each $i \in I$, K_i is a non-empty valued upper semicontinuously differentiable coordinate correspondence on X to X_i.
(4) For each $i \in I$, set $\{x \in X | P_i(x) \cap K_i(x) \neq \emptyset\}$ is open.

PROOF: Assume no equilibrium point. Under condition (4), if $x = (x_j)_{j \in I} \in X$ is not an equilibrium point, at least one $i \in I$ exists such that (i) $x_i \notin K_i(x)$, or (ii) neighborhood $U(x)$ of x in X satisfies that for every $z = (z^j)_{j \in I} \in U(x)$, $P_i(z) \cap K_i(z) \neq \emptyset$. Then, since X is compact, we have finite points $x^1, \ldots, x^k$ of X, open covering $\{U(x^1), \ldots, U(x^k)\}$ of X, finite sequence of players $i^1, \ldots, i^k \in I$, and upper semicontinuous correspondences, $w_{i^1}^1, \ldots, w_{i^k}^k, w_{i^{k+1}}^{k+1}, \ldots, w_{i^\ell}^\ell$ on X to $W_{i^1}, \ldots, W_{i^\ell}$, respectively, satisfying (i), $(1, \ldots, k)$, or (ii), $(k+1, \ldots, \ell)$, of the condition stated above and the differentiability condition for K_i's and P_i's. Moreover, we have by Theorem 3.1.3 open covering $\{V(x^1), \ldots, V(x^k)\}$ of X such that for each $t = 1, \ldots, k$, $x^t \in V(x^t)$ and closure $\bar{V}(x^t)$ is included in $U(x^t)$. Let J be set $\{i^1, \ldots, i^\ell\}$ and, for each $x \in X$, $J(x) = \{i^m | x \in V(x^m)\} \subset I$, $N(x) = \{n | x \in V(x^n)\} \subset \{1, \ldots, \ell\}$, and $\bar{N}(x) = \{n | x \in \bar{V}(x^n)\} \subset \{1, \ldots, \ell\}$. Define for each $x \in X$, $\Psi(x) \subset X$ as $\Psi(x) = \bigcap_{t \in \bar{N}(x)}(V_{i^t}(x_{i^t}, w_{i^t}^t(x)) \times \prod_{i \in I, i \neq i^t} X_i)$. $\Psi(x)$ may be considered a convex subset of X under the same product convex structure on X given by coordinates in J defined in the proof of Theorem 3.2.1. By considering condition $\bar{V}(x^t) \subset U(x^t)$ for each $t = 1, \ldots, \ell$, we can also verify that $\Psi(x)$ is not empty for each $x \in X$. (Note that $V_{i^t}(x_{i^t}, w_{i^t}^t(x))$ includes $K_{i^t}(x) \neq \emptyset$ for $t = 1, \ldots, k$ and $K_{i^t}(x) \cap P_{i^t}(x) \neq \emptyset$ for $t = k+1, \ldots, \ell$.) Furthermore, Ψ has the open lowersection property under (V5). Then by Theorem 2.3.1, Ψ has a fixed point, which contradicts condition (V2). ∎

Last, consider a theorem based on the Kakutani-type condition. Although the condition is restrictive for preference correspondences, there is an advantage: we do not have to assume the dual-system structure on strategy spaces.

THEOREM 3.3.6: (Social Equilibrium Existence: Kakutani Type) *Abstract economy $(X^i, P^i, K^i)_{i \in I}$ has a social equilibrium if the following conditions are satisfied:*

(1) For each $i \in I$, X_i is a compact set having convex structure.
(2) For each $i \in I$, K_i is a non-empty valued correspondence on X to X_i of the (Product K)-type.

(3) For each $i \in I$, correspondence $x \mapsto P_i(x) \cap K_i(x)$ on X to X_i is of the (Product K)-type.

PROOF: For each $i \in I$ and $x = (x_j)_{j \in I} \in X$, let $Q_i(x) = K_i(x)$ if $x_i \notin K_i(x)$, let $Q_i(x) = P_i(x) \cap K_i(x)$ if $x_i \in K_i(x)$ and $P_i(x) \cap K_i(x) \neq \emptyset$, and let $Q_i(x) = \emptyset$ if otherwise. Then $x^* \in X$ is a social equilibrium point for $(X_i, P_i, K_i)_{i \in I}$ iff $x^* \in X$ is a Nash equilibrium point of non-cooperative strategic-form game $(X_i, Q_i)_{i \in I}$. One may easily check under (2) and (3) that Q_i satisfies condition (Product K), so the result directly follows from Theorem 3.2.2. ∎

Bibliographic Notes

In this chapter, the Nash equilibrium existence problem (c.f. Nash (1950), Nash (1951)) and the social equilibrium existence problem (c.f. Debreu (1952), Shafer and Sonnenschein (1975)) were reexamined. By applying the theorems of the previous chapter, we obtained general results for these problems (e.g., see Theorems 3.3.2, 3.3.3, 3.3.4, 3.3.5, and 3.3.6). From an economic viewpoint, the most interesting result may be Theorem 3.3.5 that gives a clear condition (a characterization) for the existence of economic equilibria with non-convex non-ordered preferences if we apply it to finite dimensional ordinary differentiable cases. Theorem 3.3.2 with arguments in relation to the results of Bagh (1998) is also important as a discussion that directly generalizes all recent arguments.

Notes for Section 3.1

The fundamental setting of non-cooperative game theory used in this section is quite standard in microeconomic theory today. For references, see Ichiishi (1983), Kreps (1990), Mas-Colell et al. (1995), etc.

This section gives fundamental maximal element existence theorems for multi-agent game settings with some additional basic mathematical tools and a fixed-point theorem. Theorems 3.1.2 and 3.1.5 give the general existence results of maximal elements under considerably weaker conditions. These theorems may be compared with Yannelis and Prabhakar (1983, Corollary 5.1), Toussaint (1984, Theorem 2.1), etc. (The "$\mathscr{L}$-majorized" concept is quite useful and important for generalizing the open lowersection property. With respect to the maximal element existence problem, however, the Browder-, K1-, or Kakutani-type conditions we have treated here are more general. We will discuss the "$\mathscr{L}$-majorized" condition in general spaces (without linear structures) in Chapter 8.)

Notes for Section 3.2

In this section, the fixed-point arguments in the previous sections were applied to the existence of equilibrium problems for strategic form non-cooperative games. The Nash equilibrium existence problem (from its most classical treatments like Nash (1950), Nash (1951), Nikaido (1959)) has been dealt with as closely related to fixed-point arguments for the products of mappings. Economic theorists have treated them as lemmas for the existence of equilibrium proofs (e.g., see Gale and Mas-Colell (1975, Section 2, Theorem)). The fixed-point arguments in the previous sections are classified into four types of Nash-equilibria or fixed-point theorems: Theorems 3.2.1, 3.2.2, 3.2.3, and 3.2.4. These also cover and extend my earlier results, including Urai (2000), Urai and Hayashi (2000), etc. I should admit, however, even though such arguments are indispensable to all equilibrium existence discussions, they are by no means as comprehensive as suggested by the additional mixture and differentiability conditions in the next section (as well as discussions for cases with $\mathscr{L}$-majorized mappings in Chapter 8). From the view of fixed-point theory, Park (2005) treated part of such problems as "collectively fixed-point theorems" for a family of Browder-type mappings under the framework of the product of G-convex spaces.

Theorems 3.2.3 and 3.2.4 may also be considered generalizations of the result of Nishimura and Friedman (1981) since the best response correspondences under continuous ordered preferences P^i, $i \in I$, may typically be considered an example of φ^i's satisfying K1-type conditions. For Nash equilibrium existence arguments, see also Luo (2001) (under Δ-convexity) and Park (2001) (under G-convexity).

Notes for Section 3.3

Section 3.3 treats the existence of equilibrium problems for an abstract economy, generalized non-cooperative strategic form games (c.f. Debreu (1952), Shafer and Sonnenschein (1975)), etc. Theorem 3.3.2 gives a direct extension of such classical results.

As discussed before, Theorems 3.3.2–3.3.6 are also intended to give an extension of more recent researches such as Yannelis and Prabhakar

(1983), Toussaint (1984), Tan and Yuan (1994), and Bagh (1998). Consider the following: (1) in a locally convex space, compact convex valued upper semi-continuous correspondence K^i satisfies a K1-type condition; (2) in a metrizable space, for compact convex valued upper semi-continuous correspondence K^i, we can easily construct convex valued correspondence $\hat{K}^i : X \to X^i$ with open graph such that condition (Product K*) for K^i is satisfied; and (3) $\mathscr{L}$-majorized correspondences (in the sense of Bagh (1998); for a normal definition see Yannelis and Prabhakar (1983) and Tan and Yuan (1994)), P^i, $i \in I$, satisfies the condition of the Browder type, and we may arrange these theorems to include the results of Tan and Yuan (1994) and Bagh (1998) in metrizable locally convex cases. (Of course, in such a case, the condition of the Browder type, in this book, is much more general than the concept of $\mathscr{L}$-majorized maps. For non-metrizable cases, however, we cannot assure that Browder- or K1-type mappings include $\mathscr{L}$-majorized mappings.)

For the existence of Nash and generalized Nash equilibria, see also Kim and Yuan (2001) (as a survey for vector-space arguments including the social system coordination model), Park (2004), etc.

Chapter 4
Gale–Nikaido–Debreu's Theorem

4.1 Gale–Nikaido–Debreu's Theorem

The topics treated in this chapter are related to the excess demand approach for market equilibrium arguments. The mathematical essence of this framework is known as Gale–Nikaido–Debreu's abstract economy (c.f. Gale (1955), Nikaido (1956a), Debreu (1956)) and the existence of equilibrium theorem for it is called Gale–Nikaido–Debreu's theorem, which is also famous (under its finite dimensional form) as one of the mathematical equivalents of Brouwer fixed-point theorem. Hence, since we have developed new conditions for fixed-point theorems in Chapter 2, we can also develop alternative weak requirements for the existence of market equilibrium to generalize Gale–Nikaido–Debreu's theorem.

We begin with the market equilibrium existence problem in finite dimensional spaces, as in Debreu (1956). Let Δ be the $\ell-1$ dimensional unit simplex in R^ℓ, and let $\zeta : \Delta \to R^\ell$ be a non-empty valued correspondence (an excess demand correspondence) satisfying Walras' Law,[1]

(W) $\forall p \in \Delta$, $\forall z \in \zeta(p)$, $p \cdot z \leq 0$,

and the following local definiteness condition for directions of excess demands:

(LDD) For each $p \in \Delta$, if $\zeta(p) \cap -R^\ell_+ = \emptyset$, vector $y(p) \in \Delta$ and open neighborhood U^p of p in Δ exist such that $y(p) \cdot z > 0$ for all $z \in \zeta(q)$ for all $q \in U^p$.

Note that if $\zeta(p) \cap -R^\ell_+ = \emptyset$, and if $\zeta(p)$ is convex and closed, then $\zeta(p)$ and $-R^\ell_{++}$ may be strictly separated by a hyper-plane normal to vector $y(p)$ in Δ. If ζ is also upper semi-continuous, then we have open neighborhood U^p of p

---

[1] Throughout this section, the inner product of two vectors, x, y, is denoted by $x \cdot y$.

such that $\zeta(q)$ is a subset of the open half space defined by the hyper-plane for all $q \in U^p$. Hence, one of the sufficient conditions for (LDD) is upper semi-continuity with the closed convex valuedness for ζ. Note, however, that in (LDD) the excess demand correspondence ζ is not assumed to be closed convex valued or upper semi-continuous. The condition merely says that excess demands under non-equilibrium prices have *a locally common direction*. That is, if p is not an equilibrium price vector, then at least one direction $y(p)$ exists such that all excess demands under prices near p belong to the same side of the hyperplane defined by $y(p)$.

The first result in this chapter is the existence of market equilibrium price, $p^* \in \Delta$, $\zeta(p^*) \cap -R_+^\ell \neq \emptyset$, under conditions (W) and (LDD). Such an equilibrium existence theorem is called a Gale–Nikaido–Debreu Theorem, and its framework is used in various contexts in economic equilibrium theory. Though the result may be generalized to cases with infinite dimensional commodity spaces (Theorem 4.2.4) in the next section, the following version of the Gale–Nikaido–Debreu theorem may not be covered in the existing literature, even as a result for finite dimensional commodity spaces (see, e.g., Mehta and Tarafdar (1987)).

THEOREM 4.1.1: *For* $\zeta : \Delta \to R^\ell$ *satisfying conditions (W) and (LDD), equilibrium price* p^*, $\zeta(p^*) \cap -R_+^\ell \neq \emptyset$.

PROOF: Suppose no equilibrium price vector. Then under (LDD), we have, for all $p \in \Delta$, vector $y(p) \in \Delta$, $\epsilon(p) > 0$, and open neighborhood $V^p = \{q \in \Delta | \|q - p\| < \epsilon(p)\}$ of p such that for all $q \in V^q$ and $z \in \zeta(q)$, $y(p) \cdot z > 0$. Since Δ is compact, finite set $\{p^1, \ldots, p^m\} \subset \Delta$ exists such that $\Delta \subset \bigcup_{t=1}^m V^{p^t}$. For each $p \in \Delta$ and for each $t = 1, \ldots, m$, let $\theta_t(p) = \max\{\epsilon(p^t) - \|p - p^t\|, 0\}$, where $\|\cdot\|$ denotes the Euclidean norm. Define for each $t = 1, \ldots, m$, a function $\beta_t : \Delta \to [0,1]$ as $\beta_t(p) = \theta_t(p) / \sum_{s=1}^m \theta_s(p)$, (the partition of unity subordinate to covering $\{V^{p^t}\}_{t=1}^m$ of Δ). Then, mapping $f : \Delta \to \Delta$,

$$f : \Delta \ni p \mapsto \sum_{t=1}^m \beta_t(p) y(p^t) \in \Delta,$$

is easily seen to be continuous. Since $\forall p \in \Delta$, $\beta_t(p) > 0$ iff $p \in V^{p^t}$ and since for all t, $p \in V^{p^t}$, and $z \in \zeta(p)$, $y(p^t) \cdot z > 0$, we have $f(p) \cdot z > 0$ for all $p \in \Delta$ and $z \in \zeta(p)$. Then, since we have for all $p \in \Delta$ and $z \in \zeta(p)$, $p \cdot z \leq 0$ and $f(p) \cdot z > 0$, it follows that $p \neq f(p)$ for all $p \in \Delta$. That is, f has no fixed point, which contradicts Brouwer's fixed-point theorem. ∎

4: Gale–Nikaido–Debreu's Theorem

Besides the closed convex valuedness with upper semi-continuity, one sufficient condition for (LDD) is the following local definiteness condition on the *sign* of excess demands:

(LDS) For all $p \in \Delta$, if $\zeta(p) \cap -R^\ell_+ = \emptyset$, then coordinate $i^p \in \{1, 2, \ldots, \ell\}$ and open neighborhood U^p of p in Δ exist such that the i^p-th coordinate correspondence ζ_{i^p} of ζ satisfies $\zeta_{i^p}(q) \subset R_{++}$ for all $q \in U^p$.

That is, if p is not an equilibrium price vector, there exists at least one commodity such that for all prices near p the sign of excess demands for the commodity is always positive. When ζ is single valued, condition (LDS) is weaker than the continuity of ζ although it is still stronger than condition (LDD).[2] Unfortunately, however, when ζ is multi-valued, condition (LDS) may not be satisfied even if ζ is upper semicontinuous and closed convex valued.

The concept underlying the condition (LDD) as well as the method for proving Theorem 4.1.1 may be directly applicable to fixed-point arguments for multi-valued mappings. Let X be a non-empty compact convex subset of R^ℓ and let φ be a non-empty closed convex valued upper semi-continuous correspondence on X to itself. (Hence, the situation is that for which we may apply the fixed-point theorem of Kakutani (1941).) If $x \in X$ is not a fixed point of φ, then x does not belong to the non-empty closed convex set $\varphi(x)$, so hyperplane H_x strictly separates $\{x\}$ and $\varphi(x)$. Since φ is upper semi-continuous, open neighborhood $U(x)$ of x exists such that H_x also strictly separates z and $\varphi(z)$ for all $z \in U(x)$; i.e., all elements $v \in \varphi(z) - z$ for all z near x may be evaluated positively by the inner product with a vector $p(x)$ that is normal to H_x. In other words, the set of variation $\varphi(z) - z$ from z near x has a *locally common direction* represented by vector $p(x)$.

The next theorem shows that the above condition on variations of the correspondence is indeed a *sufficient* condition for the existence of fixed points.

THEOREM 4.1.2: (Extension of Kakutani's Fixed-Point Theorem) *Let φ be a non-empty valued correspondence on a compact convex*

---

[2]Hayashi (1997) shows the existence of equilibrium price p^*, $\zeta(p^*) \cap R^\ell_- \neq \emptyset$, for a single valued ζ under (W), (LDS), and a certain kind of boundary condition by using Eaves' theorem (Eaves 1974).

subset X of R^ℓ to X satisfying one of the following two conditions:

(LDV1) For each $x \in X$ such that $x \notin \varphi(x)$, vector $p(x) \in R^\ell$ and open neighborhood U^x of x exist such that $p(x) \cdot (w - z) > 0$ for all $z \in U^x$ and $w \in \varphi(z)$.

(LDV2) For each $x \in X$ such that $x \notin \varphi(x)$, vector $y(x) \in X$ and open neighborhood U^x of x exist such that $(y(x) - z) \cdot (w - z) > 0$ for all $z \in U^x$ and $w \in \varphi(z)$.

Then φ has fixed point $p^* \in \varphi(p^*)$.

PROOF: The theorem under (LDV1) will be generalized in the next section.

(Case: LDV1) Suppose $x \notin \varphi(x)$ for all $x \in X$. Then by (LDV1), for each $x \in X$, $p(x) \in X$ and an open neighborhood U^x of x in X exist such that $p(x) \cdot (w - z) > 0$ for all $z \in U^x$ and $w \in \varphi(z)$. Since $X \subset \bigcup_{x \in X} U^x$ and since X is compact, the covering $\{U^x \mid x \in X\}$ has finite subcovering $\{U^{x^t}\}_{t=1}^m$. Let $\beta_t : X \to [0, 1]$, $t = 1, \ldots, m$, be the partition of unity subordinate to $\{U^{x^t}\}_{t=1}^m$. Let us consider a function $f : X \to R^\ell$ such that $f(x) = \sum_{t=1}^m \beta_t(x) p(x^t)$. Moreover, let $\Psi : R^\ell \to X$ be the correspondence defined as $\Psi(v) = \{x \in X \mid v \cdot x = \max_{z \in X} v \cdot z\}$. Clearly, f is continuous and Ψ is a non-empty closed convex valued correspondence having a closed graph, so that we have a fixed point $\hat{x}$ for $\Psi \circ f$ by Kakutani's fixed-point theorem. By definitions of f and Ψ, we have $\sum_{t=1}^m \beta_t(\hat{x}) p(x^t) \cdot \hat{x} \geq \sum_{t=1}^m \beta_t(\hat{x}) p(x^t) \cdot z$ for all $z \in X$. On the other hand, since $\hat{x} \in U^{x^t}$ for at least one $t \in \{1, \ldots, m\}$, we have also $\sum_{t=1}^m \beta_t(\hat{x}) p(x^t) \cdot (w - \hat{x}) > 0$ for all $w \in \varphi(\hat{x})$, i.e., $\sum_{t=1}^m \beta_t(\hat{x}) p(x^t) \cdot w > \sum_{t=1}^m \beta_t(\hat{x}) p(x^t) \cdot \hat{x}$ for all $w \in \varphi(\hat{x})$: a contradiction.

(Case: LDV2) Suppose that φ does not have a fixed point. Then under (LDV2) for each $x \in X$, we have vector $y(x) \in X$ and open neighborhood U^x of x such that for all $z \in U^x$ and $v \in \varphi(z) - z$, $(y(x) - z) \cdot v > 0$. Since X is compact, we have finite set $\{x^1, \ldots, x^m\} \subset X$ such that $X \subset \bigcup_{t=1}^m U^{x^t}$. Let $\beta_t : X \to [0, 1]$, $t = 1, \ldots, m$ be the partition of unity subordinate to $\{U^{x^t}\}_{t=1}^m$. Then, mapping $f : X \to X$ defined as $f : X \ni x \mapsto x + \sum_{t=1}^m \beta_t(x)(y(x^t) - x) \in X$ is continuous. Since $\forall x \in X$, $\beta_t(x) > 0$ iff $x \in V^{x^t}$, and since for each t, $x \in V^{x^t}$ and $v \in \varphi(x) - x$ implies $(y(x^t) - x) \cdot v > 0$, we have for all $x \in X$, $(f(x) - x) \cdot v > 0$ for all $v \in \varphi(x) - x$. It follows that $f(x) \neq x$ for all $x \in X$, which contradicts that f has a fixed point by Brouwer's fixed-point theorem. ∎

4.2 Market Equilibria in General Vector Spaces

The purpose of this section is to apply our fixed-point results for general spaces in previous sections to the market equilibrium existence problem of the Gale–Nikaido–Debreu type. We can find one of the most general forms of results for this problem in Nikaido (1956b), Nikaido (1957), or Nikaido (1959). After the 1980s, the same problem (with some varieties in topologies, boundary conditions, and so on) has been treated by many authors (e.g., Aliprantis and Brown (1983), Florenzano (1983), and Mehta and Tarafdar (1987)).

Let E be a vector space, and assume duality $\langle E, F \rangle$ between E and certain vector space F.[3] Denote by $P \subset E$ a non-empty $\sigma(E, F)$-closed convex cone with vertex 0 such that $P \cap -P \neq P$, and by P^* the polar cone of $-P$ with respect to duality $\langle E, F \rangle$. Moreover, denote by P_0^* set $P^* \setminus \{0\}$. At first we apply Theorem 2.1.6 to the setting given in Nikaido (1959). The basic problem is given a correspondence $\zeta : P_0^* \to E$ (an excess demand correspondence) satisfying Walras' Law (see (D3) below) to find p^* such that $\zeta(p^*) \cap -P \neq \emptyset$ (a market equilibrium price).

In many settings, upper semi-continuity with the non-empty compact convex valuedness of the excess demand correspondence is known to be sufficient for the existence of such a p^*. Since our methods used in previous sections may directly be applicable to the market equilibrium existence problem, we may replace the upper semi-continuity and the convexity assumptions for weaker conditions on the local direction of mappings. Urai and Hayashi (1997) show that when ζ is defined on $\sigma(F, E)$-compact base Δ of P, Walras' law (see (D3) below) and the following condition (LD) on the local direction of excess demands are sufficient for the existence of an equilibrium price.

(LD) For each $p \in \Delta$ such that $\zeta(p) \cap -P = \emptyset$, vector $\bar{p}$ and neighborhood $U(p)$ of p in $\Delta \subset (F, \sigma(F, E))$ exist such that $\forall q \in U(p)$, $\forall z \in \zeta(q)$, $(\zeta(q) \cap -P = \emptyset) \Rightarrow \langle \hat{p}, z \rangle > 0$.

That is, if p is not an equilibrium price, all excess demands under price q near p are positively evaluated by an appropriately chosen direction $\bar{p}$ (depending on p).

---

[3]Precisely, we suppose a bilinear form $F : E \times F \ni (x, f) \mapsto F(x, f) \in R$ such that E separates points in F and F separates points in E, where a set of linear functionals A on X *separates points* in X if for all $x, y \in X$, $x - y \neq 0$ implies that element $f \in A$ exists such that $f(x - y) \neq 0$.

The result gives one of the most general forms of the Gale–Nikaido–Debreu theorem in the literature (see, e.g., Mehta and Tarafdar (1987)). It is more desirable, however, to present a condition like (LD) as a property with an appropriate microeconomic foundation. When the commodity space is infinite dimensional, we cannot generally expect excess demand correspondence (defined as individual maximal choices based on preferences and budget constraints) to be upper semi-continuous on the whole domain, though showing the upper semi-continuity on each finite dimensional subspace of the domain is not difficult. Hence, some of the most general forms of the Gale–Nikaido–Debreu theorem are given by several authors based on the continuity condition on every finite dimensional subspace of the domain (see Nikaido (1959); Florenzano (1983)).

At first, we apply the results and concepts in Chapter 2 to the basic setting given in Nikaido (1957) and (1959). The Gale–Nikaido–Debreu lemma in Florenzano (1983) also has such settings. We integrate and generalize their results for cases that (1) cone P possibly has no interior points (so P_0^* may not have a $\sigma(F, E)$ compact base Δ), (2) the range of ζ is not necessarily compact (assumptions in Nikaido (1957) and (1959)), and (3) the condition of compact convex valuedness and upper semi-continuity on each finite dimensional subspace for ζ is weakened through conditions like (LD) (see (D1-1) and (D1-2) in the following lemmas and theorems).[4]

For the results in this chapter, especially for intuitive or natural expressions among prices and excess demands in economic arguments, we use the following notational conventions. When p is a price and $\zeta(q)$ is a value of excess demand correspondence (under a price q), we mean by $\langle p, \zeta(q) \rangle > 0$ or by $p(\zeta(q)) > 0$ the fact that $\langle p, z \rangle > 0$ for any $z \in \zeta(q)$. The same conventions will be used for cases with $=$ and $\leq$ as far as the evaluation of excess demands under prices are concerned; i.e., we denote by $p(\zeta(p)) = 0$ that $p(z) = 0$ for all $z \in \zeta(p)$ (Walras' Law in the strict sense), $\langle p, \zeta(p) \rangle \leq 0$ to mean that $\langle p, z \rangle \leq 0$ for all $z \in \zeta(p)$ (Walras' Law in the wide sense), etc. Of course, the usage is compatible with those in cases where ζ is single valued and is identified with a function (an excess-demand function). For finite subset A of a vector space, we denote by L_A the convex cone with vertex 0 spanned by all elements in A and by $L(A)$ the finite dimensional affine subspace spanned by elements in A under

---

[4]As a matter of fact, the result is not a complete generalization of Nikaido (1959, Section 5, Theorem 5), since in Nikaido (1959), the value of the excess demand correspondence is generally allowed to be acyclic.

4: Gale–Nikaido–Debreu's Theorem

the topology of finite dimensional affine subspaces. Given duality $\langle E, F \rangle$ between two vector spaces, E and F, and cone L in E (or F), denote by L° the polar cone of L defined as $L^\circ = \{y \in F | \forall x \in L, \langle x, y \rangle \leq 0\}$ (resp., $L^\circ = \{x \in E | \forall y \in L, \langle x, y \rangle \leq 0\}$).

THEOREM 4.2.1: (Market Equilibrium Lemma) *Suppose a non-empty valued correspondence ζ defined on convex $\sigma(F, E)$-dense subset D of P_0^* to E that satisfies the following conditions:*

(D1-1: Compact Convex Valuedness) For each finite subset $A \in \mathscr{F}(D)$ and $p \in \operatorname{co} A$ such that $\zeta(p) \cap L_A^\circ = \emptyset$, element $\bar{p} \in \operatorname{co} A$ exists such that $\langle \bar{p}, \zeta(p) \rangle > 0$.

(D1-2: Upper Semicontinuity on Finite Dimensional Subspaces) For each $p^* \in D$ and $q \in D$ such that $\langle q, \zeta(p^*) \rangle > 0$ and for each finite dimensional affine subspace L of F such that $p^* \in L$, we have open neighborhood U of p^* in L satisfying each $p \in U \cap D$, $\langle q, \zeta(p) \rangle > 0$.

(D2-1) Relatively Compact Range: The range of ζ, $\bigcup_{p \in D} \zeta(p)$, is relatively $\sigma(E, F)$-compact.

(D3) Walras' Law: $\forall p \in D, \langle p, \zeta(p) \rangle \leq 0$.

Then, net $\{(p^\nu, z^\nu \in \zeta(p^\nu)), \nu \in \mathscr{N}\}$ in $D \times E$ exists such that $\lim_{\nu \in \mathscr{N}} z^\nu \in -P$ under $\sigma(E, F)$-topology.

PROOF: Let us divide the proof into three steps.

(STEP1: We use only (D1) and (D3).) For each $p \in \operatorname{co} A$, $\zeta(p) \cap L_A^\circ = \emptyset$ means, by (D1-1) and (D1-2), neighborhood $U(p) \subset L(A)$ of p and point $\bar{p}$ in $\operatorname{co} A$ exist such that $\forall q \in \operatorname{co} A \cap U(p)$, $\langle \bar{p}, \zeta(q) \rangle > 0$. Since $\operatorname{co} A$ is a compact subset of $L(A)$, by letting $K = \{p \in \operatorname{co} A | \zeta(p) \cap L_A^\circ = \emptyset\}$, $\varphi(p) = \{q \in \operatorname{co} A | \langle q, \zeta(p) \rangle > 0\}$ for $p \in K$, and $\varphi(p) = \operatorname{co} A$ for $p \notin K$, we can see that $K = \{p \in \operatorname{co} A | p \notin \varphi(p)\} = \operatorname{co} A \backslash \mathcal{F}ix(\varphi)$ by (D3) and that $\varphi : \operatorname{co} A \to \operatorname{co} A$ satisfies the condition (K*) in Theorem 2.1.6, so φ has a fixed point p_A. By the definition of φ, we have $\zeta(p_A) \cap L_A^\circ \neq \emptyset$.

(STEP2: We use only (D2-1) and the definition of p_A.) Denote by $\mathscr{A}$ the set of all convex hulls of finite subset of D directed by the inclusion. By (D2-1), arbitrarily fixed net $\{z_A \in \zeta(p_A) \cap L_A^\circ, A \in \mathscr{A}\}$ has subnet $\{z_{A_\mu} \in \zeta(p_{A_\mu}) \cap L_A^\circ, \mu \in \mathscr{M}\}$ converging to point z_* in the range of ζ under topology $\sigma(E, F)$.

(STEP3: We use the definition of $D \subset P_0^*$.) Now, assume that $z_* \notin -P$. Then, since P is closed, vector $\bar{p} \in D$ exists such that $\langle \bar{p}, z_* \rangle > 0$. On the other hand, since for all $\mu \in \mathscr{M}$ sufficiently large we have $\bar{p} \in A_\mu$, we have $\langle \bar{p}, z_{A_\mu} \rangle \leq 0$ for all $\mu \in \mathscr{M}$ sufficiently large, so we have $\langle \bar{p}, z_* \rangle \leq 0$, a contradiction. Hence, $z_* \in -P$. ∎

Nikaido (1957) and (1959) assume that the range of ζ is $\sigma(E, F)$ compact (compact-range condition). In such a case, z^* in the above lemma is obtained as an element of the range of ζ, so the lemma assures the existence of the equilibrium price. Florenzano (1983) assumes that cone P has non-empty interiors, so P_0^* has $\sigma(F, E)$ compact base Δ and we may take D as a dense subset of Δ under $\sigma(F, E)$ to ensure the existence of limit $p^* \in \Delta$ for the net of prices, $\{p^\nu, \nu \in \mathscr{N}\}$.

Note that the compact-range condition is automatically satisfied when D is compact and ζ is $\sigma(E, F)$-closed valued and upper semi-continuous on whole domain D, so with the compact-range condition the lemma includes the case in which P_0^* has a weakly compact base D and ζ is upper semi-continuous on D. (This is an ordinary situation for finite dimensional commodity space E.)

To assure the existence of equilibrium price (i.e., $p^* \in D$ such that $\zeta(p^*) \cap -P \neq \emptyset$), an alternative method to the above compact-range argument is the following approach based on a certain kind of global continuity condition with the boundary condition of excess demands. The next theorem is an extension of the result in Aliprantis and Brown (1983) that is also known as one of the most general infinite dimensional Gale–Nikaido–Debreu theorem with a boundary condition of the Grandmont (1977) type. In Aliprantis and Brown (1983), the upper semi-continuity of ζ on whole domain D under $\sigma(F, E)$-topology is assumed. Since such a global continuity is too strong for standard infinite dimensional general equilibrium settings, we use (as in the previous lemma) a limit argument for excess demands with the relatively compact range assumption (D2-1). Furthermore, their boundary condition is slightly stronger than (D2-2).

In the next theorem, we suppose the existence of a compact set that spans dual cone P_0^*. (If we assume degree 0 homogeneity of ζ, it would be appropriate to consider cases when Δ can be chosen as a subset of P_0^*, which is typically the case when P_0^* has a weakly compact base under the assumption that P has an interior point.)

THEOREM 4.2.2: (Market Equilibrium Lemma with Boundary Condition) *Suppose that P^* is spanned by $\sigma(F, E)$-compact subset Δ*

4: Gale–Nikaido–Debreu's Theorem

of P^* and that there is a non-empty valued correspondence ζ defined on a convex $\sigma(F, E)$-dense subset D of $\Delta\setminus\{0\}$ to E satisfying the following conditions:

(D1-1: Compact Convex Valuedness) For each finite subset $A \in \mathscr{F}(D)$ and $p \in \operatorname{co} A$ such that $\zeta(p) \cap L_A^\circ = \emptyset$, element $\bar{p} \in \operatorname{co} A$ exists such that $\langle \bar{p}, \zeta(p) \rangle > 0$.

(D1-2: Upper Semicontinuity on Finite Dimensional Subspaces) For each $p^* \in D$ and $q \in D$ such that $\langle q, \zeta(p^*) \rangle > 0$ and for each finite dimensional affine subspace L of F such that $p^* \in L$, we have open neighborhood U of p^* in L satisfying for each $p \in U \cap D$, $\langle q, \zeta(p) \rangle > 0$.

(D2-1) Relatively Compact Range: The range of ζ, $\bigcup_{p \in D} \zeta(p)$, is relatively $\sigma(E, F)$-compact.

(D2-2) Boundary Condition: For each net $\{p^\nu, \nu \in \mathscr{N}\}$ in D converging to point $\hat{p} \in \Delta\setminus D$, vector $\bar{\bar{p}} \in D$ exists such that for a certain subnet $\{p^{\nu(\mu)}, \mu \in \mathscr{M}\}$ of $\{p^\nu, \nu \in \mathscr{N}\}$, $\langle \bar{\bar{p}}, z \rangle > 0$ for all $z \in \zeta(p^{\nu(\mu)})$ for all $\mu \in \mathscr{M}$.

(D3) Walras' Law: $\forall p \in D, \langle p, z \rangle \leq 0$ for all $z \in \zeta(p)$.

Then, there is net $\{(p^\nu, z^\nu \in \zeta(p^\nu)), \nu \in \mathscr{N}\}$ in $D \times E$ and $p^* \in D$ such that $p^* = \lim_{\nu \in \mathscr{N}} p^\nu$ under $\sigma(F, E)$-topology and $z^* = \lim_{\nu \in \mathscr{N}} z^\nu \in -P$ under $\sigma(E, F)$-topology.

PROOF: By the previous lemma with the existence of compact $\Delta \supset D$, we obtain net $((p^\nu, z^\nu) \in D \times E)_{\nu \in \mathscr{N}}$ that satisfies that $z^* = \lim_{\nu \in \mathscr{N}} z^\nu \in -P$ under $\sigma(E, F)$-topology and $p^* = \lim_{\nu \in \mathscr{N}} p^\nu \in \Delta$ in Δ under $\sigma(F, E)$-topology. If $p^* \in \Delta\setminus D$, then by (D2-2), element $\bar{p}^* \in D$ and subnet $\{p^{\nu(\mu)}, \mu \in \mathscr{M}\}$ of $\{p^\nu, \nu \in \mathscr{N}\}$ exist such that $\langle \bar{p}^*, z \rangle > 0$ for all $z \in \zeta(p_{A_{\nu(\mu)}})$ for all $\mu \in \mathscr{M}$, which is impossible since for any $A \in \mathscr{F}(D)$ sufficiently large, we have $\bar{p}_* \in A$, and for such A and p_A (where p_A is the fixed-point price given in (STEP1) in the proof of the previous lemma and may be considered equal to a $p^{\nu(\mu)}$ for a μ sufficiently large) satisfies $\zeta(p_A) \cap L_A^\circ \neq \emptyset$; i.e., $\exists z \in \zeta(p_{A_{\mu(\nu)}})$ such that $\langle \bar{p}^*, z \rangle \leq 0$. Therefore, we have $p^* \in D$. ∎

Obviously if we assume the global continuity (upper semicontinuity) of ζ under $\sigma(F, E)$-topology on D and $\sigma(E, F)$-topology on E, we may ensure that $z^* \in \zeta(p^*)$, so we have equilibrium price p^*. As stated above, however, since we cannot expect such global continuity for standard general equilibrium settings with infinite dimensional commodity spaces, here we present a minimal requirement for this argument to assure the existence of an equilibrium price. To show $\zeta(p^*) \cap -P \neq \emptyset$, it will be sufficient that condition $\zeta(p^*) \cap -P = \emptyset$ is contradictory to the fact that $p^* = \lim_{\nu \in \mathcal{N}} p^\nu$ and $z^* = \lim_{\nu \in \mathcal{N}} z^\nu \in -P$. The next condition is sufficient for this contradiction.

(D1-3: Global Upper Semicontinuity with Respect to Set $-P$) For each $p \in D$ such that $\zeta(p) \cap -P = \emptyset$, neighborhood $U(p)$ of p in $(F, \sigma(F, E))$ and vector $\bar{p} \in D$ exist such that $\forall q \in U(p) \cap D$, $\zeta(q) \cap -P = \emptyset \implies \langle \bar{p}, \zeta(q) \rangle > 0$.

Indeed, (D1-3) assures that for any net $p^\nu \to p^*$, $\langle \bar{p}, \zeta(p^\nu) \rangle > 0$ for all ν sufficiently large, which contradicts the definition of p_A in the (STEP1) for sufficiently large $A \ni \bar{p}$.

Note that for excess-demand arguments, condition $\langle p, \zeta(q) \rangle > 0$ is extremely important in the boundary as well as the continuity conditions. It is useful to define for two prices p, q in D, relation "$p \succ q$ (p dominates q)" as

$$p \succ q \Leftrightarrow \langle p, \zeta(q) \rangle > 0. \qquad (4.1)$$

Relation $\succ$ on D is called the *revealed preference relation*.[5] Notice also that the equilibrium price may be characterized as a maximal element under the revealed preference relation.

LEMMA 4.2.3: (Market Equilibrium and Revealed Preference Relation) *Equilibrium price $p^* \in D$, $\zeta(p^*) \cap -P \neq \emptyset$, is a maximal element of revealed preference relation. The converse is also true if $\zeta(p) \cap -P = \emptyset \implies \exists q \in D, \langle q, \zeta(p) \rangle > 0$.*

PROOF: Since δ is a subset of the polar of $-P$, no element of D can dominate p^* such that $\zeta(p^*) \cap -P \neq \emptyset$. The converse is obvious. ∎

If we define correspondence $\psi : D \to D$ as $\psi(q) = \{p \in D | p \succ q\}$, it is easy to see that ψ is convex valued. When Walras' Law ($p(\zeta(p)) \leqq 0$ for all $p \in D$) holds, ψ is irreflexive. Moreover, some upper semicontinuity

---

[5]The notion was developed by Nikaido (1968, p. 325) and Aliprantis and Brown (1983).

4: Gale–Nikaido–Debreu's Theorem

arguments may be utilized to assure the open lower-section property for ψ, so we can repeat the argument to obtain p_A in co A in (STEP1) through the maximal element existence theorem like Theorem 3.1.2.

The previous lemma and two limit theorems, Theorems 4.2.1 and 4.2.2, are sufficient to show the existence of a general equilibrium in many cases with infinite dimensional commodity spaces. We sum up this section by considering the next theorem with strong assumptions that ensure (1) that the existence of equilibrium price is in D, and (2) that domain D is not restricted as a dense subset of a compact base Δ of P_0^*.

In the next theorem, we use a global continuity condition for interior points (D1-4). As (D1-3), we cannot expect such a global continuity unless we restrict our discussion on finite dimensional commodity spaces. For finite dimensional cases, however, the condition is automatically satisfied when ζ is compact convex valued and Walras' Law is satisfied in the strict sense, $\langle p, \zeta(p) \rangle = 0$ for all $p \in D$. We use (D1-4) instead of (D1-3) since we can assure the existence of equilibrium even for cases in which D is not a dense subset of Δ. Since the validity of condition (D1-4) is not certain when Walras' Law is given in the wide sense, we use the strict one with the strict equilibrium condition, $0 \in \zeta(p^*)$.

THEOREM 4.2.4: (Market Equilibrium Existence) *Suppose that P^* is spanned by a $\sigma(F, E)$-compact subset Δ of P_0^* and that there is a non-empty valued correspondence ζ defined on a convex subset D of Δ to E satisfying the following conditions:*

(D1-1: Compact Convex Valuedness) For each finite subset $A \in \mathscr{F}(D)$ and $p \in \text{co } A$ such that $\zeta(p) \cap L_A^\circ = \emptyset$, element $\bar{p} \in \text{co } A$ exists such that $\langle \bar{p}, \zeta(p) \rangle > 0$.

(D1-2: Upper Semicontinuity on Finite Dimensional Subspaces) For each $p^* \in D$ and $q \in D$ such that $\langle q, \zeta(p^*) \rangle > 0$ and for each finite dimensional affine subspace L of F such that $p^* \in L$, we have open neighborhood U of p^* in L satisfying for each $p \in U \cap D$, $\langle q, \zeta(p) \rangle > 0$.

(D1-4) Global Upper Semicontinuity for Interior Points: For each p in int $D \subset \Delta$ and int $P_0^* \subset F$ such that $0 \notin \zeta(p)$, neighborhood $U(p)$ of p in $(F, \sigma(F, E))$ and vector $\bar{p} \in D$ exist such that $\forall q \in U(p) \cap D$, $0 \notin \zeta(q) \Rightarrow \langle \bar{p}, \zeta(q) \rangle > 0$.

(D2-3) Boundary Condition: For each net $\{p^\nu, \nu \in \mathcal{N}\}$ in D converging to point $\hat{p} \in \partial P_0^*$ in F or $\Delta \backslash D$, vector $\bar{\bar{p}} \in D$ exists such that for a certain subnet $\{p^{\nu(\mu)}, \mu \in \mathcal{M}\}$ of $\{p^\nu, \nu \in \mathcal{N}\}$, $\langle \bar{\bar{p}}, z \rangle > 0$ for all $z \in \zeta(p^{\nu(\mu)})$ for all $\mu \in \mathcal{M}$.

(D3) Walras' Law in the Strict Sense: $\forall p \in D, \langle p, z \rangle = 0$ for all $z \in \zeta(p)$.

Then, $\exists p^* \in D, 0 \in \zeta(p^*)$.

PROOF: The argument in (STEP1) in the proof of Theorem 4.2.1 may completely be repeated here, so we have for each $A \in \mathscr{F}(D)$, $p_A \in \text{co } A$ such that $\zeta(p_A) \cap L_A^\circ \neq \emptyset$. Let us consider net $\{p_A, A \in \mathscr{F}(D)\}$, where $\mathscr{F}(D)$ is directed by set-theoretical inclusion. By the existence of compact $\Delta \supset D$, we obtain subnet $\{p_{A(\nu)}, \nu \in \mathcal{N}\}$ of $\{p_A, A \in \mathscr{F}(D)\}$, satisfying that $p^* = \lim_{\nu \in \mathcal{N}} p_{A(\nu)} \in \Delta$ exists in Δ under $\sigma(F, E)$-topology. If p^* is not in $\text{int } D \subset \Delta$ and $\text{int } P_0^* \subset F$, then by (D2-3), element $\bar{p}^* \in D$ and subnet $\{p_{A(\nu(\mu))}, \mu \in \mathcal{M}\}$ of $\{p_{A(\nu)}, \nu \in \mathcal{N}\}$ exist such that $\langle \bar{p}^*, z \rangle > 0$ for all $z \in \zeta(p_{A_{\nu(\mu)}})$ for all $\mu \in \mathcal{M}$, which is impossible since for any $A(\nu(\mu)) \in \mathscr{F}(D)$ sufficiently large, we have $\bar{p}_* \in A(\nu(\mu))$, and for such $A(\nu(\mu))$ and $p_{A(\nu(\mu))}$, we have $\zeta(p_{A(\mu(\nu))}) \cap L_A^\circ \neq \emptyset$, i.e. $\exists z \in \zeta(p_{A(\nu(\mu))})$ such that $\langle \bar{p}^*, z \rangle \leq 0$. Therefore, we have p^* in $\text{int } D \subset \Delta$ and $\text{int } P_0^* \subset F$. Last, we show that $0 \in \zeta(p^*)$. If not, then by (D1-4), neighborhood $U(p^*)$ of p^* in $(F, \sigma(F, E))$ and vector $\bar{p} \in D$ exist such that $\forall q \in U(p^*) \cap D, 0 \notin \zeta(q) \implies \langle \bar{p}, \zeta(q) \rangle > 0$, which is impossible since for all sufficiently large $A(\nu(\mu))$ such that $p_{A(\nu(\mu))} \in U(p^*)$ and $A(\nu(\mu)) \ni \bar{p}$, we have $\zeta(p_{A(\nu(\mu))}) \cap L_{A(\nu(\mu))}^\circ \neq \emptyset$, i.e., element $z \in \zeta(p_{A(\nu(\mu))})$ exists such that $\langle \bar{p}, z \rangle \leq 0$. ∎

In Theorem 4.2.4, if we consider the case that E and F are finite dimensional, all (D1)-type conditions are completely natural and condition (D2-3) is almost equal to (D2-2), the weakest one of the Grandmont (1977) type. It follows that the result generalizes the argument in Neuefeind (1980) allowing for the possibility that D is not a dense subset of Δ under the weaker boundary condition (D2-3). Even with (D1-3) or (D1-4), the results in this section give the most general forms of the Gale–Nikaido–Debreu Theorem. (See, e.g., Mehta and Tarafdar (1987, Theorem 8). Note also that we do not assume the value of ζ to be compact and/or convex like our fixed-point theorems in Chapters 2 and 3.)

In all preceding theorems of this section, the condition (D3: Walras' Law) may be replaced by the following weak version of Walras' Law (used

4: Gale–Nikaido–Debreu's Theorem

in Yannelis (1985), Mehta and Tarafdar (1987)), without any changes in the proofs.

(D3-1) Weak Walras' Law: $\forall p \in D, \langle p, z \rangle \leq 0$ for a certain $z \in \zeta(p)$.

It is difficult to find any positive implication from this generalization, however, since Walras' Law from the economic viewpoint has an important meaning representing that the circulation of income is closed in a model.

4.3 Demand-Supply Coincidence in General Spaces

We end this chapter with a corollary to the theorem on the coincidence of two mappings. The result may be interpreted as the coincidence of *demand* and *supply* correspondences in economic equilibrium theory, i.e., a sort of Gale–Nikaido–Debreu's theorem. Mathematically, the result may also be classified in a generalized form of the variational inequality problem under a generalized dual-system structure.

THEOREM 4.3.1: (**Gale–Nikaido–Debreu's Theorem under a Generalized Dual-System Structure**) *Let X be a compact Hausdorff space having convex structure $\{f_A | A \in \mathscr{F}(X)\}$ and topological dual-system structure (W, V) on X. Assume that W is also a compact Hausdorff space having convex structure $\{g_A | A \in \mathscr{F}(W)\}$. Let $D : W \to 2^X$ and $S : W \to 2^X$ be two non-empty multi-valued mappings such that if $D(w) \cap S(w) = \emptyset$, open neighborhood U^w of w and point $\theta(w) \in W$ exist that satisfy for all $w' \in U^w$ and $s \in S(w')$,*

$$D(w') \subset V(s, \theta(w)). \quad (Continuity)$$

Moreover, suppose that for all $w \in W$, $\exists s \in S(w)$,

$$D(w) \subset X \backslash V(s, w). \quad (Weak\ Walras'\ Law)$$

Then there is at least one $w^* \in W$ such that $D(w^*) \cap S(w^*) \neq \emptyset$.

PROOF: Assume the contrary, i.e., for all $w \in W$, $D(w) \cap S(w) = \emptyset$. Then since W is compact, there are finite points $w^1, \ldots, w^n$ and covering $U^1 = U^{w^1}, \ldots, U^n = U^{w^n}$ of W satisfying the condition stated in the theorem. Let us consider the partition of unity subordinate to $U^1, \ldots, U^n$,

$\beta^1 : U^1 \to [0,1], \ldots, \beta^n : U^n \to [0,1]$. Define multi-valued mapping φ on W to itself as

$$\varphi : W \ni w \mapsto \{w' \in W | \forall s \in S(w), D(w) \subset V(s,w')\} \in 2^W.$$

Since $w \in U^t$ means that $\theta(w^t) \in \varphi(w)$, φ is a non-empty valued correspondence. It is convex valued by condition (V3) for V. It is also clear that for all $w \in W \backslash \mathcal{F}\!ix(\varphi)$, point $y^w \in \varphi(w)$ and open neighborhood U^w of w exist such that for all $z \in U^w \backslash \mathcal{F}\!ix(\varphi)$, we have $y^w \in \varphi(z)$. (Indeed, if $w \in U^t$, let y^w be element $\theta(w^t)$ and U^w be U^t.) Therefore, φ is a mapping that satisfies the condition in Theorem 2.3.1. Let w^* be a fixed point of φ. Then we have $\forall s \in S(w^*), D(w^*) \subset V(s,w^*)$, which contradicts Walras' Law. ∎

Bibliographic Notes

The arguments in Section 4.1 are based on Urai and Hayashi (2000).

The results in Section 4.2 are a generalization of Gale–Nikaido–Debreu's theorem in infinite dimensional vector spaces. Such arguments were first formulated by Nikaido (1959) and extended in the 1980s by Aliprantis and Brown (1983), Florenzano (1983), and Mehta and Tarafdar (1987), etc. Usually, Gale–Nikaido–Debreu's theorem is used as a lemma for equilibrium existence theorems, and for this purpose, it is sometimes too strong to assume the global continuity of excess demand mappings.

From a purely mathematical viewpoint, Gale–Nikaido–Debreu's theorem (if we neglect boundary arguments and the global continuity problems stemming from the economic equilibrium framework) may be classified as a sort of *variational inequality problem* (for a generalization of the concept in this book, see Chapter 8). Since the discussion in Section 4.3 is stated under a general topological framework (without using the vector space structure), it would be interesting to relate Theorem 4.3.1 with a variational inclusion problem like Kristály and Varga (2003).

Chapter 5
General Economic Equilibrium

5.1 General Preferences and Basic Existence Theorems

In this chapter, we apply the theorems from previous chapters to problems in the general economic equilibrium setting, especially to arguments on the existence and optimality of the equilibrium.

As in Chapter 3, let us start from arguments on a general preference relation on choice set X. Let $P : X \to X$ be a (possibly empty valued) correspondence on subset X of Hausdorff space E to itself. Assume that P satisfies

$$\text{(Irreflexivity)} \quad \forall x \in X, x \notin P(x).$$

In the following, we regard X as an individual choice set and $P(x) \subset X$ as the set of points preferred to x for each $x \in X$. Then element $x^* \in X$ may be interpreted as a *maximal element* for preference correspondence P if $P(x^*) = \emptyset$. It is sometimes beneficial to write $x \succ y$ instead of $y \in P(x)$ and call $\succ$ the *preference relation based on* $P : X \to X$.

In Chapter 3, we saw in Theorems 3.1.2 and 3.1.5 that fixed-point theorems may be easily modified to existence theorems on maximal elements. The existence of maximal elements for possibly empty-valued irreflexive mappings (preference relations) may be considered a contra-positive assertion to the existence of fixed points for non-empty valued correspondences. Since in the maximal-element existence problem, mapping P directly represents an individual preference, the importance of our generalization of mappings in fixed-point theorems (in Chapters 2 and 3) should be reconsidered in view of the generality for a representation of our general preferences.

Note that as a condition for individual preferences, local-intersection or fixed-direction properties like (K*) and (K1) for the fixed-point theorems in

Chapters 2 and 3 are not only mathematically general but also intuitively natural for describing our ordinary preference attitudes. It is, indeed, far more natural than continuity. Moreover, many concrete examples may not be treated in the standard argument but may be treated in our scope. For example, we may consider an important sort of ordered preferences that are *not continuous* and fail to have open lower sections. The most important kind is preferences based on *lexicographic ordering* (Figure 27).[1] We may also consider many kinds of ordered preferences that are *not complete* and fail to have open lower sections. (For example, consider relation $\leq$ in R^n representing strict monotonicity at each point. See Figure 28. The left picture shows a type of such orderings where $(y_1, y_2) \succ (x_1, x_2)$ if and only if ($y_1 > x_1$ and $y_2 \geq x_2$). In the right picture, $(y_1, y_2) \succ (x_1, x_2)$ if and only if ($y_1 \geq x_1$ and $y_2 \geq x_2$ and $(y_1, y_2) \neq (x_1, x_2)$). In all cases, the better sets at x are denoted by the shaded areas. Both fail to have the open lower section property at every point.)

Note that in such cases with orderings in vector spaces (including the case with lexicographic ordering), preference relations may not have open lower sections but have the local intersection property to satisfy property (K*). Hence, our treatments of preference correspondences, mainly as mappings of the Browder-type (as (P1)–(P3) in Chapter 3), are sufficient to include many natural (possibly non-continuous) preferences. (All cases

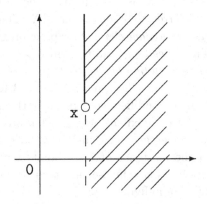

Figure 27: Lexicographic ordering

---

[1] In this case, the preference correspondence cannot even be an $\mathscr{L}$-majorized (see Chapter 8) mapping.

5: General Economic Equilibrium

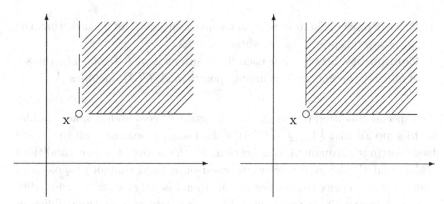

Figure 28: Orderings in vector spaces

in Figures 27 and 28 satisfy condition (K*) or (P1)–(P3). The cases in Figure 28 are also covered by (K#) and all (K1)-type conditions.)

Of course, in the following, we can see that all the generalizations in this book (not only the preference conditions but also the convex and dual-system structures on commodity and price spaces) may be utilized to construct the general equilibrium argument. As we see below, we may reconstruct the consumption, production, demand-supply arguments, etc., so that they are free from convexity and dual-system structures. By doing so, we can utilize developments in abstract convexity theory or changes in our general definitions for dual-systems, directly and independently, in general equilibrium theory without reconstructing models and arguments.

To begin with, I restate here the basic maximal-element existence result, Theorem 3.1.5, in the special settings of this chapter with alternative conditions taken as weakly as possible in view of the general equilibrium argument. Preceding the theorem, I describe several basic (minimal) settings for our economic equilibrium arguments.

(B0) [Commodity and Price Spaces]

1: **Convex Structures** Let E and F be Hausdorff spaces having convex structures $\{f_A \,|\, A \in \mathscr{F}(E)\}$ and $\{g_A \,|\, A \in \mathscr{F}(F)\}$.

2: **Dual-System Structure** Function $\langle \cdot, \cdot \rangle$ on $E \times F$ into R exists such that for each $(x,p) \in E \times F$, $\langle \cdot, p \rangle : E \to R$ and $\langle x, \cdot \rangle : F \to R$ are continuous and $V_E : E \times F \ni (x,p) \mapsto \{y \in E \,|\, \langle y, p \rangle < \langle x, p \rangle\} \subset E$ forms a dual-system structure $(E, F, V_E : E \times F \to E)$ on E satisfying (V1)–(V3). Moreover, suppose that for each $x, y \in E$ and $\alpha \in R$, the set of all $p \in F$ such that $\langle x, p \rangle > \langle y, p \rangle + \alpha$ is convex.

3: **Price Space** There is compact convex subset Δ of F such that the prices are taken in a convex subset D of Δ.
4: **Commodity Space** For each finite subset A of E (resp., D), the set of all convex combinations among points in A, $C(A)$, includes A.

The special condition on the convex structure of commodity space enables us to suppose that $\bigcup_{A \in \mathscr{F}(E)} C(A) = E$. I shall repeatedly use the above basic settings throughout this section. In the above, I wrote conditions (B0-1) and (B0-3) as if they were based on a fixed (unique) topology of commodity and price spaces. For infinite dimensional cases with commodity spaces, however, it is more desirable to prepare two or three different topologies on the commodity and price spaces. In such cases, condition (B0) merely states one such alternative topology.[2]

(B1) [Consumers and Technologies]

1: **Consumption Sets** Let I be a non-empty index set of consumers, and for each $i \in I$, let X_i be non-empty closed convex subsets of E representing the set of all consumption plans of consumer i.
2: **Technology Sets** Let J be a non-empty index set of technologies in this economy. For each $j \in J$, technology Y_j is assumed to be a non-empty closed convex subset of E.
3: **Aggregation Process for Technologies** For a list of the production plans of $j \in J$, $(y_j \in Y_j)_{j \in J}$, with the social initial endowments of this economy, $\omega \in E$, we have element $\omega + \sum_{j \in J} y_j$ of E, an aggregate production including all initial endowments, where $\omega + \sum_{j \in J}$ is supposed to be a continuous operator on $\prod_{j \in J} Y_j$ to E. Denote by $Y \subset E$ the set representing the aggregate technology (all aggregate production plans) in this economy including all initial endowments. Also assume that Y is non-empty, closed, and convex.

---

[2] For example, given duality $\langle L_\infty, ba \rangle$ chosen by T. F. Bewley's seminal paper (1972), we may identify Δ with a $\sigma(ba, L_\infty)$ compact base of non-negative cone in ba and D with $L_1 \cap \Delta$. Topology on $E = L_\infty$ is taken as $\sigma(L_\infty, L_1)$ to guarantee the relative compactness of attainable sets and is taken as $\tau(L_\infty, L_1)$ for preferences. As an alternative approach for the same duality, it is even possible to choose Mackey topology $\tau(L_\infty, ba)$ (which equals the norm and strong topology and enables us to use the joint continuity of the duality bilinear form) on L_∞ as long as the compactness of consumption and the production sets are assumed.

5: General Economic Equilibrium

4: **Aggregation Process for Consumption** For a list of all consumption plans of $i \in I$, $(x_i)_{i \in I}$, we have element $\sum_{i \in I} x_i$ of E, where $\sum_{i \in I}$ is supposed to be a continuous operator on $\prod_{i \in I} X_i$ to E.

5: **Attainable Set** We call a list of consumption plans, $(x_i)_{i \in I}$, *feasible (attainable)* if $\sum_{i \in I} x_i \in Y$. It is assumed that for each $i \in I$, the set of all i-th components of feasible (attainable) list of consumption plans, i-th *feasible components* of the economy, $\hat{X}_i$, is non-empty, convex, and relatively compact.

We call set $\mathcal{Alloc} = \prod_{i=1}^{m} X_i$ the *set of allocations* (for consumers) and set $\mathcal{Falloc} = \{((x^i)_{i=1}^{m}) \in \mathcal{Alloc} \mid \sum_{i \in I} x^i \in Y\}$ the *set of feasible (attainable) allocations*. Under (B1-4), we assume that $\mathcal{Falloc}$ is a non-empty relatively compact convex subset of $\mathcal{Alloc}$.[3]

THEOREM 5.1.1: (Maximal-Element Existence) *Under* (B0) *and* (B1), *assume that preference correspondence* $P_i : X = \prod_{j \in I} X_j \to X_i$ *satisfies one of the following conditions:* (Product B), (Product K), (Product K1-V4*), (Product K1-V4), (USC Differentiable under (V5)) *in Chapter 3, Section 3.2 (p. 67) and Section 3.3 (p. 81). For any $x \in X = \prod_{j \in I} X_j$, and a non-empty compact convex subset A_i of X_i, element $x_i^* \in A_i$ exists such that $P_i(x_i^*; x_{\hat{i}}) \cap A_i = \emptyset$, i.e., a maximal element in A_i for better-set relation $\succ_i$ defined by $P_i(\cdot ; x_{\hat{i}}) : X_i \to X_i$ on X_i.*

PROOF: The result immediately follows from Theorem 3.1.5 by restricting the duality on A_i and D instead of X_i and W_i. For the upper semi-continuously differentiable case, use Theorem 3.3.5. ■

(B2) [Income Circulation]

1: **Wealth Function** For each $i \in I$, function $W_i : \prod_{j \in J} Y_j \times D \to R$ exists such that (1) for each $(y_j)_{j \in J} \in \prod_{j \in J} Y_j$ and $p \in D$, $\sum_{j \in I} W_j(y, p) = \langle \omega + \sum_{j \in J} y_j, p \rangle$, (2) continuous function $g_i : R \to R$ exists such that $W_i(y, p) = g_i(\langle \omega + \sum_{j \in J} y_j, p \rangle)$, and (3) for each $(z_j)_{j \in J} \in \prod_{j \in J} Y_j$, $\langle z_j, p \rangle \geqq \langle y_j, p \rangle$ for all $j \in J$ implies that $W_i((z_j)_{j \in J}, p) \geqq W_i((y_j)_{j \in J}, p)$.

---

[3] As in Chapter 3 (3.1, Multi-Agent Game Setting and Product of Mappings), for allocation $x \in \prod_{i=1}^{m} X_i$, we often denote by x_i the i-th coordinate of x and by $x_{\hat{i}}$ the vector obtained by eliminating the i-th coordinate x_i of x as long as there is no fear of misunderstanding or confusion.

2: **Minimum Wealth Condition** For each $i \in I$, there are $\bar{x}_i \in X_i$ and $(y^i_j)_{j \in J} \in \prod_{j \in J} Y_j$ such that for all $p \in D$, $\langle \bar{x}_i, p \rangle < W_i((y^i_j)_{j \in J}, p)$.

3: **Aggregation of Consumption Values** For each $p \in D$ and $(x_i)_{i \in I} \in \mathcal{A}lloc$, we have $\langle \sum_{i \in I}(x_i)_{i \in I}, p \rangle \leqq \sum_{i \in I} \langle x_i, p \rangle$.

Under settings (B0), (B1), and (B2) with preference correspondence $P_i : X = \prod_{j \in I} X_j \to X_i$ for each $i \in I$, we call $\mathcal{E} = ((X_i, P_i, W_i)_{i \in I}, (Y_j)_{j \in J})$ an *economy*. An *equilibrium* of economy $\mathcal{E}$ is $(p^*, (x^*_i)_{i \in I}, (y^*_j)_{j \in J}) \in D \times \prod_{i \in I} X_i \times \prod_{j \in J} Y_j$ such that (1) for each $i \in I$, x^*_i is a $P_i(\cdot, x^*_i)$-maximal element of $B_i((y^*_j)_{j \in J}, p^*) = \{x_i \in X_i \mid \langle x_i, p^* \rangle \leqq W_i((y^*_j)_{j \in J}, p^*)\} \subset X_i$, (2) for each $j \in J$, $y^*_j \in Y_j$ satisfies $\langle y^*_j, p^* \rangle \geqq \langle y_j, p^* \rangle$ for all $y_j \in Y_j$, and (3) $(\exists y \in Y, \sum_{i \in I} x^*_i = y)$ or $((x^*_i)_{i \in I} \in \mathcal{F}alloc)$. The third condition, (3), is sometimes weakened to incorporate a systematic free disposability in the model,

$$(3') \quad \exists y \in Y, \forall p \in \Delta, \langle \sum_{i \in I} x^*_i, p \rangle \leqq \langle y, p \rangle.$$

The situation satisfying (1), (2), and (3′) is called a *free disposal equilibrium*.[4]

Note that under (B2-1) and (B2-3), we obtained Walras' Law, the closed income circulation in the model.

Walras' Law For all $p \in D$, $(y_j)_{j \in J} \in \prod_{j \in J} Y_j$, and $(x_i)_{i \in I} \in \prod_{i \in I} X_i$ such that $(x_i)_{i \in I} \in B_i((y_j)_{j \in J}, p) = \{x_i \in X_i \mid \langle x_i, p \rangle \leqq W_i((y_j)_{j \in J}, p)\}$, we have $\langle \sum_{i \in I}(x_i)_{i \in I}, p \rangle \leqq \langle \omega + \sum_{j \in J} y_j, p \rangle$.

Now we can prove several fundamental equilibrium existence theorems, which are based on the same simplified conditions on individual choice sets and duality structure such that all consumption sets X_i ($i \in I$) and production sets Y_j ($j \in J$) are compact and that duality operator $\langle \cdot, \cdot \rangle$ restricted on those sets and Δ are jointly continuous. By these circumstances, note that by Theorem 5.1.1, given any price $p \in D$, every technology $j \in J$ has at least one profit-maximizing production plan $y^*_j(p)$, $\langle p, y^*_j(p) \rangle \geqq \langle p, y_j \rangle$ for all $y_j \in Y_j$, and that every consumer $i \in I$ has at least one maximal consumption $x^*_i(p)$ with respect to his/her preference in budget set $B_i((y^*_j(p))_{j \in J}, p) = \{x_i \in X_i \mid \langle x_i, p \rangle \leqq W_i((y^*_j(p))_{j \in J}, p)\}$, where $W_i((y^*_j(p))_{j \in J}, p)$ does not

---

[4] For infinite dimensional topological vector spaces with positive cones having no interior points, the set appropriate for defining the free disposability as Δ in (3′) fails to be compact and convex. Even in such cases, if the positive cone can be approximated as a limit of cones having non-empty interiors, we can use a family of increasing compact convex sets whose totality defines the free disposability, as is treated in the next section.

5: General Economic Equilibrium

depend on the choice of special profit-maximizing plans by (B2-1). Moreover, the value of wealth $W_i((y_j^*(p))_{j\in J}, p)$ is easily seen to be a continuous function with respect to price p.[5]

THEOREM 5.1.2: (Existence of Economic Equilibrium: Compact Domains with Browder-Type Preferences) *Under* (B0), (B1), *and* (B2), *assume* (1) *that for each* $i \in I$ *and* $j \in J$, X_i *and* Y_j *are compact*, (2) *that the restrictions of* $\langle \cdot, \cdot \rangle$ *on* $\Delta \times X_i$ *and* $\Delta \times Y_j$ *are jointly continuous, and* (3) *that* $D = \Delta$. *Moreover, suppose the following conditions:*

(i) *For each* $i \in I$, P_i *is the (Product BK1)-type.*
(ii) *For each* $i \in I$, *mapping* $(x, y, p) \mapsto P_i(x) \cap B_i(y, p)$ *is the (Product BK1)-type.*

Then $\mathcal{E}$ *has free disposal equilibrium* $(p^*, (x_i^*)_{i\in I}, (y_j^*)_{j\in J}) \in D \times \prod_{i\in I} X_i \times \prod_{j\in J} Y_j$. (*When each* P_i *has open values and the open lowersection property,* (ii) *is automatically satisfied since for each* $p \in D$ *and* $\alpha \in R$, $\{x \in X \mid \langle x, p \rangle < \alpha\}$ *is dense in* $\{x \in X \mid \langle x, p \rangle \leqq \alpha\}$.)

PROOF: Consider an abstract economy that consists of consumers I, producers J and an auctioneer. The producer's behavior (better-set correspondence) is to take better (more profitable) production plans, $(x, y, p) \mapsto \{y_j' \in Y_j \mid \langle y_j', p \rangle > \langle y_j, p \rangle\}]$, and the auctioneer's better-set correspondence is to give prices evaluating excess demand more highly, $(x, y, p) \mapsto \{q \in D \mid \langle \sum_{i\in I} x_i, q \rangle - \langle \omega + \sum_{j\in J} y_j, q \rangle > \langle \sum_{i\in I} x_i, p \rangle - \langle \omega + \sum_{j\in J} y_j, p \rangle\}$. By the joint continuity of $\langle \cdot, \cdot \rangle$, the dual-system structure satisfies (V4*), so budget correspondence B_i is the (Product BK1)-type under (V4*) such that $(x, y, p) \mapsto \text{int } B_i(y, p)$ has the open lowersection property for all $i \in I$. Hence, by Theorem 3.3.3, the abstract economy has an equilibrium. Note that the set of interior points of $B_i(y, p)$, which is non-empty by (B2-2), is always dense in itself. Therefore, if each P_i has open values and open lowersection property, $P_i(x) \cap B_i(y, p)$ is the (Product B) type. It is a routine task to check whether the equilibrium of the abstract economy is a free disposal equilibrium of the economy. ∎

---

[5]For this assertion, by (B2-1), it is sufficient to show that maximum value $\langle y_j^*(p), p \rangle$ is continuous with respect to p. Consider net $p^\nu \to p \in D$. Since for all $\epsilon > 0$, $\langle y_j^*(p), p \rangle + \epsilon > \langle y_j^*(p^\nu), p \rangle + \epsilon/2 > \langle y_j^*(p^\nu), p^\nu \rangle$ for all sufficiently large ν, the limit supremum of $\langle y_j^*(p^\nu), p^\nu \rangle$ is less than or equal to $\langle y_j^*(p), p \rangle$. On the other hand, since $\langle y_j^*(p^\nu), p^\nu \rangle \geqq \langle y_j^*(p), p^\nu \rangle$ for all ν, the limit infimum of $\langle y_j^*(p^\nu), p^\nu \rangle$ is greater than or equal to $\langle y_j^*(p), p \rangle$.

As noted in the previous proof, by jointly continuous duality operator $\langle \cdot, \cdot \rangle$, the dual-system structure satisfies all topological conditions $(V4)$, $(V4^*)$, and $(V5)$ in Chapter 3. Since in such contexts, the continuous differentiability of preferences (see Chapter 3) is more general than the K1-type condition, we can summarize such K1-type arguments by the following theorem:

THEOREM 5.1.3: (Existence of Economic Equilibrium: Compact Domains with USC Differentiable Preferences) *Under* (B0), (B1), *and* (B2), *assume* (1) *that for each* $i \in I$ *and* $j \in J$, X_i *and* Y_j *are compact*, (2) *that the restrictions of* $\langle \cdot, \cdot \rangle$ *on* $\Delta \times X_i$ *and* $\Delta \times Y_j$ *are jointly continuous, and* (3) *that* $D = \Delta$. *Moreover, suppose the following conditions*:

(i) *For each* $i \in I$, P_i *is upper semi-continuously differentiable.*
(ii) *For each* $i \in I$, $\{(x, y, p) \mid P_i(x) \cap B_i(y, p) \neq \emptyset\} \subset \prod_{i \in I} X_i \times \prod_{j \in J} Y_j \times D$ *is open.*

Then $\mathscr{E}$ *has free disposal equilibrium* $(p^*, (x_i^*)_{i \in I}, (y_j^*)_{j \in J}) \in D \times \prod_{i \in I} X_i \times \prod_{j \in J} Y_j$. *If each* P_i *has open values and the open lowersection property, and if for each* $p \in D$ *and* $\alpha \in R$, $\{x \in X \mid \langle x, p \rangle < \alpha\}$ *is dense in* $\{x \in X \mid \langle x, p \rangle \leq \alpha\}$, *condition* (ii) *is unnecessary.*

PROOF: Note that the budget correspondences are also upper semi-continuously differentiable. The proof is essentially the same as that of the previous theorem except for using Theorem 3.3.5. The last assertion is also obvious as before since by (B2-2), the budget correspondences always have interior points at which, by the joint continuity of $\langle \cdot, \cdot \rangle$, the local intersection property holds. ■

As Browder- and K1-type preferences, the same result holds for Kakutani-type preferences.

THEOREM 5.1.4: (Existence of Economic Equilibrium: Compact Domains with Kakutani-Type Preferences) *Under* (B0), (B1), *and* (B2), *assume* (1) *that for each* $i \in I$ *and* $j \in J$, X_i *and* Y_j *are compact*, (2) *that the restrictions of* $\langle \cdot, \cdot \rangle$ *on* $\Delta \times X_i$ *and* $\Delta \times Y_j$ *are jointly continuous, and* (3) *that* $D = \Delta$. *Moreover, suppose the following conditions*:

(i) *For each* $i \in I$, P_i *satisfies Condition* (*Product K*) *with* (*Closedness*).
(ii) *For each* $i \in I$, *mapping* $(x, y, p) \mapsto P_i(x) \cap B_i(y, p)$ *satisfies Condition* (*Product K*) *with* (*Closedness*).

Then $\mathscr{E}$ has free disposal equilibrium $(p^*, (x_i^*)_{i \in I}, (y_j^*)_{j \in J}) \in D \times \prod_{i \in I} X_i \times \prod_{j \in J} Y_j$. (When each P_i has open values and the open lowersection property, condition (ii) is unnecessary since for each $p \in D$ and $\alpha \in R$, $\{x \in X \mid \langle x, p \rangle < \alpha\}$ is dense in $\{x \in X \mid \langle x, p \rangle \leqq \alpha\}$.)

PROOF: Note that the budget correspondences are also the (Product K) type. The proof is essentially the same as that of Theorem 5.1.2 except for using Theorem 3.3.6. ∎

5.2 Pareto Optimal Allocations

We next consider the problem of the existence of maximal elements among many agents, i.e., social optima. As in the standard arguments, under our general settings of the spaces and preference conditions to obtain a relation between social optima and equilibria, however, it is desirable to characterize our preferences based not only on better sets but also on indifferent points.[6] Once we have a correspondence representing (possibly non-ordered) the relation "preferred or indifferent to" for each i, however, it is not difficult to obtain the *first fundamental theorem of welfare economics* (every price equilibrium allocation is Pareto optimal), merely by assuming the next condition:

(C) For each $i \in I$ and maximal consumption plan $x_i^*(p)$ under $p \in D$, if x_i' is preferred or indifferent to $x_i^*(p)$, then $\langle x_i^*(p), p \rangle \leqq \langle x_i', p \rangle$.

Indeed, if we assume (C) with an additional assumption that the inequality in condition (B2-3) always holds with equality, then we may confirm it merely through a simple aggregation of allocations and the evaluation under p.[7]

With (B0) and (B1), let us assume that for each $i = 1, \ldots, m$, two correspondences $P_i : \textbf{\textit{Alloc}} \to X_i$ and $\tilde{P}_i : \textbf{\textit{Alloc}} \to X_i$, the *strong* and *weak*

---

[6] For weak Pareto optima (characterized only by strict preferences) to be Pareto optima, conditions like open valuedness with local non-satiation would be necessary. When we consider better-set correspondences that may not have such open graph properties, as in cases with lexicographic preferences, therefore, there is no warranty for equilibrium allocations (characterized by strict preferences) to be Pareto optimal. (Recall that the existence of "demand saturation" gives an example such that price equilibrium allocations fail to be Pareto optimal.)

[7] In our general setting, the *second welfare theorem* (every Pareto optimal allocation can be regarded as a general equilibrium allocation by changing the wealth functions) is rather appropriate to be classified as a requirement for the system (in some cases, we can obtain the assertion as a theorem).

preference correspondences, respectively, satisfy that $\forall x = (x^1, \ldots, x^m) \in$ ***Alloc***, $x^i \notin P_i(x)$ (irreflexivity), $x^i \in \tilde{P}_i(x)$ (reflexivity), and $P_i(x) \subset \tilde{P}_i(x)$. Then, define correspondence, $P : $ ***Alloc*** $\to$ ***Alloc***, as follows:

$$P(x) = \{w = (w^1, \ldots, w^m) \mid \forall i, w^i \in \tilde{P}_i(x) \text{ and } \exists i, w^i \in P_i(x)\}.$$

For allocations $x, y \in $ ***Alloc***, y is said to be *Pareto superior* to x if and only if condition $y \in P(x)$ is satisfied. It is easy to check that for all $z \in $ ***Alloc***, $z \notin P(z)$, i.e., $P : $ ***Alloc*** $\to$ ***Alloc*** is an irreflexive correspondence. A feasible allocation $x \in $ ***Falloc*** is said to be *Pareto optimal* if $P(x) \cap $ ***Falloc*** $= \emptyset$. The next theorem shows that the existence of Pareto optimal allocations may be assured even for cases with non-ordered, non-convex, and non-continuous individual preferences. Note that the necessary condition for individual preferences to assure the existence of social optima (Pareto optimal allocations), (D), is essentially the same as the necessary condition for the existence of individual optima (maximal elements) in Theorem 5.1.1.

THEOREM 5.2.1: (Existence of Pareto Optimal Allocations)
Assume that for each $i = 1, \ldots, m$, P_i and $\tilde{P}_i$ satisfy the following conditions for directions of mappings:

(D) *For each $x \in $ **Alloc**, (for each $x \in $ **Alloc** such that $P_i(x) \neq \emptyset$, resp.), Neighborhood U_x of x and upper semicontinuous compact valued correspondence $\theta^i_x : U_x \to E'_X$ exist such that $\langle p, w_i - z_i \rangle \geq 0$ ($\langle p, w_i - z_i \rangle > 0$, resp.) for all $p \in \theta^i_x(z)$, $w_i \in \tilde{P}_i(z)$, ($w_i \in P_i(z)$, resp.), and $z = (z_1, \ldots, z_m) \in U_x$.*

Then there is a Pareto optimal allocation.

PROOF: Assume the contrary. Then correspondence $P_F : $ ***Falloc*** $\ni x \mapsto P(x) \cap $ ***Falloc*** is non-empty valued. By condition (D) and the compactness of ***Alloc***, through the argument using the partition of unity, we obtain an upper semicontinuous compact valued correspondence $\theta^i : $ ***Alloc*** $\to E'_X$ for each i such that $\forall z \in $ ***Alloc*** and $\forall p \in \theta^i(z)$, $(w_i \in \tilde{P}_i(z)) \to (\langle p, w_i - z_i \rangle \geq 0)$ and $(w_i \in P_i(z)) \to (\langle p, w_i - z_i \rangle \geq 0)$. For each i and $z \in $ ***Alloc***, denote by $\hat{P}_i(z)$, (resp., $\hat{\tilde{P}}_i(z)$), the set $\{w_i \in X_i \mid \forall p \in \theta^i(z), \langle p, w_i - z_i \rangle > 0\}$, (resp., set $\{w_i \in X_1 \mid \forall p \in \theta^i(z), \langle p, w_i - z_i \rangle \geq 0\}$). Clearly, for each i and $z \in $ ***Alloc***, $\hat{P}_i(z) \supset P_i(z)$ and $\hat{\tilde{P}}_i(z) \supset \tilde{P}_i(z)$. Since $\hat{P}_i(z)$ is open, and since the sum operation is continuous, the non-emptiness of P_F means that for each $w \in $ ***Falloc***, neighborhood O_w in ***Falloc*** and index of consumer $i(w)$ exist such that for each $z \in O_w$, there is an element $y = (y_1, \ldots, y_m) \in $ ***Falloc***

5: General Economic Equilibrium

such that $y_{i(w)} \in \hat{P}_{i(w)}(z)$. Since $\mathcal{F}\!alloc$ is compact, there is a finite covering $O_{w^1}, \ldots, O_{w^k}$ of $\mathcal{F}\!alloc$ and partition of unity $\alpha_t : O_{w^t} \to [0,1], t = 1, \ldots, k$, subordinate to it. Let us define correspondence Φ on $\mathcal{F}\!alloc$ to itself as

$$\Phi(z) = z + \alpha^t(z)(\Phi(z)^t - z),$$

where $\Phi^t(z)$ equals

$$\hat{\hat{P}}_1(z) \times \cdots \times \hat{\hat{P}}_{i(w^t)-1}(z) \times \hat{\hat{P}}_{i(w^t)}(z) \times \hat{\hat{P}}_{i(w^t)+1} \times \cdots \times \hat{\hat{P}}_m(z) \cap \mathcal{F}\!alloc.$$

Φ is convex valued since each Φ^t is. Φ has non-empty valued since each Φ^t is non-empty valued as long as P_F is. Moreover, by defining $\theta(z) : \mathcal{A}\!lloc \to \mathcal{A}\!lloc$ and $\Psi : \mathcal{F}\!alloc \to \mathcal{F}\!alloc$, respectively, as

$$\theta(z) = (\theta^1(z), \ldots, \theta^m(z)), \quad \text{and}$$

$$\Psi(z) = \{w \in \mathcal{F}\!alloc \mid \forall p \in \theta(z), \langle p, w \rangle > 0\},$$

we have $\Phi(z) \subset \Psi(z)$, so that correspondence Ψ is a non-empty convex valued on $\mathcal{F}\!alloc$ to $\mathcal{F}\!alloc$ having no fixed point. Note that θ is also a compact valued upper semicontinuous correspondence on $\mathcal{A}\!lloc$ to $E'^{(m)}_X$, where $E'^{(m)}_X$ denotes the m-th product of E'_X. Then for all $w^* \in \Psi(z^*)$, there is an $\epsilon^* > 0$ such that $\langle p, w^* \rangle > \epsilon^* > 0$ by the compactness of $\theta(z^*)$; hence, by the joint continuity of the duality operation (precisely, for each component), neighborhood O^* of z^* exists such that for all $z \in O^*$, $w^* \in \Psi(z)$. Therefore, Ψ has the local intersection property at each z, so that by Theorem 2.3.1, Ψ has a fixed point, and we have a contradiction. ∎

5.3 Existence of General Equilibrium

In Section 5.1, we treated the existence of general economic equilibria for cases with compact domains and the joint continuity of dual-system operator $\langle \cdot, \cdot \rangle$. We cannot generally expect, however, the joint continuity of the dual-system operator unless our attention is restricted on finite dimensional vector spaces. It is also restrictive to assume that consumption and production sets are compact even in cases with finite dimensional vector spaces. Moreover, a compact domain may not exist for prices that define appropriately the free disposability, especially when an infinite dimensional vector space fails to have the positive cone with interior points.

In this section, we extend the basic existence theorems to cases with (1) consumption and production sets that are not assumed to be compact without changing the assumptions for price spaces and dual-system operators (including standard arguments in the finite dimensional Arrow–Debreu economy), (2) possibly non-compact consumption and production sets with a dual-system operator that is not assumed to be jointly continuous while we leave the condition for the price space unchanged (e.g., typical infinite dimensional settings like Bewley's $\langle L_\infty, ba \rangle$), and, furthermore, (3) possibly non-compact consumption and production sets, possibly non-jointly-continuous dual-system operators, and the price space with no appropriate compact convex set defining the free disposability in the economy (e.g., cases with commodity spaces whose positive cones may not have interior points, as treated by Nikaido (1959) and Mas-Colell (1986)).

5.3.1 Non-compact consumption and production sets

Even for classical general equilibrium settings like Arrow–Debreu (1954) and Nikaido (1956a), the consumption and production sets are usually not restricted to be compact. In such cases, however, they are instead based on sufficient conditions for assuring the compactness of attainable (feasible) allocations in the economy. For example, if $E = R^\ell$ and all X_i are R^ℓ_+, it is often assumed that the intersection of Y (the set of all aggregate production plans including initial endowments) and R^ℓ_+ is bounded, hence it is compact. Usually, this automatically implies the boundedness for set $\hat{X}_i$ of all i-th components of feasible list of consumption plans.[8] For the boundedness of the set of all j-th components of feasible production plans, $\hat{Y}_j$, however, there seem to be no suitable conditions except for assuming the boundedness (compactness) directly for each $j \in J$. We often use the next assumption:

(B3) [Boundedness Assumption] Topology τ_w exists under which $\langle \cdot, q \rangle$ is continuous for every price $q \in D$ satisfying the following conditions:

1: **Attainable Consumption Plans** For each $j \in I$, $\hat{X}_j = \{x_j \in X_j \,|\, \forall i \in I, i \neq j, \exists x_i \in X_i, \sum_{i \in I} x_i \in Y\}$ is relatively compact under τ_w.

---

[8]Even for infinite dimensional vector spaces, norm or order boundedness may possibly be utilized for ensuring the compactness for attainable sets as long as we appropriately choose the topology for commodity spaces. For example, in AL- (abstract L-) spaces, order bounded sets are weakly relatively compact (see Schaefer (1971, p. 249, 8.6, Corollary 1)). Norm bounded sets may also be used as relatively compact sets under the weak star topology by the theorem of Alaoglu–Bourbaki (see Schaefer (1971, p. 84, 4.3, Corollary)).

5: General Economic Equilibrium

2: Attainable Production Plans For each $i \in J$, $\hat{Y}_i = \{y_i \in Y_i \mid \forall k \in I, \exists x_k \in X_k, \forall j \in J, j \neq i, \exists y_j \in Y_j, \sum_{k \in I} x_k = \omega + \sum_{j \in J} y_j \in Y\}$ is relatively compact under τ_w.

This section treats the simplest case where compact convex set K has interior points in E such that all individual feasible plans, $\hat{X}_i$, $i \in I$, and $\hat{Y}_j$, $j \in J$ are contained in int K.

THEOREM 5.3.1: (Existence of General Equilibrium I) *Under (B0), (B1), and (B2), assume that (1) compact convex set K that has interior points in E exists such that for each $i \in I$ and $j \in J$, the set of individual feasible consumption and production plans, $\hat{X}_i$ and $\hat{Y}_j$ are contained in* int K, *(2) restrictions of $\langle \cdot, \cdot \rangle$ on $\Delta \times X_i$ and $\Delta \times Y_j$ are jointly continuous, and (3) $D = \Delta$. Moreover, suppose the following conditions:*

(i) *For each $i \in I$, P_i is the (Product BK1)-type.*
(ii) *Each P_i satisfies that for all $x \in X$ and $y_i \in P_i(x)$, $C(\{x_i, y_i\}) \setminus \{x_i\}$ is included in $P_i(x)$.*
(iii) *For each $j \in J$, for all $y_j, z_j \in Y_j$, and $p \in \Delta$, $\langle z_j, p \rangle > \langle y_j, p \rangle$ implies that $\langle z'_j, p \rangle > \langle y_j, p \rangle$ for all $z'_j \in C(\{z_j, y_j\}) \setminus \{y_j\}$.*
(iv) *For each $i \in I$, mapping $(x, y, p) \mapsto P_i(x) \cap B_i(y, p)$ is the (Product BK1)-type.*

Then $\mathscr{E}$ has free disposal equilibrium $(p^*, (x_i^*)_{i \in I}, (y_j^*)_{j \in J}) \in D \times \prod_{i \in I} X_i \times \prod_{j \in J} Y_j$. (When each P_i has open values and the open lowersection property, condition (iv) is unnecessary since for each $p \in D$ and $\alpha \in R$, $\{x \in X \mid \langle x, p \rangle < \alpha\}$ is dense in $\{x \in X \mid \langle x, p \rangle \leq \alpha\}$.)

PROOF: Let us define $\tilde{X}_i$ as $\tilde{X}_i = X_i \cap K$ and $\tilde{Y}_j$ as $\tilde{Y}_j = Y_j \cap K$ for each $i \in I$ and $j \in J$. Restrict all P_i and W_i on $\tilde{X} = \prod_{i \in I} \tilde{X}_i$ and $\prod_{j \in J} \tilde{Y}_j \times \Delta$, respectively, and consider economy $\tilde{\mathscr{E}} = ((\tilde{X}_i, P_i, W_i)_{i \in I}, (Y_j)_{j \in J})$. By Theorem 5.1.2, economy $\tilde{\mathscr{E}}$ has equilibrium $(p^*, (x_i^*)_{i \in I}, (y_j^*)_{j \in J})$. It is a routine task to check that conditions (ii) and (iii) are sufficient for assuring that x_i^* and y_j^* are individual maxima not only in $X_i \cap K$ and $Y_j \cap K$ but also in X_i and Y_j for each i and j. ∎

In the above theorem, condition (i) for preferences may be replaced by K1- and Kakutani-types. (For proofs, use Theorems 5.1.3 and 5.1.4 instead of Theorem 5.1.2, respectively.)

5.3.2 For cases with general dual-system operators

For infinite dimensional vector spaces, we cannot generally expect the dual-system operator $\langle \cdot, \cdot \rangle$ to be jointly continuous. The following is a list of conditions that enable us to use the limit argument of the Bewley type (Bewley 1972) for the existence of general equilibrium.

(B4) [Miscellaneous Additional Assumptions]

1: **Convex Structure** For each non-empty finite subset $B \in \mathscr{F}(E)$ of E, set $C(B)$ of all convex combinations among points in B is compact and convex.
2: **Duality Operator** For each $B \in \mathscr{F}(E)$, duality operator $\langle \cdot, \cdot \rangle$ is jointly continuous on $C(B) \times F$.
3: **Aggregation of Production Values** For each $(z_j)_{j \in J} \in \prod_{j \in J} Y_j$ such that $\langle z_j, p \rangle \geqq \langle y_j, p \rangle$ for all $j \in J$ and for at least one $j \in J$, $\langle z_j, p \rangle > \langle y_j, p \rangle$, there is at least one $i \in J$ such that $W_i((z_j)_{j \in J}, p) > W_i((y_j)_{j \in J}, p)$.
4: **Aggregation of Consumption Values** For each $x \in \mathbf{\mathit{Alloc}}$ and $p \in D$, we have $\langle \sum_{i \in I} x_i, p \rangle = \sum_{i \in I} \langle x_i, p \rangle$.
5: **Continuity of Wealth Function** For each $i \in I$ and $(y_j)_{j \in J} \in \prod_{j \in J} Y_j$, $W_i((y_j)_{j \in J}, \cdot)$ is continuous on D.
6: **Attainable Allocations** The boundedness assumption (B3) for attainable allocations holds even when we replace Y with $Y - C$, where $Y - C$ denotes the set defining free disposability, i.e., $Y - C = \{z \in E \mid \exists y \in Y, \forall p \in \Delta, \langle z, p \rangle \leqq \langle y, p \rangle\}$.

Conditions are quite standard (indeed, except for the last one, they are unnecessary) for ordinary arguments under vector space settings. Condition (B4-6) is also standard as requirements for the existence of free disposal equilibria (see, e.g., Toussaint 1984).

THEOREM 5.3.2: (Existence of General Equilibrium II) *Under* (B0), (B1), (B2), *and* (B4), *with conditions* $I = \{1, 2, \ldots, m\}$, $J = \{1, 2, \ldots, n\}$ *(the sets of consumers and producers are finite), and that* $D = \Delta$, *suppose the following*:

(i) *For each* $i \in I$, P_i *is the (Product BK1) type.*
(ii) *For any allocation x such that for some $y \in Y$, $\langle \sum_{i \in I} x_i, p \rangle \leqq \langle y, p \rangle$ for all $p \in \Delta$, set $P_i(x)$ is non-empty for all $i \in I$ (non-satiation property).*

5: General Economic Equilibrium

(iii) *Each P_i satisfies for all $x \in X$ and $y_i \in P_i(x)$, that $C(\{x_i, y_i\}) \setminus \{x_i\}$ is included in $P_i(x)$ (convexity and local non-satiation properties) and that $P_i^{-1}(y_i)$ includes a neighborhood of x in E under τ_w (open lowersection property under τ_w).*

(iv) *Each P_i has open values, and for each $p \in D$ and two points x and x' in X_i such that $\langle x', p \rangle < \langle x, p \rangle$, we can take $z \in C(\{x', x\})$ satisfying $\langle z, p \rangle < \langle x, p \rangle$ arbitrarily near to x.*

Then $\mathscr{E}$ has a free disposal equilibrium $(p^*, (x_i^*)_{i \in I}, (y_j^*)_{j \in J}) \in D \times \prod_{i \in I} X_i \times \prod_{j \in J} Y_j$.

PROOF: Let $\tilde{B} \in \mathscr{F}(E)$ be a finite set including all points, $\bar{x}_i \in X_i$, $i \in I$, $y_j^i \in Y_j$, $j \in J$, stated in (B2-2) ensuring the minimum wealth condition. For each $B \in \mathscr{F}(E)$ such that $B \supset \tilde{B}$, define an economy $\mathscr{E}^B$ as $\mathscr{E}^B = ((X_i^B, P_i, W_i)_{i \in I}, (Y_j)_{j \in J})$, where $X_i^B = C(B) \cap X_i$ for each $i \in I$, $Y_j^B = C(B) \cap Y_j$ for each $j \in J$, and P_i and W_i are restrictions on them for each $i \in I$. Then by (B4-1) and (B4-2), economy $\mathscr{E}^B$ satisfies all conditions in Theorem 5.1.2, so $\mathscr{E}^B$ has an equilibrium $(p^B, (x_i^B)_{i \in I}, (y_j^B)_{j \in J}) \in D \times \prod_{i \in I} X_i^B \times \prod_{j \in J} Y_j^B$. By (B0-3) and (B4-6), there are converging subnets $(p^{B(\nu)})_{\nu \in \mathcal{N}}$, $(x_i^{B(\nu)})_{\nu \in \mathcal{N}}$, $i \in I$, and $(y_j^{B(\nu)})_{\nu \in \mathcal{N}}$, $j \in J$, with their limits, $p^* \in D = \Delta$, $x_i^* \in \hat{X}_i$, $i \in I$, and $y_j^* \in \hat{Y}_j$, $j \in J$, where $\hat{X}_i$ and $\hat{Y}_i$ are defined to be compact through set $Y - C$, as stated in (B4-6). In the following, we see that for each $i \in I$ if $x_i \in P_i(x^*)$, where $x^* = (x_1^*, x_2^*, \ldots, x_m^*)$, then for any $y_j \in Y_j$, $j \in J$, we have

$$\langle x_i, p^* \rangle \geqq W_i((y_j)_{j \in J}, p^*). \tag{5.1}$$

Indeed, if we take sufficiently large $B(\nu) \supset \tilde{B}$ having x_i and all y_j, $j \in J$ as its elements, and if we consider that $x^{B(\nu)} = (x_1^{B(\nu)}, \ldots, x_m^{B(\nu)})$ converges to x^* under τ_w with the open lowersection property in (iii), we may suppose $x_i \in P_i(x^{B(\nu)})$, so we have

$$\langle x_i, p^{B(\nu)} \rangle > \langle x_i^{B(\nu)}, p^{B(\nu)} \rangle = W_i((y_j^{B(\nu)})_{j \in J}, p^{B(\nu)}) \geqq W_i((y_j)_{j \in J}, p^{B(\nu)}).$$

The equality holds by (ii) and the local non-satiation property in (iii) with the joint continuity on $C(B(\nu)) \times D$ of duality function $\langle \cdot, \cdot \rangle$. The last $\geqq$ holds by (B2-1). Hence, by taking limit $p^{B(\nu)} \to p^*$, by (B4-5), we have Equation (5.1). Since $\langle \sum_{i \in I}(x_i^{B(\nu)})_{i \in I}, q \rangle \leqq \langle \omega + \sum_{j \in J}(y_j^{B(\nu)}), q \rangle$ for all q (condition of free disposal equilibrium) for all ν, we have $\langle \sum_{i \in I}(x_i^*)_{i \in I}, q \rangle \leqq \langle \omega + \sum_{j \in J}(y_j^*), q \rangle$ in the limit (under τ_w) for all q, so x^* and y^* satisfy the free disposal market equilibrium condition. Note, especially that $\langle\sum_{i\in I}(x_i^*)_{i\in I}, p^*\rangle \leqq \langle\omega + \sum_{j\in J}(y_j^*), p^*\rangle$ holds. By the local non-satiation property in (iii), we may choose $x_i \in P_i(x^*)$ arbitrarily near to x_i^*, so Equation (5.1) gives

$$\langle x_i^*, p^*\rangle \geqq W_i((y_j)_{j\in J}, p^*)$$

for all $(y_j)_{j\in J} \in \prod_{j\in J} Y_j$. Under (B4-4), by summing up the above relation for $i \in I$, we obtain

$$\sum_{i\in I} W_i((y_j^*)_{j\in J}, p^*) = \left\langle \omega + \sum_{j\in J}(y_j^*), p^* \right\rangle \geqq \left\langle \sum_{i\in I}(x_i^*)_{i\in I}, p^* \right\rangle$$
$$= \sum_{i\in I} \langle x_i^*, p^*\rangle \geqq \sum_{i\in I} W_i((y_j)_{j\in J}, p^*) \qquad (5.2)$$

for all $(y_j)_{j\in J} \in \prod_{j\in J} Y_j$. This means, by (B4-3), that for each $j \in J$, y_j^* is a profit maximizing production plan under p^* for each $j \in J$. Equation (5.2) also shows that

$$\langle x_i^*, p^*\rangle = W_i((y_j^*)_{j\in J}, p^*)$$

for all $i \in I$. Hence, x_i^* satisfies the budget constraint under p^* for each $i \in I$. Equation (5.1) and the previous one assert that every x_i such that $\langle x_i, p^*\rangle < \langle x_i^*, p^*\rangle$ does not belong to $P_i(x^*)$, especially point $\bar{x}_i$ assuring the minimum wealth condition for i is not an element of $P_i(x^*)$. If $P_i(x^*) \cap B_i((y_j^*)_{j\in J}, p^*) \neq \emptyset$, in the convex combination between two points x_i and $\bar{x}_i$, $C(\{x_i, \bar{x}_i\})$, we can take z_i satisfying $\langle z_i, p^*\rangle < \langle x_i, p^*\rangle$ arbitrarily near to x_i by condition (iv), which contradicts the fact that $x_i \in P_i(x^*)$ and $P_i(x^*)$ is open. Hence, we have $P_i(x^*) \cap B_i((y_j^*)_{j\in J}, p^*) = \emptyset$. ∎

5.3.3 Price domains having no compact bases

In this subsection, we replace the condition that $D = \Delta$, the existence of compact convex set Δ appropriately defining the notion of free disposability. This is a typical situation for infinite dimensional vector spaces with positive cones having no interior points since dual cone C' of C (the polar of $-C$) with respect to the duality $\langle E, E'\rangle$ may fail to have a compact base in such cases. In view of economics, however, equilibrium problems that must be treated in such a large (infinite dimensional) space under a weak topology (such that the positive cone has no interior points) does not seem to occur so often. Of course, we may consider many natural situations in commodity spaces like R^∞, where positive cones have no interior points. But in many

5: General Economic Equilibrium

cases it is also possible to transform such arguments into smaller commodity spaces such as $\ell^\infty \subset R^\infty$, where we may be allowed to define stronger (e.g., norm) topologies under which the same positive cones have interior points. Note that the problem of cones having no interior points is also typical for cases with ordinary finite dimensional vector commodity spaces with possibly "negative" prices defining the free disposability through the polar cones of prices. Since problems deeply depend on vector space structures, we assume on E and F vector space structures and the duality structure separating points in each of the other space throughout this section. On E, we also consider Hausdorff vector space topology $\mathscr{T}_E$ which is at least as strong as $\sigma(E, F)$.

Suppose that we have a set $Y - C \subset E$ which is defined by the set Y (social production possibility including endowments) and a non-empty closed convex cone (with vertex 0) $P \subset F$ of values on E as

$$Y - C = Y - \{x \in E \mid \forall p \in P, \langle x, p \rangle \geq 0\}.$$

Intuitively, $P \subset F$ defines the *free disposability* condition in this model as $Y - C \subset E$. The problem (in cases allowing for negative prices or for some infinite dimensional vector spaces) is that when the cone

$$-C = \{x \in E \mid \forall p \in P, \langle x, p \rangle \leq 0\},$$

characterizing the free disposability under the vector-space structure, fails to have interior points, we cannot expect P to have an appropriate compact convex subset on which we may restrict our attention to seek an equilibrium price.

If cone $-C$ has interior point $-u$, there is a 0-neighborhood U such that $-u + U \subset -C$. Define non-empty convex set $P_u = \{p \in P \mid \langle u, p \rangle = 1\}$. Since $-u$ is an interior point of $-C$, we have for all $p \in P$, $\langle -u, p \rangle < 0$, so for all $p \in P$, there is a $\lambda \in R_{++}$ such that $\lambda p \in P_u$. Moreover, P_u is $\sigma(F, E)$-compact as a closed subset of set $\{q \in F \mid \forall x \in U, \langle x, q \rangle \leq 1\}$, which is compact under the $\sigma(F, E)$-topology by the theorem of Alaoglu.[9] When set $-C$ has no interior points, we cannot use all the above arguments to define appropriate compact base P_u for the domain of equilibrium prices.

---

[9]Set $H = \{q \in F \mid \forall x \in U, \langle x, q \rangle \leq 1\} \subset F$ under $\sigma(F, E)$-topology can, by the definition of weak star topology, be identified with a subset of R^E under the product topology, through mapping $h : q \mapsto (q(x))_{x \in E}$. Image $h(H)$ is contained in compact set $[-1, 1,]^E$ and is shown to be closed since $f(F) = f(E')$ is closed in R^E. (Alaoglu's Theorem)

In the following, we suppose a pair of sets $Y - C \subset E$ and $P \subset F$, where $-C$ is the polar cone of P. Cone C does not necessarily have an interior point, so P may not have an appropriate compact base. We use the condition that enables us to approximate $Y - C$ and P through a pair of sets, $Y - C_V \subset E$ and $P_V \subset F$, defined for each V in a neighborhood base $\mathscr{V}$ of a certain element v of C such that C_V has v as an interior point and P_V has a non-empty compact convex subset having $-C_V$ as its polar cone.

(B5) [Commodity Space Having Cone without Interior Points]

1: **Vector Space Structure** E and F are vector spaces with standard convex structures and a system of duality $\langle f, E, F \rangle$ (i.e., E and F separate each other's points).

2: **Topology**   On E consider a Hausdorff topological vector space topology $\mathscr{T}_E$ that is at least as strong as $\sigma(E, F)$ and satisfies that all convex $\mathscr{T}_E$ closed sets are $\sigma(E, F)$ closed.

3: **Conditions (B1) and (B2)**   Under the topological vector-space structures stated above, conditions (B1) and (B2) are satisfied, where the convexity and the aggregation process (summation) for consumers and producers are interpreted as those derived from the vector-space structures.

4: **Cone Defining Free Disposability**   Sets $Y - C \subset E$ and $P \subset F$ exist, where Y is the aggregate production set including social endowments, and P is a non-empty closed cone with vertex 0 such that $-C \neq \emptyset$ is the polar cone of P.

5: **Approximation of the Cone without Interior Points**   An element v of C exists such that (1) for family $\mathscr{V}$ of neighborhood base at v (directed by the inclusion), set P_V defined as $P_V = \{p \in F \,|\, \forall z \in V \cup C, \forall \lambda \in R_{++}, \langle \lambda z, p \rangle \geq 0\}$ is not an empty set for each $V \in \mathscr{V}$, and (2) for each $V' \in \mathscr{V}$, $\bigcup_{V \in \mathscr{V}, V \subset V'} P_V$ is $\sigma(F, E)$-dense in P.

Condition (B5-5) is automatically satisfied when E under $\mathscr{T}_E$) is a locally convex topological vector space, F is its topological dual, and C is a closed convex cone with vertex 0 with element v such that $-v \notin C$.[10] Note that under (B5-5)-(2), we have for each $V' \in \mathscr{V}$,

$$\bigcap_{V \in \mathscr{V}, V \subset V'} C_V = C, \tag{5.3}$$

---

[10] See Nikaido (1959), where this property is used with the strong assumption of compact range (see the discussion after Theorem 4.2.1 in Chapter 4).

5: General Economic Equilibrium

where C_V is the polar of P_V for each $V \in \mathscr{V}$. (Indeed, since each C_V includes C, $C \subset \bigcap_{V \in \mathscr{V}, V \subset V'} C_V$ is clear. On the other hand, assume that $x \notin C$ and for some $V' \in \mathscr{V}$, $x \in \bigcap_{V \in \mathscr{V}, V \subset V'} C_V$. By definition of C, there is element $p \in P$ such that $\langle x, p \rangle = \epsilon > 0$, so there is $\sigma(F, E)$-open neighborhood $O(p)$ of p such that $\forall q \in O(p)$, $\langle x, q \rangle > \epsilon/2$. Since $\bigcup_{V \in \mathscr{V}, V \subset V'} P_V$ is $\sigma(F, E)$-dense in P, $V \in \mathscr{V}$ exists such that $V \subset V'$ and $P_V \cap O(p) \neq \emptyset$. Then, since $x \in C_V$, the polar of P_V, an element $q \in P_V \cap O(p)$ satisfies $\langle x, q \rangle \leqq 0$, which is a contradiction.)

The third existence theorem of the general equilibrium is that under the approximation setting of (B5-5) for commodity and price spaces, we may characterize the existence of equilibrium through the following limit condition for commodity and price spaces:

(B*) [Limit Conditions] A sufficiently small $\tilde{V} \in \mathscr{V}$ exists as a neighborhood of v satisfying the following conditions:

1: **For Attainable Allocations** The boundedness assumption (B3) for attainable allocations holds even when we replace for each $i \in I$, X_i with $X_i + C_{\tilde{V}}$, Y with $Y - C_{\tilde{V}}$, and τ_w with $\sigma(E, F)$.
2: **For Prices** For all $x \in \hat{X} = \prod_{i \in I} \hat{X}_i$, $y \in \hat{Y} = \prod_{j \in J} \hat{Y}_j$, and $p \in P \setminus P_{\tilde{V}}$, there is at least one consumer i or producer j with vector $z^p \in C_{\tilde{V}}$ such that $\langle z^p, p \rangle < 0$ together with a $\sigma(E, F) \times \sigma(E, F) \times \sigma(F, E)$ open neighborhood U of $(x, y, p) \in X \times Y \times P$ such that for i (or j, resp.,) $C(\{x'_i, x'_i + z^p\}) \cap P_i(x'_i) \neq \emptyset$ $(C(\{y'_j - z^p, y'_j\}) \cap Y_j \neq \emptyset$, resp.,) for each $(x', y', q) \in U$.

Condition (B*) is closely related to popular conditions in locally convex commodity spaces like "production sets having interior points" (see, e.g., Toussaint 1984), "uniformly properness for preferences" (Mas-Colell 1986).

THEOREM 5.3.3: (Existence of General Equilibrium III) *Under assumption* (B5) *with conditions* $I = \{1, 2, \ldots, m\}$ *and* $J = \{1, 2, \ldots, n\}$ *(the sets of consumers and producers are finite), suppose the following:*

(i) *For each* $i \in I$, P_i *is the (Product BK1)-type.*
(ii) *For any allocation* x *such that for some* $y \in Y$, $\langle \sum_{i \in I} x_i, p \rangle \leqq \langle y, p \rangle$ *for all* $p \in P$, *set* $P_i(x)$ *is non-empty for all* $i \in I$ *(non-satiation property).*
(iii) *Each* P_i *satisfies for all* $x \in X$ *and* $y_i \in P_i(x)$, *that* $C(\{x_i, y_i\}) \setminus \{x_i\}$ *is included in* $P_i(x)$ *(convexity and local non-satiation properties) and that* $P_i^{-1}(y_i)$ *includes a* $\sigma(E, F)$ *neighborhood of* x *in* E *(open lowersection property).*

(iv) *Each P_i has open values, and for each $p \in D$ and two points x and x' in X_i such that $\langle x',p\rangle < \langle x,p\rangle$, we can take $z \in C(\{x',x\})$ satisfying $\langle z,p\rangle < \langle x,p\rangle$ arbitrarily near to x.*

Then, condition (B*) is sufficient for assuring that economy $\mathscr{E}$ has a free disposal equilibrium $(p^*,(x_i^*)_{i\in I},(y_j^*)_{j\in J}) \in (P\backslash\{0\}) \times \prod_{i\in I} X_i \times \prod_{j\in J} Y_j$.

PROOF: As in the proof of the previous theorem, let $\tilde{B} \in \mathscr{F}(E)$ be a finite set including all points, $\bar{x}_i \in X_i$, $i \in I$, $y_j^i \in Y_j$, $i \in I$, $j \in J$, stated in (B2-2) and for each $B \in \mathscr{F}(E)$ and $V \in \mathscr{V}$ such that $B \supset \tilde{B}$ and $V \subset \tilde{V}$, define an economy $\mathscr{E}^{BV}$ as $\mathscr{E}^{BV} = ((X_i^B, P_i, W_i)_{i\in I}, (Y_j)j \in J)$, where $X_i^B = C(B) \cap X_i$ for each $i \in I$, $Y_j^B = C(B) \cap Y_j$ for each $j \in J$, the price space is P_V, and P_i and W_i are restrictions on them for each $i \in I$. By (B5-1) and (B5-2), all X_i^B and Y_j^B are compact and economy $\mathscr{E}^{BV}$ satisfies all conditions in Theorem 5.1.2, so $\mathscr{E}^{BV}$ has a free disposal equilibrium $(p^{BV},(x_i^{BV})_{i\in I},(y_j^{BV})_{j\in J}) \in D^V \times \prod_{i\in I} X_i^B \times \prod_{j\in J} Y_j^B$, $\sum_{i\in I} x_i^{BV} \in (\omega + \sum_{j\in J} y_j^{BV}) - C_V$, where D^V is the compact base $\{p \in P_V\,|\,\langle u,p\rangle = 1\}$. Condition (B*-1) assures that such equilibrium vectors, x_i^{BV}, $i \in I$, and y_j^{BV}, $j \in J$, belong to $\sigma(E,F)$ relatively compact sets, $\hat{X}_i^{\tilde{V}} = \{x_i \in X_i\,|\,\forall j \in I, j \neq i, \exists x_j \in X_j, \sum_{j\in I} x_j \in Y - C_{\tilde{V}}\}$, $i \in I$ and $\hat{Y}_j^{\tilde{V}} = \{y_j \in Y_j\,|\,\forall i \in I, \exists x_i \in X_i, \forall k \in J, k \neq j, \exists y_k \in Y_k, \sum_{i\in I} x_i - \sum_{k\in J} y_k - \omega \in Y - C_{\tilde{V}}\}$, $j \in J$, respectively. Then for each V, there is a $\sigma(E,F)$ converging subnet $(p^{B(\mu)V},x^{B(\mu)V},y^{B(\mu)V})_{\mu\in\mathscr{M}_V}$ of $(p^{BV},x^{BV},y^{BV})_{B\in\mathscr{F}(E),B\supset\tilde{B}}$, whose limit $(p^V,x^V,y^V) \in D^V \times \prod_{i\in I} X_i \times \prod_{j\in J} Y_j$ satisfies $\sum_{i\in I} x_i^V \in (\omega + \sum_{j\in J} y_j^V) - C_V$. (Note that by (B5-2), $\mathscr{T}_E$-closed convex sets are $\sigma(E,F)$-closed.) We can also assure that p^V never falls in $D^V\backslash D^{\tilde{V}}$, since for such p^V, by (B*-2), there is vector $z^{p^V} \in C_{\tilde{V}}$ preventing all $x_i^{B(\mu)V}$ (or $y_j^{B(\mu)V}$) sufficiently large from individual maximal (equilibrium) points. Since $D^{\tilde{V}}$, $\hat{X}_i^{\tilde{V}}$, and $\hat{Y}_j^{\tilde{V}}$ are compact (under $\sigma(F,E)$, and $\sigma(E,F)$'s, respectively), we can further take a subnet $(p^{V(\nu)},x^{V(\nu)},y^{V(\nu)})_{\nu\in\mathscr{N}}$ of $(p^V,x^V,y^V)_{V\in\mathscr{V},V\subset\tilde{V}}$ such that $\lim_{\nu\in\mathscr{N}}(p^{V(\nu)},x^{V(\nu)},y^{V(\nu)}) = (p^*,x^*,y^*) \in D^{\tilde{V}} \times \prod_{i\in I} \hat{X}_i^{\tilde{V}} \times \prod_{j\in J} \hat{Y}_j^{\tilde{V}}$. It is also easy to check that

$$\sum_{i\in I} x_i^* \in \left(\omega + \sum_{j\in J} y_j^*\right) - C,$$

the free disposal market equilibrium condition, since $\bigcap_{V\in\mathscr{V}} C_V = C$. By considering product directed set $\mathscr{L} = \mathscr{N} \times \prod_{\nu\in\mathscr{N}} \mathscr{M}_{V(\nu)}$, we obtain net

5: General Economic Equilibrium

$S : \mathscr{L} \ni (\nu, f) \mapsto (p^{B(f(\nu))V(\nu)}, x^{B(f(\nu))V(\nu)}, y^{B(f(\nu))V(\nu)})$ converging to (p^*, x^*, y^*).[11]

It remains to be shown that (x^*, y^*, p^*) is a free disposal equilibrium. As stated above, the free disposal market equilibrium condition is satisfied. To check individual maximization conditions for consumers and producers, we can follow the same argument in the proof of the previous theorem merely by replacing the converging net of points $(p^{B(\nu)}, x^{B(\nu)}, y^{B(\nu)})$'s in the proof of the previous theorem with points $(p^{B(f(\nu))V(\nu)}, x^{B(f(\nu))V(\nu)}, y^{B(f(\nu))V(\nu)})$'s in this proof. Though the following lines include simple repetition of the arguments in the proof of the previous theorem, I write them for the sake of the completeness of the proof.

Let us consider converging nets $(p^{B(f(\nu))V(\nu)})_{(\nu,f)\in\mathscr{L}}$, $(x_i^{B(f(\nu))V(\nu)})_{(\nu,f)\in\mathscr{L}}$, $i \in I$, and $(y_j^{B(f(\nu))V(\nu)})_{(\nu,f)\in\mathscr{L}}$, $j \in J$, with limits $p^* \in D^{\tilde{V}}$, $x_i^* \in \hat{X}_i^{\tilde{V}}$, $i \in I$ and $y_j^* \in \hat{Y}_j^{\tilde{V}}$, $j \in J$. As before, let us see that for each $i \in I$, if x_i is an element of $P_i(x^*)$, where $x^* = (x_1^*, x_2^*, \ldots, x_m^*)$, then for any $y_j \in Y_j$, $j \in J$, we have

$$\langle x_i, p^* \rangle \geq W_i((y_j)_{j \in J}, p^*). \tag{5.4}$$

Indeed, if we take $B(f(\nu)) \supset \tilde{B}$ sufficiently large so that $B(f(\nu))$ has x_i and y_j for all $j \in J$ as elements, and if we consider the fact that $x^{B(f(\nu))V(\nu)} = (x_1^{B(f(\nu))V(\nu)}, \ldots, x_m^{B(f(\nu))V(\nu)})$ converges to x^* together with the local intersection property in (iii), we may suppose that $x_i \in P_i(x^{B(f(\nu))V(\nu)})$ for all $(\nu, f) \in \mathscr{L}$ sufficiently large, so we have

$$\langle x_i, p^{B(f(\nu))V(\nu)} \rangle > \langle x_i^{B(f(\nu))V(\nu)}, p^{B(f(\nu))V(\nu)} \rangle$$
$$= W_i((y_j^{B(f(\nu))V(\nu)})_{j \in J}, p^{B(f(\nu))V(\nu)})$$
$$\geq W_i((y_j)_{j \in J}, p^{B(f(\nu))V(\nu)}),$$

for all sufficiently large $(\nu, f) \in \mathscr{L}$. The equality holds by (ii) and the local non-satiation property in (iii) with the joint continuity on $C(B(f(\nu))) \times D^V$ of $\langle \cdot, \cdot \rangle$. The last $\geq$ holds by (B2-1). Hence, by taking limit $p^{B(f(\nu))V(\nu)} \to p^*$, (use (B2-1)-(2)), we have Equation (5.4). In the above, we have already seen that $\sum_{i \in I} x_i^* \in (\omega + \sum_{j \in J} y_j^*) - C$, so we have $\langle \sum_{i \in I}(x_i^*)_{i \in I}, q \rangle \leq \langle \omega + \sum_{j \in J}(y_j^*), q \rangle$ for all $q \in P$ (x^* and y^* satisfies the free disposal market equilibrium condition). Note

---

[11] Use the definition of product preordering on $\mathscr{L}$. The non-emptiness is assured by the axiom of choice. (See also the theorem on iterated limits in Kelley (1955, p. 69).)

that $\langle \sum_{i \in I}(x_i^*)_{i \in I}, p^* \rangle \leq \langle \omega + \sum_{j \in J}(y_j^*), p^* \rangle$ holds. By (ii) and the local non-satiation property in (iii), we may choose $x_i \in P_i(x^*)$ arbitrarily near to x_i^*, so Equation (5.4) gives

$$\langle x_i^*, p^* \rangle \geq W_i((y_j)_{j \in J}, p^*)$$

for all $(y_j)_{j \in J} \in \prod_{j \in J} Y_j$. By summing up the above relation for $i \in I$, we obtain

$$\sum_{i \in I} W_i((y_j^*)_{j \in J}, p^*) = \left\langle \omega + \sum_{j \in J}(y_j^*), p^* \right\rangle$$

$$\geq \left\langle \sum_{i \in I}(x_i^*)_{i \in I}, p^* \right\rangle$$

$$= \sum_{i \in I} \langle x_i^*, p^* \rangle$$

$$\geq \sum_{i \in I} W_i((y_j)_{j \in J}, p^*) \qquad (5.5)$$

for all $(y_j)_{j \in J} \in \prod_{j \in J} Y_j$. This means by (B2-1), that for each $j \in J$, y_j^* is a profit maximizing production plan under p^* for each $j \in J$. Equation (5.5) also shows that

$$\langle x_i^*, p^* \rangle = W_i((y_j^*)_{j \in J}, p^*)$$

for all $i \in I$. Hence, x_i^* satisfies the budget constraint under p^* for each $i \in I$. Equation (5.4) and the previous equation assert that every x_i such that $\langle x_i, p^* \rangle < \langle x_i^*, p^* \rangle$ does not belong to $P_i(x^*)$, especially point $\bar{x}_i$ that assures the minimum wealth condition for i is not an element of $P_i(x^*)$. If $P_i(x^*) \cap B_i((y_j^*)_{j \in J}, p^*) \neq \emptyset$, in the convex combination between two points x_i and $\bar{x}_i$, $C(\{x_i, \bar{x}_i\})$, we can take z_i satisfying $\langle z_i, p^* \rangle < \langle x_i, p^* \rangle$ arbitrarily near to x_i by condition (iv), which contradicts that $x_i \in P_i(x^*)$ and $P_i(x^*)$ is open. Hence, we have $P_i(x^*) \cap B_i((y_j^*)_{j \in J}, p^*) = \emptyset$. ∎

Bibliographic Notes

The basic settings used in this chapter crucially depend on the classical formulation of Debreu (1959). I also borrow several concepts from the settings of Gale and Mas-Colell (1975). The argument in this chapter is not comprehensive but is devoted to one area of general equilibrium theory: equilibrium in abstract commodity spaces. The complete discussion is designed to give a unified perspective on the basic results for spaces with positive cones that have interior points like Bewley (1972), without interior points like Mas-Colell (1986), and cases with production technologies with interior points like Toussaint (1984), etc. As Kajii (1988) pointed out, methods like the uniformly properness condition for preferences and the interior condition for production technology are essentially intended to solve the same problem. Nikaido (1957) (published as a part of Nikaido (1959)) also used the same method for commodity space.

Chapter 6

The Čech Type Homology Theory and Fixed Points

6.1 Basic Concepts in Algebraic Topology

6.1.1 Structures and categories

Strictly speaking, a mathematical theory is constructed by a formal language that describes (*mathematical*) *objects* as its *terms* and (*mathematical*) *relations* as its *formulas*. Since our arguments are based on set theory, all the mathematical objects we treat in our formal theory are sets. Note, however, that many rigorously described "properties" of objects are useful that cannot be treated by themselves as objects in the domain of discourse described under formal language. In such cases, we (informally) use the notion "class" instead of "set" to describe the fact that we are concerned with any object having such a property, even though the collection or totality of such objects may not be treated as a set.

In mathematics, "structure" has a rigorous special meaning. A *species of mathematical structures* is a list of rules $\mathfrak{S}$, defining a certain class of mathematical objects, $\mathfrak{Obj}(\mathfrak{S})$. The rules consist of: (1) a specification of the *principal base set* on which a structure of the species is defined, (2) a specification of the *auxiliary base set* fixed to characterize the species of structures, (3) a *typical characterization* of a *generic structure* as an element of a set constructed by fundamental finite procedures (including taking subsets and products) among base sets, and (4) a relation called the *axiom* of the species of structures specifying the properties of a generic structure.

A class of *morphisms* $\mathfrak{M}$ for a species of structures $\mathfrak{S}$ is a list of rules that define the following: (1) for each $X, Y \in \mathfrak{Obj}(\mathfrak{S})$, a set $\hom(X, Y)$ such that $\hom(X,Y) \cap \hom(X', Y') = \emptyset$, unless $X = X'$ and $Y = Y'$, (2) for each $f \in \hom(X, Y)$ and $g \in \hom(Y, Z)$, element $g \circ f \in \hom(X, Z)$ called the *composition* of f and g that satisfies associativity ($(h \circ g) \circ f = h \circ (g \circ f)$ as long as they are defined), and (3) for each $X \in \mathfrak{Obj}(\mathfrak{S})$, *identity element*

$1_X \in \hom(X,X)$ such that $1_X \circ f = f \circ 1_X$ for each $f \in \hom(X,Y)$. An element of $\hom(X,Y)$ is called a *morphism* (or an $\mathfrak{M}$-morphism).

When class $\mathfrak{M}$ of morphisms is associated with a species of structures $\mathfrak{S}$, *category* $\mathfrak{C}$ (the list of rules $\mathfrak{S}$ together with $\mathfrak{M}$) is defined. An element of $\mathfrak{Obj}(\mathfrak{S})$ is also said to be an object in the category of $\mathfrak{C}$, and an element of $\hom(X,Y)$ defined under $\mathfrak{M}$ is also called a morphism in $\mathfrak{C}$. The class of all objects in $\mathfrak{C}$ is denoted by $\mathfrak{Obj}(\mathfrak{C})$ ($=\mathfrak{Obj}(\mathfrak{S})$), and the class of all morphisms in $\mathfrak{C}$ is denoted by $\mathfrak{Mor}(\mathfrak{C})$. In category $\mathfrak{C}$, two objects, $X, Y \in \mathfrak{Obj}(\mathfrak{C})$, are *isomorphic* if $\exists f \in \hom(X,Y)$ and $\exists g \in \hom(Y,X)$ such that $f \circ g = 1_Y$ and $g \circ f = 1_X$. Such a morphism (f or g) is called an *isomorphism* in category $\mathfrak{C}$.

Given two categories $\mathfrak{C}$ and $\mathfrak{D}$, we say that a list of rules $\mathfrak{T}$ is a *covariant* (*contravariant*, resp.,) *functor* on $\mathfrak{C}$ to $\mathfrak{D}$ if for each $A \in \mathfrak{Obj}(\mathfrak{C})$ it defines an element $\mathfrak{T}(A) \in \mathfrak{Obj}(\mathfrak{D})$ and for each $f \in \mathfrak{Mor}(\mathfrak{C})$ an element $\mathfrak{T}(f) \in \mathfrak{Mor}(\mathfrak{D})$, such that: (1) for each $f \in \hom(X,Y)$, $\mathfrak{T}(f) \in \hom(\mathfrak{T}(X), \mathfrak{T}(Y))$, ($\mathfrak{T}(f) \in \hom(\mathfrak{T}(Y), \mathfrak{T}(X))$, resp.), (2) for each $1_X \in \hom(X,X)$, $\mathfrak{T}(1_X) = 1_{\mathfrak{T}(X)}$, and (3) for each $f \in \hom(X,Y)$ and $g \in \hom(Y,Z)$, $\mathfrak{T}(g \circ f) = \mathfrak{T}(g) \circ \mathfrak{T}(f)$, ($\mathfrak{T}(g \circ f) = \mathfrak{T}(f) \circ \mathfrak{T}(g)$, resp.).

6.1.2 Inverse and direct limits

Let I be a directed set. Suppose that for each $i \in I$ there is set E_i and for each $i, j \in I$, $i \leq j$, there is mapping $h_{ij} : E_j \to E_i$ ($h_{ji} : E_i \to E_j$, resp.,) such that for each $i, j, k \in I$, $i \leq j \leq k$, $h_{ij} \circ h_{jk} = h_{ik}$ ($h_{kj} \circ h_{ji} = h_{ki}$, resp.). The system of sets and mappings $(E_i, h_{ij})_{i,j \in I, i \leq j}$ (($E_i, h_{ji})_{i,j \in I, i \leq j}$, resp.,) is called an *inverse system* (*direct system*, resp.,) of sets and mappings over directed set I.

Let $(E_i, h_{ij})_{i,j \in I, i \leq j}$ be an inverse system over directed set I and let us consider direct product $\prod_{i \in I} E_i$. Subset $E \subset \prod_{i \in I} E_i$ such that $x \in E$ if and only if $\mathrm{pr}_i(x) = h_{ij} \circ \mathrm{pr}_j(x)$, where pr_i denotes the projection onto the i-th component of the direct product, is called the *inverse limit* of the inverse system and is denoted by $E = \varprojlim E_i$. An *essential element* of E_i is an element x_i such that $(i \leq j) \Rightarrow (x_i \in h_{ij}(E_j))$. It is clear that for each $x \in E = \varprojlim E_i$, $pr_j(x) \in E_j$ is an essential element of E_j for each $j \in I$.

For direct system $(E_i, h_{ji})_{i,j \in I, i \leq j}$ over directed set I, let us consider direct sum $\bigcup_{i \in I} E_i \times \{i\}$ and quotient set $F = (\bigcup_{i \in I} E_i \times \{i\})/\sim$, where $\sim$ denotes an equivalence relation on $\bigcup_{i \in I} E_i \times \{i\}$ such that $x \sim y$ ($x \in E_i$, $y \in E_j$) if and only if $\exists k$, $i \leq k$, $j \leq k$, $h_{ki}(x) = h_{kj}(y)$. Set F is called the *direct limit* of a direct system and is denoted by $\varinjlim E_i$.

6.1.3 Simplex

In an $n+1$-dimensional Euclidean space, R^{n+1}, $(n = 1, 2, \ldots,)$ we denote by Δ^n the set $\{(x_1, \ldots, x_{n+1}) \in R^{n+1} | \sum_{i=1}^{n+1} x_i = 1, x_1 \in R_+, \ldots, x_{n+1} \in R_+\}$. Δ^n is called the *standard n-simplex* (or the *n-dimensional standard simplex*). We denote by $A = \{e^i | i = 1, \ldots, n+1\}$ the *standard basis* of vector space R^{n+1}, i.e., e^i denotes the vector of R^{n+1} whose j-th coordinate is 0 if $j \neq i$ and is 1 if $j = i$. Set Δ^n may be identified with the set of all *convex (linear) combinations* among points in $A = \{e^1, \ldots, e^{n+1}\}$, i.e., $\{\sum_{i=1}^{n+1} \alpha_i e^i | \sum_{i=1}^{n+1} \alpha_i = 1, \alpha_1 \in R_+, \ldots, \alpha_{n+1} \in R_+\}$. Let B be a non-empty subset of A, and let $\sharp B = k+1$. The set of convex combinations of points in $B = \{e^{i_0}, \ldots, e^{i_k}\}$, $\{\sum_{j=0}^{k} \alpha_j e^{i_j} | \sum_{j=0}^{k} \alpha_j = 1, \alpha_0 \in R_+, \ldots, \alpha_k \in R_+\}$, is called a *$k$-dimensional face* (or *$k$-face*) of Δ^n. We say that an *orientation* for standard simplex Δ^n is determined if a total ordering on the set of $n+1$ points $\{e^1, \ldots, e^{n+1}\}$ is chosen up to even permutations. Δ^n has *standard orientation* such that $e^1 < e^2 < \cdots < e^{n+1}$. Standard orientation *induces* for each face $\{\sum_{j=0}^{k} \alpha_j e^{i_j} | \sum_{j=0}^{k} \alpha_j = 1, \alpha_0 \in R_+, \ldots, \alpha_k \in R_+\}$ an orientation equivalent to the restriction of $e^1 < e^2 < \cdots < e^{n+1}$ on $\{e^{i_0}, \ldots, e^{i_k}\}$. We denote standard n-simplex and its k-face by $e^1 \cdots e^{n+1}$ and $e^{i_0} \cdots e^{i_k}$, respectively. To fix an orientation for each simplex or face, we use $\langle \cdot \rangle$ to denote by $\langle e^1 \cdots e^{n+1} \rangle$ and $\langle e^{i_0} \cdots e^{i_k} \rangle$ the oriented simplex and its face.

In R^n, consider a set of $m+1$ points $\{x^0, \ldots, x^m\}$ such that $x^1 - x^0, x^2 - x^0, \ldots, x^m - x^0$ are linearly independent. (Such $m+1$ points are called *affine independent*.) Set $\{\sum_{i=1}^{m} a_i x^i | \sum_{i=1}^{m} a_i = 1, a_1 \in R_+, \ldots, a_m \in R_+\}$ is called the *m-(dimensional) simplex* defined by points $x^0, \ldots, x^m$ in R^n. It is easy to check that an m-simplex in R^n is topologically isomorphic (homeomorphic) to standard m-simplex. As standard simplices, we denote an m-simplex in R^n by $x^1 \cdots x^m$ or $\langle x^1 \cdots x^m \rangle$ (with a fixed orientation) as long as there is no fear of confusion. As before, a *k-face* of m-simplex $x^1 \cdots x^m$ is the set of all convex combinations among points in a certain subset $\{x^{i_0}, \ldots, x^{i_k}\}$ of $\{x^1, \ldots, x^m\}$ and is denoted by $x^{i_0} \cdots x^{i_k}$ (or $\langle x^{i_0} \cdots x^{i_k} \rangle$ with fixed orientation). A *(simplicial) complex* K is a family of simplices such that: (1) for all $\sigma \in K$, all faces of σ are also elements of K, and (2) if σ and σ' are elements of K, the intersection $\sigma \cap \sigma'$ (in the simple set-theoretic sense) is also an element of K. For complex K, we denote by $|K|$ its *underlying space* $\bigcup_{\sigma \in K} \sigma$. The underlying space of a complex in R^n is called the *polyhedron*. The maximal dimension of simplexes in K is called the *dimension* of complex K or the polyhedron $|K|$. Simplicial complex K

is said to be *oriented* if with each $\sigma \in K$ an arbitrary fixed orientation is associated. Simplex σ having such an orientation is denoted by $\langle \sigma \rangle$.

More generally, let A be an arbitrary finite set and let K be a family of subsets of A satisfying:

(1) $\forall a \in A$, $\{a\} \in K$, and
(2) $(s \in K) \Rightarrow ((s' \subset s) \Rightarrow s' \in K)$.

We call K an *abstract simplicial complex*. Set A is called the *vertex set* of K and is denoted by **Vert**(K). Element s of K such that $\sharp s = q + 1$ is called a *q-simplex* of K, and elements of s are called the *vertices* of s. q-simplex s is said to be *oriented* if the vertices of s are totally ordered up to even permutations. Abstract simplicial complex K is said to be oriented if every $s \in K$ has an arbitrarily fixed orientation. Simplex s with such a fixed orientation is denoted by $\langle s \rangle$. In the following, we treat a simplicial complex in R^n as a special case of abstract simplicial complex.

Let K and L be two abstract simplicial complexes. Mapping f : **Vert**$(K) \to$ **Vert**(L) is a *simplicial map* if for each simplex $\{v_0, \ldots, v_q\}$ in K, the set of vertices $\{f(v_0), \ldots, f(v_q)\}$ is a simplex in L. The class of abstract simplicial complexes with simplicial mappings determines a category.

Given finite open covering $\{U^i\}_{i \in I}$ of X, consider family X_I of subsets of index set I such that $J \in X_I$ if and only if $\bigcap_{j \in J} U^j \neq \emptyset$. Family X_I is called the *nerve* of the covering $\{U^i\}_{i \in I}$. It is easy to check that family X_I satisfies: (1) that for each member $i \in I$, $\{i\} \in X_I$, and (2) that every non-empty subset of a member of X_I is a member of X_I. Hence, by identifying each member J with an abstract $(\sharp J - 1)$-simplex, we may identify X_I with an abstract simplicial complex. Čech (1932) first defined the nerve of a finite open covering and used such complexes as approximations to a space.

6.1.4 Homology theory

Homology theory associates an *algebraic structure* (e.g., groups, modules, vector spaces, etc.) with a topological space. Such an algebraic object represents the connectivity among points in each dimension of the topological space. In this book, though we use the word "group," all homology groups we treat are vector spaces over real field R.

Denote by **Tops** the category whose objects $X, Y \in \mathfrak{Obj}($**Tops**$)$ are topological spaces and hom(X, Y) is the set of all continuous functions on X to Y. In general, a homology theory may be identified with a functor on a certain admissible subcategory of *Tops* or a pair of such subcategories to a category of certain algebraic objects (in this book, mainly graded vector spaces over R). The basic homology theory we use is *simplicial homology theory* that associates a graded sequence of vector spaces over R with each topological space in the form of finite dimensional *polyhedron* $|K|$ (where K is an oriented simplicial complex) as follows. For each $q = 0, 1, \ldots, n$, where n is the dimension of K, define vector-space $C_q(K)$, the set of all *q-chains* on K, as $C_q(K) = R^{(A)}$, where A denotes the set of all q-simplex in K and $R^{(A)}$ is the vector space of all functions on A to R having finite *supports* (i.e., the set of functions $f : A \to R$ such that $\{\sigma \in A | f(\sigma) \neq 0\}$ is a finite set). By formally identifying element $f \in R^{(A)}$ with the finite *abstract summation* $\sum_{\sigma \in A} f(\sigma)\sigma$, $C_q(K)$ may be considered a vector space over R. It is a good idea to interpret $1\sigma = 1_\sigma \in R^{(A)}$ ($1_\sigma(\sigma) = 1$ and $1_\sigma(\sigma') = 0$ for each $\sigma' \neq \sigma$) as $\langle\sigma\rangle$ (simplex $\sigma \in K$ with a certain fixed orientation) and $-1\sigma = -1_\sigma \in R^{(A)}$ as the same simplex with reverse orientation. In this sense we denote by $-\langle\sigma\rangle$ the oriented simplex having reverse orientation to $\langle\sigma\rangle$. Set $\{\langle\sigma\rangle | \sigma \in A\}$ may be considered the base of $C_q(K)$. For each $q < 0$ and $q > n$, define $C_q(K)$ as $C_q(K) = 0$.

Given oriented simplicial complex K, we define for each pair of two simplexes of K, a q-simplex $\langle\sigma^q\rangle = \langle v_0, v_1, \ldots, v_q\rangle$ and $(q-1)$-simplex $\langle\sigma^{q-1}\rangle = \langle u_0, \ldots, u_{q-1}\rangle$, the *incidence number* $[\langle\sigma^q\rangle, \langle\sigma^{q-1}\rangle]$ as $[\langle\sigma^q\rangle, \langle\sigma^{q-1}\rangle] = 0$ if σ^{q-1} is not a face of σ^q, $[\langle\sigma^q\rangle, \langle\sigma^{q-1}\rangle] = 1$ if σ^{q-1} is a face of σ^q such that $\{v_0, v_1, \ldots, v_q\} \setminus \{u_0, \ldots, u_{q-1}\} = \{v_i\}$ and $\langle\sigma^q\rangle = \langle v_i, u_0, \ldots, u_{q-1}\rangle$, and $[\langle\sigma^q\rangle, \langle\sigma^{q-1}\rangle] = -1$ if σ^{q-1} is a face of σ^q such that $\{v_0, v_1, \ldots, v_q\} \setminus \{u_0, \ldots, u_{q-1}\} = \{v_i\}$ and $\langle\sigma^q\rangle = -\langle v_i, u_0, \ldots, u_{q-1}\rangle$. In the following we denote that σ^{q-1} is a $(q-1)$-face of $\sigma^q = \langle v_0, \ldots, v_q\rangle$ such that the vertex set of σ^{q-1} is $\{v_0, \ldots, v_q\} \setminus \{v_i\}$ by $\sigma^{q-1} = \langle v_0, \ldots, \hat{v}_i, \ldots, v_q\rangle$. Now, we define for each q, $1 \leq q \leq n$, where n is the dimension of K, *boundary operator* $\partial_q : C_q(K) \to C_{q-1}(K)$ as a linear extension of

$$\partial_q(\langle v_0, \ldots, v_q\rangle) = \sum_{i=0}^{q}(-1)^i\langle v_0, \ldots, \hat{v}_i, \ldots, v_q\rangle.$$

By using the incidence number, it is also possible to write $\partial_q(\sigma^q)$ as

$$\partial_q(\sigma^q) = \sum_{\sigma^{q-1} \subset \sigma^q} [\langle\sigma^q\rangle, \langle\sigma^{q-1}\rangle]\langle\sigma^{q-1}\rangle,$$

where $\sigma^{q-1} \subset \sigma^q$ represents all $(q-1)$-faces of q-simplex σ^q. For each $q < 1$ and $q > n$, define ∂_q as the linear mapping whose values constantly equal 0 to obtain the following sequence of graded vector spaces:

$$\cdots \to 0 \xrightarrow{\partial_{n+1}} C_n(K) \xrightarrow{\partial_n} \cdots \to C_2(K) \xrightarrow{\partial_2} C_1(K) \xrightarrow{\partial_1} C_0(K) \to 0 \to \cdots$$

The basic property of boundary operators is that $\partial_{q-1} \circ \partial_q = 0$ for all q.[1] Hence, it is always the case that for each q, the image of $C_{q+1}(K)$ under ∂_{q+1}, $B_q(K)$, is a subset of the kernel of ∂_q, $Z_q(K) = \{\sigma \in C_q(K) | \partial_q(\sigma) = 0\}$. Such a sequence $\{C_q\}$ with boundary operators $\{\partial_q\}$ is called a *chain complex*. Since every ∂_q is a linear operator, it is easy to check that for each q, $B_q(K)$ and $Z_q(K)$ are vector subspaces of $C_q(K)$. An element b of $B_q(K) \subset C_q(K)$ is called a *q-boundary* (or is said to *bound* a chain in C_{q+1}), and an element of $Z_q(K)$ is called a *q-cycle*. The q-th *homology group* $H_q(K)$ is the quotient vector space $Z_q(K)/B_q(K)$, i.e., we identify two elements z and z' in $Z_q(K)$ if and only if $z' - z \in B_q(K)$. We denote by $[z]$ the equivalence class of $z \in Z_q(K)$ under $B_q(K)$. See Figure 29. Two cycles, $\langle ab \rangle + \langle bc \rangle + \langle cd \rangle + \cdots + \langle ma \rangle$ and $\langle ab \rangle + \langle bx \rangle + \langle xc \rangle + \langle cd \rangle + \cdots + \langle ma \rangle$ are identified since their difference $\langle bc \rangle - \langle bx \rangle - \langle xc \rangle = \langle bc \rangle + \langle cx \rangle + \langle xb \rangle$ is the boundary of $\langle bcx \rangle$.

It is clear that we may define for abstract simplicial complex K such concepts as the vector space of *chains, abstract summations, incidence*

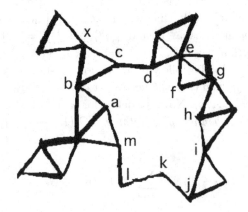

Figure 29: Equivalence class of cycles under $B_q(K)$

---

[1]One may prove it through simple calculation by applying the definition twice on a certain simplex. See, e.g., Hocking and Young (1961, p. 224).

numbers, *boundary operators, boundaries, cycles,* and *homology groups* in the same way. Let K and K' be two abstract simplicial complexes, and let $f : \mathbf{Vert}(K) \to \mathbf{Vert}(K')$ be a simplicial map. Then for each q, f *induces* linear mapping $f_q : C_q(K) \to C_q(K')$ such that for each q-simplex in K, a q-simplex in K' is associated as $f_q(\langle v_0, \ldots, v_q \rangle) = \langle f(v_0), \ldots, f(v_q) \rangle$ if $\langle f(v_0), \ldots, f(v_q) \rangle$ is a q-simplex in K' and $f_q(\langle v_0, \ldots, v_q \rangle) = 0$, if otherwise. One can prove (e.g., see Hocking and Young (1961, p. 249)) that $f_{q-1} \circ \partial_q = \partial'_q \circ f_q$, where ∂_q and ∂'_q are boundary operators relative to K and K', respectively. More generally, a sequence of homomorphisms (in our case, linear mappings), $\varphi_q : C_q(K) \to C_q(K')$, $q = \ldots, -1, 0, 1, 2, \ldots$, is called the *chain map* (or *chain homomorphism*) if it satisfies $\partial'_q \circ \varphi_q = \varphi_{q-1} \circ \partial_q$. It is easy to verify that a chain map $\{\varphi_q\}$ *induces* a sequence of homomorphisms (linear maps) $\varphi_{*q} : H_q(K) \to H_q(K')$ as $\varphi_{*q}([\sigma]) = [\varphi_q([\sigma])]$.

Two chain homomorphisms $\{\psi_q : C_q(K) \to C_q(K')\}$ and $\{\psi'_q : C_q(K) \to C_q(K')\}$ are called *chain homotopic* if there is a sequence of homomorphism $\{\Phi_q : C_q(K) \to C_{q+1}(K')\}$ such that $\partial_{q+1} \circ \Phi_q = \psi_q - \psi'_q - \Phi_{q-1} \circ \partial_q$. If z^q is a cycle of $C_q(K)$, this implies $\partial_{q+1}(\Phi_q(z^q)) = \psi_q(z^q) - \psi'_q(z^q)$, so that $\psi_q(z^q) - \psi'_q(z^q)$ is a boundary. Hence, two chain homotopic maps induces the same homomorphisms between homology groups. $\{\Phi_q\}$ is called a *chain homotopy*. If for a chain map $\{\varphi_q : C_q(K) \to C_q(K')\}$, there is a chain map $\{\eta_q : C_q(K') \to C_q(K)\}$ such that $\{\eta_q \circ \varphi_q : C_q(K) \to C_q(K)\}$ is homotopic to identity chain map $\{Id_q^K : C_q(K) \to C_q(K)\}$ and $\{\varphi_q \circ \eta_q : C_q(K') \to C_q(K')\}$ is homotopic to identity chain map $\{Id_q^{K'} : C_q(K') \to C_q(K')\}$, then $\{\varphi_q\}$ is a *chain equivalence*. In this case, two chain complexes are said to be *chain equivalent*.[2]

6.1.5 Star refinement of covering

Let X be a compact Hausdorff space and let $\mathfrak{M}$ be a finite open covering of X. If $\mathfrak{N}$ is a finite open covering that is a refinement of $\mathfrak{M}$, we write $\mathfrak{N} \preccurlyeq \mathfrak{M}$.

---

[2]Intuitively, chain homotopy and chain equivalence are the algebraic counterparts of the concept of *homotopy* $F : X \times [0, 1] \to Y$ (a continuous map for two maps $f_0(x) = F(x, 0)$ and $f_1(x) = F(x, 1)$), and the concept of *homotopy equivalence* (a continuous $f : X \to Y$ having continuous $g : Y \to X$ such that $f \circ g$ is homotopic to id^X and $g \circ f$ is homotopic to id^Y), respectively. When $f : X \to Y$ is a homotopy equivalence, X and Y are of the same *homotopy type*. Homology theory is constructed to satisfy that if X and Y have the same homotopy type, then chain complexes constructed on X and Y are chain equivalent.

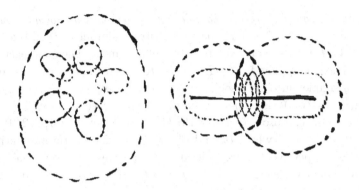

Figure 30: Star refinements

If $\mathfrak{N} \preccurlyeq \mathfrak{M}$ and for each $N \in \mathfrak{N}$, set $St(N;\mathfrak{N}) = \bigcup\{N'| N \cap N' \neq \emptyset, N' \in \mathfrak{N}\}$ is a subset of an element of $\mathfrak{M}$, covering $\mathfrak{N}$ is a *star refinement* of $\mathfrak{M}$, and write $\mathfrak{N} \preccurlyeq^* \mathfrak{M}$ (Figure 30).

The concept of the finite coverings of compact Hausdorff space X plays essential roles in this book. We denote by ***Cover***(X) the set of all finite coverings of compact Hausdorff space X. Relation $\preccurlyeq$ is clearly reflexive and transitive on ***Cover***(X). Since for each $\mathfrak{M}, \mathfrak{N} \in$ ***Cover***(X), *intersection covering* $\{M \cap N | M \in \mathfrak{M}, N \in \mathfrak{N}\}$ is a refinement of both $\mathfrak{M}$ and $\mathfrak{N}$, it is also clear that (***Cover***$(X), \preccurlyeq$) is a directed set. Note also that for each covering $\mathfrak{M} \in$ ***Cover***(X), covering $\mathfrak{N} \in$ ***Cover***(X) such that $\mathfrak{N} \preccurlyeq^* \mathfrak{M}$ exists. Since this is a crucial property, I will simply sketch a direct proof for our special case (for a reference, see, e.g., Tukey (1940, p. 47)).

LEMMA 6.1.1: *Let X be a compact Hausdorff space. For each covering $\mathfrak{M} \in$ **Cover**(X), a star refinement $\mathfrak{N} \in$ **Cover**(X) of $\mathfrak{M}$, $\mathfrak{N} \preccurlyeq^* \mathfrak{M}$, exists.*

PROOF: Binary covering is a covering that consists of two sets. It is easy to verify that (1) for each finite covering, there is a family of binary coverings whose intersection covering (the set of all finite intersections among elements of the binary coverings) refines the original covering, and (2) if $\mathfrak{M}_2 \preccurlyeq^* \mathfrak{M}_1$ and $\mathfrak{N}_2 \preccurlyeq^* \mathfrak{N}_1$, then the intersection covering $\{M_2 \cap N_2 | M_2 \in \mathfrak{M}_2$ and $N_2 \in \mathfrak{N}_2\}$ is a star refinement of the intersection covering $\{M_1 \cap N_1 | M_1 \in \mathfrak{M}_1$ and $N_1 \in \mathfrak{N}_1\}$. Hence, for our purpose, it is sufficient to show the result when $\mathfrak{M}$ is a binary covering. Assume that X is covered by family $\mathfrak{U} = \{U_1, U_2\}$. By Theorem 3.1.3, there is a refinement $\mathfrak{V} = \{V_1, V_2\}$ of $\mathfrak{U}$ such that two closed sets C_1 and C_2 satisfy $V_1 \subset C_1 \subset U_1$ and $V_2 \subset C_2 \subset U_2$ (Figure 31). Then we can see that the

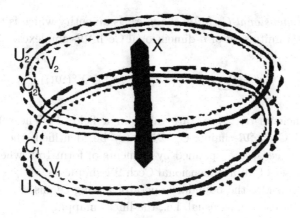

Figure 31: Construction of a star refinement

family $\mathfrak{N} = \{X\backslash C_1, U_1 \cap V_2, V_1 \cap V_2, U_2 \cap V_1, X\backslash C_1\}$ is a star refinement of $\mathfrak{U}$. ∎

6.1.6 Čech homology

Let X be a compact Hausdorff space. The *nerve* of covering $\mathfrak{M}$ of X, $X^c(\mathfrak{M})$, is an abstract complex such that the set of vertices of $X^c(\mathfrak{M})$ is $\mathfrak{M}$ and n-dimensional simplex $\sigma^n = M_0 M_1 \cdots M_n$ belongs to $X^c(\mathfrak{M})$ if and only if $\bigcap_{i=0}^{n} M_i \neq \emptyset$. We call an n-dimensional simplex σ^n in $X^c(\mathfrak{M})$ an *n-dimensional Čech $\mathfrak{M}$-simplex* (or simply, Čech simplex, n-dimensional Čech simplex, Čech $\mathfrak{M}$-simplex, etc., as long as there is no fear of confusion). $X^c(\mathfrak{M})$ is also called the Čech $\mathfrak{M}$-complex. In the following, we assume that every Čech $\mathfrak{M}$-complex is oriented. Since $\mathfrak{M}$ is a finite covering, we may identify $X^c(\mathfrak{M})$ with a polyhedron in a finite dimensional Euclidean space.[3]

If $p: \mathfrak{N} \to \mathfrak{M}$ is a mapping such that for all $N \in \mathfrak{N}$, $N \subset p(N) \in \mathfrak{M}$, we say that p is a *projection*. It is clear that if $\mathfrak{N}$ is a refinement of $\mathfrak{M}$, then for each $N_1, N_2 \in \mathfrak{N}$, $N_1 \cap N_2 \neq \emptyset$ implies that $p(N_1) \cap p(N_2) \neq \emptyset$. Hence, vertex mapping, projection p, uniquely induces simplicial map $X^c(\mathfrak{N}) \ni N_1 N_2 \cdots N_k \mapsto p(N_1)p(N_2)\cdots p(N_k) \in X^c(\mathfrak{M})$ that is also denoted by p and called a projection.

---

[3] An identification mapping from an abstract simplicial complex to the Euclidean space having standard simplexes is called a *geometric realization* (Rotman 1988, p. 142).

An n-dimensional Čech $\mathfrak{M}$-chain, c^n, is an entity which is represented uniquely as a finite sum of n-dimensional Čech $\mathfrak{M}$-simplexes,

$$c^n = \sum_{i=1}^{k} \alpha_i \sigma_i^n, \quad (\sigma_1^n, \ldots, \sigma_k^n \in X^c(\mathfrak{M})),$$

where coefficients $\alpha_1, \ldots, \alpha_k$ are taken in a field, F. The set of all n-dimensional Čech $\mathfrak{M}$-chains, $C_n^c(\mathfrak{M})$, may be identified, therefore, with the vector space over F spanned by elements of form $1\sigma^n$, where σ^n runs through the set of all n-dimensional Čech $\mathfrak{M}$-simplexes.

Let us consider the boundary operator among chains, $\partial_n : C_n^c(\mathfrak{M}) \to C_{n-1}^c(\mathfrak{M})$, for each n, as usual, i.e., the linear mapping,

$$\partial_n : M_0 M_1 \cdots M_n \to \sum_{i=0}^{n} (-1)^i M_0 M_1 \cdots \hat{M}_i \cdots M_n,$$

where the series of vertices with a circumflex over a vertex means the ordered array obtained from the original array by deleting the vertex with the circumflex and for all n such that there are no n-dimensional Čech $\mathfrak{M}$-simplexes, $C_n^c(\mathfrak{M})$ is supposed to be 0 with corresponding 0-mappings (boundary operators).[4] Then the set of all n-dimensional Čech $\mathfrak{M}$-cycles, $Z_n^c(\mathfrak{M})$, and the set of n-dimensional Čech $\mathfrak{M}$-boundaries, $B_n^c(\mathfrak{M})$, may be defined as usual, so that we obtain the n-th Čech $\mathfrak{M}$-homology group, $H_n^c(\mathfrak{M})$, for each n.

For each $\mathfrak{N} \preccurlyeq \mathfrak{M}$ and dimension n, simplicial map p induces chain homomorphism $p_n^{\mathfrak{M}\mathfrak{N}}$. Note that if $\mathfrak{N} \preccurlyeq \mathfrak{M}$, and if $p : \mathfrak{N} \to \mathfrak{M}$ and $p' : \mathfrak{N} \to \mathfrak{M}$ are projections, two simplicial maps, p and p', are *contiguous*, i.e., for each Čech $\mathfrak{N}$-simplex, $N_0 N_1 \cdots N_k$, images $p(N_0) p(N_1) \cdots p(N_k)$ and $p'(N_0) p'(N_1) \cdots p'(N_k)$ are faces of a single simplex.[5] Since two contiguous simplicial maps are *chain homotopic*,[6] p and p' induce the same

---

[4] More precisely, such $C_n^c(\mathfrak{M})$'s are defined as vector spaces having empty basis, so by definition, are equal to 0.

[5] Indeed, it is clear that the intersection $(\bigcap_{i=0}^{k} p(N_i)) \cap (\bigcap_{i=0}^{k} p'(N_i)) \supset \bigcap_{i=1}^{k} N_i \neq \emptyset$. Hence, $p(N_0) p(N_1) \cdots p(N_k) p'(N_0) p'(N_1) \cdots p'(N_k)$ is a Čech $\mathfrak{M}$-simplex.

[6] See, for example, Eilenberg and Steenrod (1952, p. 164). If we are allowed to define piecewise linear extensions $\bar{p}$ and $\bar{p}'$ of p and p', respectively, it may also be easy to find a homotopy bridge between $\bar{p}$ and $\bar{p}'$. One can also prove the assertion by directly defining the chain homotopy $\Phi_q : C_q^c(\mathfrak{N}) \to C_{q+1}^c(\mathfrak{M})$ as $\Phi_q(\langle N_0 \cdots N_k \rangle) = \sum_{j=0}^{k} (-1)^j \langle p(N_0) \cdots p(N_j) p'(N_j) \cdots p'(N_k) \rangle$. (Compare this with the mapping defined in Equation (6.1).)

homomorphism, $p_{*n}^{\mathfrak{M}\mathfrak{N}} : H_n^c(\mathfrak{N}) \to H_n^c(\mathfrak{M})$ for each n. The limit for the inverse system, $(H_n^c(\mathfrak{M}), p_{*n}^{\mathfrak{M}\mathfrak{N}})$, on the preordered family, $(\mathbf{Cover}(X), \preccurlyeq)$,

$$H_n^c(X) = \varprojlim_{\mathfrak{M}} H_n^c(\mathfrak{M}),$$

is the n-dimensional Čech Homology group.

Under the definitions of the homology group and the inverse limit, an element of $H_n^c(X)$ may be considered, intuitively, an equivalence class of a sequence of Čech cycles, $\{z^n(\mathfrak{M}) \in Z_n^c(\mathfrak{M}) : \mathfrak{M} \in \mathbf{Cover}(X)\}$, such that for each $\mathfrak{M}, \mathfrak{N} \in \mathbf{Cover}(X)$ satisfying that $\mathfrak{N} \preccurlyeq \mathfrak{M}$, we have $z^n(\mathfrak{M}) \sim p_n^{\mathfrak{M}\mathfrak{N}}(z^n(\mathfrak{N}))$, where the equivalence relation is defined relative to the class of Čech boundaries, i.e., $z^n(\mathfrak{M}) - p_n^{\mathfrak{M}\mathfrak{N}}(z^n(\mathfrak{N})) \in B_n^c(\mathfrak{M})$.[7]

6.1.7 Vietoris homology

An n-*dimensional Vietoris simplex* is a collection of $n+1$ points of compact Hausdorff space X, $x_0 x_1 \cdots x_n$. Vietoris simplex $\sigma = x_0 x_1 \cdots x_n$ is an $\mathfrak{M}$-*simplex* if the set of vertices, $\{x_0, x_1, \ldots, x_n\}$, is a subset of an element of $\mathfrak{M}$. The set of all Vietoris $\mathfrak{M}$-simplexes forms a simplicial (infinite) complex (Vietoris $\mathfrak{M}$-complex) and is denoted by $X^v(\mathfrak{M})$. An *orientation* for n-dimensional Vietoris simplex $x_0 x_1 \cdots x_n$ is a total ordering on $\{x_0, x_1, \ldots, x_n\}$ up to even permutations. In the following, we suppose that every Vietoris $\mathfrak{M}$-complex is oriented.

The set of all n-*dimensional Vietoris* $\mathfrak{M}$-*chain*, $C_n^v(\mathfrak{M})$, is the vector space whose elements are uniquely represented as a finite sum of n-dimensional Vietoris $\mathfrak{M}$-simplexes,

$$c^n = \sum_{i=1}^{k} \alpha_i \sigma_i^n, \quad (\sigma_1^n, \ldots, \sigma_k^n \in X^v(\mathfrak{M})),$$

where coefficients $\alpha_1, \ldots, \alpha_k$ are taken in field F. We may also consider the boundary operator among chains, $\partial_n : C_n^v(\mathfrak{M}) \to C_{n-1}^v(\mathfrak{M})$, for each n, as the linear map satisfying,

$$\partial_n : x_0 x_1 \cdots x_n \to \sum_{i=0}^{n} (-1)^i x_0 x_1 \cdots \hat{x}_i \cdots x_n,$$

---

[7] For more details of the Čech homology theory, see Eilenberg and Steenrod (1952). For more introductory arguments, Hocking and Young (1961, Chapter 8) is also recommended.

where the circumflex over a vertex means elimination as before, and it is supposed that $C_n^v(\mathfrak{M}) = 0$ for all $n < 0$.[8] The set of all n-dimensional Vietoris $\mathfrak{M}$-cycles, $Z_n^v(\mathfrak{M})$, and the set of n-dimensional Vietoris $\mathfrak{M}$-boundaries, $B_n^v(\mathfrak{M})$, may also be defined as usual, so that we obtain the n-th Vietoris $\mathfrak{M}$-homology group, $H_n^v(\mathfrak{M})$, for each n.

For coverings $\mathfrak{M}, \mathfrak{N} \in \text{Cover}(X)$, it is clear that $(\mathfrak{N} \preccurlyeq \mathfrak{M}) \Rightarrow (X^v(\mathfrak{N}) \subset X^v(\mathfrak{M}))$. Denote by $h_n^{\mathfrak{M}\mathfrak{N}} : C_n^v(\mathfrak{N}) \to C_n^v(\mathfrak{M})$ the chain homomorphism that is induced by the above inclusion. Then, for each n, the system of vector spaces with mappings, $(C_n^v(\mathfrak{M}), h_n^{\mathfrak{M}\mathfrak{N}})_{\mathfrak{M},\mathfrak{N} \in \text{Cover}(X)}$, their cycles, $(Z_n^v(\mathfrak{M}), h_n^{\mathfrak{M}\mathfrak{N}})_{\mathfrak{M},\mathfrak{N} \in \text{Cover}(X)}$, and boundaries, $(B_n^v(\mathfrak{M}), h_n^{\mathfrak{M}\mathfrak{N}})_{\mathfrak{M},\mathfrak{N} \in \text{Cover}(X)}$, form inverse systems. The inverse limit of the inverse system, $(Z_n^v(\mathfrak{M})/B_n^v(\mathfrak{M}), h_{*n}^{\mathfrak{M}\mathfrak{N}})_{\mathfrak{M},\mathfrak{N} \in \text{Cover}(X)}$,

$$H_n^v(X) = \varprojlim_{\mathfrak{M}} H_n^v(\mathfrak{M}),$$

is the n-dimensional (n-th) Vietoris Homology group. If W is a subset of X, the set of all Vietoris $\mathfrak{M}$ simplexes whose vertices are in W is denoted by $X^v(\mathfrak{M}) \cap W$. It is obvious that boundary operators and inclusion mappings with respect to each $X^v(\mathfrak{M}) \subset X^v(\mathfrak{N})$ are closed in $X^v(\mathfrak{M}) \cap W$, so $X^v(\mathfrak{M}) \cap W$ is a *closed subcomplex* of $X^v(\mathfrak{M})$ and for each n, the inverse system, $(Z_n^v(\mathfrak{M})/B_n^v(\mathfrak{M}))$, may adequately be restricted as relations on subcomplex $X^v(\mathfrak{M}) \cap W$. For this inverse limit, we use notation $H_n^v(W)$ even when W is not a compact subset of X.

An element of $H_n^v(X)$ may be identified with an equivalence class of a sequence of n-dimensional Vietoris $\mathfrak{M}$-cycles, $\mathfrak{M} \in \text{Cover}(X)$, (an n-dimensional Vietoris cycle), $\{z^n(\mathfrak{M}) \in Z_n^v(\mathfrak{M}) | \mathfrak{M} \in \text{Cover}(X)\}$, such that for each $\mathfrak{M}, \mathfrak{N} \in \text{Cover}(X)$ satisfying that $\mathfrak{N} \preccurlyeq \mathfrak{M}$, we have $z^n(\mathfrak{M}) \sim h_n^{\mathfrak{M}\mathfrak{N}}(z^n(\mathfrak{N}))$, where the equivalence class is taken with respect to Vietoris $\mathfrak{M}$-boundaries, i.e., $z^n(\mathfrak{M}) - h_n^{\mathfrak{M}\mathfrak{N}}(z^n(\mathfrak{N})) \in B_n^v(\mathfrak{M})$.[9]

---

[8] Precisely, as is the case with Čech cycles, by definition they equal 0 as vector spaces having an empty basis.

[9] The concept of Vietoris homology group was originally introduced by Vietoris (1927) as the first homology theory of the Čech type for metric spaces. Though the theory has been used in many researches, e.g., Eilenberg and Montgomery (1946), it has not been frequently discussed as the more general Čech theory. The theory was extended to be applicable for cases of compact Hausdorff spaces by Begle (1950a), and the result was used in Nikaido (1959) to prove an analogue of Sperner's lemma.

6.1.8 Vietoris and Čech cycles

The Čech homology theory is a powerful tool to approximate a space with groups of a finite complex. The Vietoris homology theory, on the other hand, has an intuitional advantage with which we may directly characterize the space by its elements (points). Fortunately, we may utilize both merits since the two homological concepts give the same homology groups (see Theorem 6.1.3 below).

Before proving this, note the following facts on the equivalences of two cycles on a simplicial complex. Since a homology group is nothing but a set of equivalence classes of cycles, it is not surprising that homological arguments often depend on such equivalence results. Let K be a simplicial complex. Suppose that the set of vertices of K, $\textbf{Vert}(K)$, is simply ordered in an arbitrary way, and let $\sigma^n = \langle a_0, a_1, \ldots, a_n \rangle$ be an n-simplex (oriented by simple order) in K. The *product simplicial complex* of K and the unit interval denoted by $K \times \{0, 1\}$ is the family of simplexes of the form $\langle (a_0, 0), (a_1, 0), \ldots, (a_i, 0), (a_i, 1), \ldots, (a_n, 1) \rangle$ for each $\langle a_0, a_1, \ldots, a_n \rangle \in K$ with all of their faces (Figure 32). The subcomplex of $K \times \{0, 1\}$ constructed by all simplexes of the form $\langle (a_0, 0), \ldots, (a_n, 0) \rangle$ with $\langle a_0, \ldots, a_n \rangle \in K$ may clearly be identified with K and be called the *base* of $K \times \{0, 1\}$. An isomorphism also exists between K and the subcomplex of all simplexes of form $\langle (a_0, 1), \ldots, (a_n, 1) \rangle$ with $\langle a_0, \ldots, a_n \rangle \in K$, called the *top* of $K \times \{0, 1\}$. For each n-simplex $\langle \sigma^n \rangle = \langle a_0, \ldots, a_n \rangle$ of K, define an $n+1$-chain, $\Phi_n(\sigma^n)$, on product simplicial complex $K \times \{0, 1\}$ as

$$\Phi_n(\sigma^n) = \sum_{j=0}^{n} (-1)^j \langle (a_0, 0), \ldots, (a_j, 0), (a_j, 1), \ldots, (a_n, 1) \rangle. \tag{6.1}$$

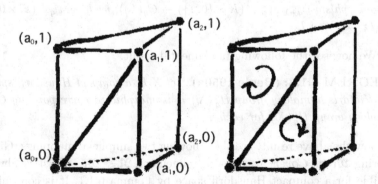

Figure 32: Prism $K \times \{0, 1\}$

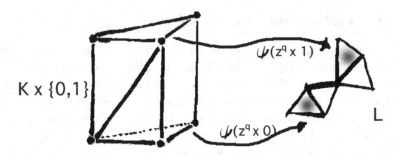

Figure 33: Mapping ψ on Prism $K \times \{0,1\}$ to L

Extend each Φ_n to a homomorphism on $C_n(K)$ to $C_n(K \times \{0,1\})$. Then we can verify through direct calculations that for each n-chain $c^n \in K$,

$$\partial_{n+1}\Phi_n(c^n) + \Phi_{n-1}\partial_n(c^n) = c^n \times 1 - c^n \times 0 \in C_{n-1}(K \times \{0,1\}), \quad (6.2)$$

where $c^n \times 1$ (resp., $c^n \times 0$) is the chain on the top (resp. base) of $K \times \{0,1\}$ formed by replacing each vertex of each simplex of c^n by the vertex of the ordered pair with 0 (resp., 1). Hence, if z^n is a cycle on K,

$$\partial_{n+1}\Phi_n(z^n) = z^n \times 1 - z^n \times 0 \in B_n(K \times \{0,1\}), \quad (6.3)$$

i.e., we have $z^n \times 0 \sim z^n \times 1$ on $K \times \{0,1\}$. Therefore, if simplicial mapping ψ on $K \times \{0,1\}$ to certain simplicial complex L exists, the next lemma holds.

LEMMA 6.1.2: *Assume the existence of simplicial mapping ψ on $K \times \{0,1\}$ to simplicial complex L. For two images, $\psi_{q+1}(z^q \times 0)$ and $\psi_{q+1}(z^q \times 1)$, in q-th chain group $C_q(L)$ of q-cycle $z^q \in C_q(K)$ (through induced homomorphism $\psi_{q+1} : C_{q+1}(K \times \{0,1\}) \to C_q(L)$), we have $\psi_{q+1}(z^q \times 0) \sim \psi_{q+1}(z^q \times 1)$ on L.*

We now see the following fundamental result.

THEOREM 6.1.3: (Begle 1950a) *Let X be a compact Hausdorff space. q-th Vietoris homology group $H_q^v(X)$ is isomorphic to corresponding Čech homology group $H_q^c(X)$ for each q.*

To show the above result, use the following two simplicial mappings. Given covering $\mathfrak{M}$ in $\mathbf{Cover}(X)$, choose refinement $\mathfrak{N} \preccurlyeq^* \mathfrak{M}$, which is always possible for a compact Hausdorff space by Lemma 6.1.1. It is convenient for the discussion below to denote one such selection for each $\mathfrak{M}$ by a

6: The Čech Type Homology Theory and Fixed Points

fixed operator on $\mathbf{Cover}(X)$ as $\mathfrak{N} = {}^*\mathfrak{M}$.[10] For each $\mathfrak{M} \in \mathbf{Cover}(X)$ and for each $x \in X$, there are $N_x \in {}^*\mathfrak{M}$ and $M_x \in \mathfrak{M}$ such that $x \in N_x$ and $St(N_x; {}^*\mathfrak{M}) \subset M_x$. Moreover, for each $N \in {}^*\mathfrak{M}$ there is an element $x_N \in N$. Define functions $\zeta^b_{\mathfrak{M}}$ and $\varphi^b_{\mathfrak{M}}$ as

$$\zeta^b_{\mathfrak{M}} : \mathbf{Vert}(X^v({}^*\mathfrak{M})) = X \ni x \mapsto M_x \in \mathfrak{M} = \mathbf{Vert}(X^c(\mathfrak{M})) \quad (6.4)$$

$$\varphi^b_{\mathfrak{M}} : \mathbf{Vert}(X^c({}^*\mathfrak{M})) = {}^*\mathfrak{M} \ni N \mapsto x_N \in X = \mathbf{Vert}(X^v(\mathfrak{M})). \quad (6.5)$$

Under the definition of star refinement, it is easy to see that $\zeta^b_{\mathfrak{M}}$ and $\varphi^b_{\mathfrak{M}}$ are simplicial mappings. Hence, we obtain chain homomorphisms $\zeta^b_{\mathfrak{M}q} : C^v_q(\mathfrak{N}) \to C^c_q(\mathfrak{M})$ and $\varphi^b_{\mathfrak{M}q} : C^c_q(\mathfrak{M}) \to C^v_q(\mathfrak{P})$. As we see below, these mappings play essential roles in characterizing the relations between Vietoris and Čech homology groups. Especially, mappings $\zeta^b_{\mathfrak{M}q}$ and $\varphi^b_{\mathfrak{M}q}$ respectively induce isomorphisms $\zeta^b_{*q} : H^v_q(X) \to H^c_q(X)$ and $\varphi^b_{*q} : H^c_q(X) \to H^v_q(X)$ (Theorem 6.1.3), and $\varphi^b_{\mathfrak{M}q} \circ \zeta^b_{\mathfrak{N}q}$ ($\mathfrak{N} = {}^*\mathfrak{M}$) assures the finite dimensional character of acyclic spaces (Theorem 6.2.2) or locally connected spaces (Theorem 6.2.4).

PROOF OF THEOREM 6.1.3: Let $\gamma^q = \{\gamma^q(\mathfrak{N}) | \mathfrak{N} \in \mathbf{Cover}(X)\}$, (or simply, $\{\gamma^q(\mathfrak{N})\}$) be a q-dimensional Vietoris cycle. For each $\mathfrak{M} \in \mathbf{Cover}(X)$ and $\mathfrak{N} = {}^*\mathfrak{M}$, define $z^q(\mathfrak{M})$ as $z^q(\mathfrak{M}) = \zeta^b_{\mathfrak{M}q}(\gamma^q(\mathfrak{N}))$. We see that (1) $z^q = \{z^q(\mathfrak{M})\}$ is a Čech cycle, and that (2) mapping $\zeta^b_{*q} : \gamma^q \mapsto z^q$ is an isomorphism on $H^v_q(X)$ to $H^c_q(X)$.

(1) Since $\zeta^b_{\mathfrak{M}q} : C^v_q(\mathfrak{N}) \to C^c_q(\mathfrak{M})$ is a chain homomorphism, all $z^q(\mathfrak{M})$ ($\mathfrak{M} \in \mathbf{Cover}(X)$) are cycles in $C^c_q(\mathfrak{M})$. Hence, by definition of the inverse limit, all we have to show is $z^q(\mathfrak{M}_1) \sim p_q^{\mathfrak{M}_1 \mathfrak{M}_2}(z^q(\mathfrak{M}_2))$ for each $\mathfrak{M}_2 \preccurlyeq \mathfrak{M}_1$. Let $\mathfrak{N}_1$ and $\mathfrak{N}_2$ be refinements of $\mathfrak{M}_1$ and $\mathfrak{M}_2$, respectively, to define mappings $\zeta^b_{\mathfrak{M}_1 q}$ and $\zeta^b_{\mathfrak{M}_2 q}$. By Lemma 6.1.1, we can take $\mathfrak{P}$ as $\mathfrak{P} \preccurlyeq^* \mathfrak{N}_1$ and $\mathfrak{P} \preccurlyeq^* \mathfrak{N}_2$. Note that since $\{\gamma^q(\mathfrak{N})\}$ is a Vietoris cycle, we have $h_q^{\mathfrak{N}_1 \mathfrak{P}}(\gamma^q(\mathfrak{P})) \sim \gamma^q(\mathfrak{N}_1)$ and $h_q^{\mathfrak{N}_2 \mathfrak{P}}(\gamma^q(\mathfrak{P})) \sim \gamma^q(\mathfrak{N}_2)$. Hence, $z^q(\mathfrak{M}_1) = \zeta^b_{\mathfrak{M}_1 q}(\gamma^q(\mathfrak{N}_1)) \sim \zeta^b_{\mathfrak{M}_1 q}(h_q^{\mathfrak{N}_1 \mathfrak{P}}(\gamma^q(\mathfrak{P})))$ and $p_q^{\mathfrak{M}_1 \mathfrak{M}_2}(z^q(\mathfrak{M}_2)) = p_q^{\mathfrak{M}_1 \mathfrak{M}_2}(\zeta^b_{\mathfrak{M}_2 q}(\gamma^q(\mathfrak{N}_2))) \sim p_q^{\mathfrak{M}_1 \mathfrak{M}_2}(\zeta^b_{\mathfrak{M}_2 q}(h_q^{\mathfrak{N}_2 \mathfrak{P}}(\gamma^q(\mathfrak{P}))))$.[11] It follows that all we have to show is $\zeta^b_{\mathfrak{M}_1 q}(\gamma^q(\mathfrak{P})) \sim p_q^{\mathfrak{M}_1 \mathfrak{M}_2}(\zeta^b_{\mathfrak{M}_2 q}(\gamma^q(\mathfrak{P})))$. Let $K = K(\gamma^q(\mathfrak{P}))$ be the complex that consists of all simplexes in cycle $\gamma^q(\mathfrak{P})$ with their faces. Then by Lemma 6.1.2, it is sufficient to show the

---

[10] For this, the Axiom of Choice is needed.
[11] In the above, inclusion mappings $h_q^{\mathfrak{N}_1 \mathfrak{P}}$ and $h_q^{\mathfrak{N}_2 \mathfrak{P}}$ might be abbreviated. Since including relation $C^v_q(\mathfrak{N}') \subset C^v_q(\mathfrak{N})$ for each $\mathfrak{N}' \preccurlyeq \mathfrak{N}$ is obvious, these operators will be omitted henceforth as long as there is no fear of confusions.

existence of simplicial map ψ on $K \times \{0,1\}$ to $L = X^c(\mathfrak{M}_1)$ such that $\zeta^b_{\mathfrak{M}_1 q}(\gamma^q(\mathfrak{P}))$ and $p_q^{\mathfrak{M}_1 \mathfrak{M}_2}(\zeta^b_{\mathfrak{M}_2 q}(\gamma^q(\mathfrak{P})))$ are images through induced map $\psi_{q+1} : C_{q+1}(K \times \{0,1\}) \to X^c(\mathfrak{M}_1)$ of $\gamma^q(\mathfrak{P}) \times 0$ and $\gamma^q(\mathfrak{P}) \times 1$, respectively. For each $a \in \mathbf{Vert}(K)$, define ψ as $\psi((a,0)) = \zeta^b_{\mathfrak{M}_1}(a)$ and $\psi((a,1)) = p^{\mathfrak{M}_1 \mathfrak{M}_2} \zeta^b_{\mathfrak{M}_2}(a)$. For any simplex $\langle (a_0, 0), \ldots, (a_i, 0), (a_i, 1), \ldots, (a_k, 1) \rangle$ in $K \times \{0,1\}$, we have simplex $a_0 \cdots a_k$ of $K = K(\gamma^q(\mathfrak{P}))$, so $P \in \mathfrak{P}$ exists such that $a_0, \ldots, a_k \in P$. We have to show that $\langle \zeta^b_{\mathfrak{M}_1}(a_0), \ldots, \zeta^b_{\mathfrak{M}_1}(a_i), p^{\mathfrak{M}_1 \mathfrak{M}_2} \zeta^b_{\mathfrak{M}_2}(a_i), \ldots, p^{\mathfrak{M}_1 \mathfrak{M}_2} \zeta^b_{\mathfrak{M}_2}(a_k) \rangle$ forms a simplex in $X^c(\mathfrak{M}_1)$. For each j, $0 \leq j \leq i$, since $\mathfrak{P} \preccurlyeq^* \mathfrak{N}_1 \preccurlyeq^* \mathfrak{M}_1$, each $\zeta^b_{\mathfrak{M}_1}(a_j) = M_{1 a_j}$ ($0 \leq j \leq i$) includes $St(N_{1 a_j}, \mathfrak{N}_1)$ for a certain $N_{1 a_j} \ni a_j$. Hence, P which has a_j and satisfies $St(P, \mathfrak{P}) \subset N_1$ for a certain $N_1 \in \mathfrak{N}_1$ must be a subset of $St(N_{1 a_j}, \mathfrak{N}_1) \subset M_{1 a_j}$. For each j, $i \leq j \leq k$, since $\mathfrak{P} \preccurlyeq^* \mathfrak{N}_2 \preccurlyeq^* \mathfrak{M}_2 \preccurlyeq \mathfrak{M}_1$, each $p^{\mathfrak{M}_1 \mathfrak{M}_2} \zeta^b_{\mathfrak{M}_2}(a_j) = p^{\mathfrak{M}_1 \mathfrak{M}_2} M_{2 a_j}$ ($i \leq j \leq k$) includes $St(N_{2 a_j}, \mathfrak{N}_2)$ for a certain $N_{2 a_j} \ni a_j$. Hence P, which has a_j and satisfies $St(P, \mathfrak{P}) \subset N_2$ for a certain $N_2 \in \mathfrak{N}_2$, must be a subset of $St(N_{2 a_j}, \mathfrak{N}_2) \subset M_{2 a_j}$ and the corresponding element for $M_{2 a_j}$ under projection $p^{\mathfrak{M}_1 \mathfrak{M}_2}$ of $\mathfrak{M}_1$. Therefore, we have $\zeta^b_{\mathfrak{M}_1}(a_0) \cap \cdots \cap \zeta^b_{\mathfrak{M}_1}(a_i) \cap p^{\mathfrak{M}_1 \mathfrak{M}_2} \zeta^b_{\mathfrak{M}_2}(a_i) \cap \cdots \cap p^{\mathfrak{M}_1 \mathfrak{M}_2} \zeta^b_{\mathfrak{M}_2}(a_k) \supset P \neq \emptyset$ and $\langle \zeta^b_{\mathfrak{M}_1}(a_0), \ldots, \zeta^b_{\mathfrak{M}_1}(a_i), p^{\mathfrak{M}_1 \mathfrak{M}_2} \zeta^b_{\mathfrak{M}_2}(a_i), \ldots, p^{\mathfrak{M}_1 \mathfrak{M}_2} \zeta^b_{\mathfrak{M}_2}(a_k) \rangle \in X^c(\mathfrak{M}_1)$, i.e., ψ is a simplicial map. By considering the construction way for the induced map ψ_q, it is also clear that $\psi_{q+1}(\gamma^q \times 0) = \zeta^b_{\mathfrak{M} q}(\gamma^q)$ and $\psi_{q+1}(\gamma^q \times 1) = p_q^{\mathfrak{M} \mathfrak{N}} \zeta^b_{\mathfrak{N} q}(\gamma^q)$.

(2) We have to show that mapping $\zeta^b_{*q} : Z^v_q(X) \ni \gamma^q \mapsto z^q \in Z^c_q(X)$ is one to one and onto. For this purpose, we use three steps: (2-1) defines mapping $\varphi^b_{*q} : Z^c_q(X) \to Z^v_q(X)$, (2-2) shows that composite $\varphi^b_{*q} \circ \zeta^b_{*q}$ is the identity, and (2-3) shows that composite $\zeta^b_{*q} \circ \varphi^b_{*q}$ is the identity.

(2-1) Let us define a function that gives for each $\mathfrak{M}$ and $z^q = \{z^q(\mathfrak{M})\} \in Z^c_q(X)$, an element $\varphi^b_{\mathfrak{N} q}(z^q(\mathfrak{N})) \in Z^v_q(\mathfrak{M})$, where $\mathfrak{N} = {}^*\mathfrak{M}$. Denote the relation by $\varphi^b_{*q} : Z^c_q(X) \ni z^q \mapsto \{\varphi^b_{\mathfrak{N} q}(z^q({}^*\mathfrak{M})) | \mathfrak{M} \in \mathbf{Cover}(X)\} \in \prod_{\mathfrak{M} \in \mathbf{Cover}(X)} Z^v_q(\mathfrak{M})$. We see that for each $\mathfrak{M}_2 \preccurlyeq \mathfrak{M}_1$ with $\mathfrak{N}_1 = {}^*\mathfrak{M}_1$ and $\mathfrak{N}_2 = {}^*\mathfrak{M}_2$, $\varphi^b_{\mathfrak{N}_1 q}(z^q(\mathfrak{N}_1)) \sim h_q^{\mathfrak{M}_1 \mathfrak{M}_2} \varphi^b_{\mathfrak{N}_2 q}(z^q(\mathfrak{N}_2))$, so that sequence $\{\varphi^b_{\mathfrak{N} q}(z^q({}^*\mathfrak{M})) | \mathfrak{M} \in \mathbf{Cover}(X)\}$ is a Vietoris cycle. We can assume $\mathfrak{M}_2 \preccurlyeq^* \mathfrak{N}_1 \preccurlyeq^* \mathfrak{M}_1$ without loss of generality since the existence of a common star refinement, $\mathfrak{M}_3$, of $\mathfrak{N}_2$ and $\mathfrak{N}_1$ combined with the same assertions for $\mathfrak{M}_3 \preccurlyeq^* \mathfrak{N}_1 \preccurlyeq^* \mathfrak{M}_1$ and $\mathfrak{M}_3 \preccurlyeq^* \mathfrak{N}_2 \preccurlyeq^* \mathfrak{M}_2$ will assure the results for $\mathfrak{M}_2 \preccurlyeq \mathfrak{M}_1$ through $h_q^{\mathfrak{M}_1 \mathfrak{M}_3} \varphi^b_{\mathfrak{M}_3 q}(z^q({}^*\mathfrak{M}_3))$. Take a common star refinement $\mathfrak{P}$ of $\mathfrak{N}_1$ and $\mathfrak{N}_2$. Since $z^q = \{z^q(\mathfrak{M})\}$ is a Čech cycle, all we have to show is $\varphi^b_{\mathfrak{N}_1 q}(p_q^{\mathfrak{N}_1 \mathfrak{P}} z^q(\mathfrak{P})) \sim h_q^{\mathfrak{M}_1 \mathfrak{M}_2} \varphi^b_{\mathfrak{N}_2 q}(p_q^{\mathfrak{N}_2 \mathfrak{P}} z^q(\mathfrak{P}))$. Let $K = K(z^q(\mathfrak{P}))$ be the complex formed by all simplexes in cycle

6: The Čech Type Homology Theory and Fixed Points 141

$z^q(\mathfrak{P}) \in X_q^c(\mathfrak{P})$ with their faces. By Lemma 6.1.2, it is sufficient for our purpose to show the existence of simplicial map ψ on $K \times \{0,1\}$ to $L = X^v(\mathfrak{M}_1)$ such that $\varphi^b_{\mathfrak{M}_1 q}(p_q^{\mathfrak{N}_1 \mathfrak{P}} z^q(\mathfrak{P}))$ and $h_q^{\mathfrak{M}_1 \mathfrak{M}_2} \varphi^b_{\mathfrak{M}_2 q}(p_q^{\mathfrak{N}_2 \mathfrak{P}} z^q(\mathfrak{P}))$ are images through induced map $\psi_{q+1} : C_{q+1}(K \times \{0,1\}) \to X^v(\mathfrak{M}_1)$ of $z^q(\mathfrak{P}) \times 0$ and $z^q(\mathfrak{P}) \times 1$, respectively. For each $a \in \mathbf{Vert}(K) \subset \mathfrak{P}$, define ψ as $\psi((a,0)) = \varphi^b_{\mathfrak{M}_1}(p^{\mathfrak{N}_1 \mathfrak{P}}(a))$ and $\psi((a,1)) = \varphi^b_{\mathfrak{M}_2}(p^{\mathfrak{N}_2 \mathfrak{P}}(a))$. For any simplex $\langle (a_0,0), \ldots, (a_i,0), (a_i,1), \ldots, (a_k,1) \rangle$ in $K \times \{0,1\}$, we have a simplex $a_0 \cdots a_k$ of $K = K(z^q(\mathfrak{P}))$, so that $a_0 \cap \cdots \cap a_k \neq \emptyset$. We have to show that $\langle \varphi^b_{\mathfrak{M}_1}(p^{\mathfrak{N}_1 \mathfrak{P}}(a_0)), \ldots, \varphi^b_{\mathfrak{M}_1}(p^{\mathfrak{N}_1 \mathfrak{P}}(a_i)), \varphi^b_{\mathfrak{M}_2}(p^{\mathfrak{N}_2 \mathfrak{P}}(a_i)), \ldots, \varphi^b_{\mathfrak{M}_2}(p^{\mathfrak{N}_2 \mathfrak{P}}(a_k)) \rangle$ forms a simplex in $X^v(\mathfrak{M}_1)$. Note that for each j, $0 \leq j \leq i$, $\mathfrak{P} \preccurlyeq^* \mathfrak{N}_1 \preccurlyeq^* \mathfrak{M}_1$, and for each j, $i \leq j \leq k$, $\mathfrak{P} \preccurlyeq^* \mathfrak{N}_2 \preccurlyeq^* \mathfrak{M}_2 \preccurlyeq^* \mathfrak{N}_1 \preccurlyeq^* \mathfrak{M}_1$. Since $a_0 \cap \cdots \cap a_k \neq \emptyset$, there are $N_1 \in \mathfrak{N}_1$ and $N_2 \in \mathfrak{N}_2$ such that $a_0 \cup \cdots \cup a_k \subset N_1$ and $a_0 \cup \cdots \cup a_k \subset N_2$. By definitions of φ^b and p, $St(N_1; \mathfrak{N}_1)$ and $St(N_2; \mathfrak{N}_2)$ contain all points of the form $\varphi^b_{\mathfrak{M}_1}(p^{\mathfrak{N}_1 \mathfrak{P}}(a_j))$, ($0 \leq j \leq i$) and $\varphi^b_{\mathfrak{M}_2}(p^{\mathfrak{N}_2 \mathfrak{P}}(a_j))$, ($i \leq j \leq k$). There are $M_1 \in \mathfrak{M}_1$ and $M_2 \mathfrak{M}_2$ such that $St(N_1; \mathfrak{N}_1) \subset M_1$ and $St(N_2; \mathfrak{N}_2) \subset M_2$. The fact $\mathfrak{M}_2 \preccurlyeq^* \mathfrak{N}_1$ means, however, that $M_2 \subset N_1'$ for some N_1' in $\mathfrak{N}_1$. Since $N_1' \cap N_1 \supset a_0 \cup \cdots \cup a_k$, $N_1' \subset St(N_1; \mathfrak{N}_1)$, so that M' includes both $St(N_1; \mathfrak{N}_1)$ and $St(N_2; \mathfrak{N}_2)$. Hence, $\langle \varphi^b_{\mathfrak{M}_1}(p^{\mathfrak{N}_1 \mathfrak{P}}(a_0)), \ldots, \varphi^b_{\mathfrak{M}_1}(p^{\mathfrak{N}_1 \mathfrak{P}}(a_i)), \varphi^b_{\mathfrak{M}_2}(p^{\mathfrak{N}_2 \mathfrak{P}}(a_i)), \ldots, \varphi^b_{\mathfrak{M}_2}(p^{\mathfrak{N}_2 \mathfrak{P}}(a_k)) \rangle$ forms a simplex in $X^v(\mathfrak{M}_1)$.

(2-2) We see for each $\mathfrak{M}$, $\mathfrak{N} = {}^*\mathfrak{M}$, $\mathfrak{P} = {}^*\mathfrak{N}$, and $\gamma^q \in C^v(X)$, $\varphi^b_{\mathfrak{M} q} \circ \zeta^b_{\mathfrak{N} q}(\gamma^q(\mathfrak{P})) \sim \gamma^q(\mathfrak{P})$, which is sufficient for assertion $\zeta^b_{*q} \circ \varphi^b_{*q}(\gamma^q) = \gamma^q$. Let $K = K(\gamma^q(\mathfrak{P}))$ be the subcomplex of $X^v(\mathfrak{P})$ formed by simplexes of $\gamma^q(\mathfrak{P})$ and their faces. By Lemma 6.1.2, we may reduce the problem to show the existence of simplicial map ψ on $K \times \{0,1\}$ to $L = X^v(\mathfrak{M})$ such that $\varphi^b_{\mathfrak{M} q} \circ \zeta^b_{\mathfrak{N} q}(\gamma^q(\mathfrak{P}))$ and $\gamma^q(\mathfrak{P})$ are images under induced map $\psi_{q+1} : C_{q+1}(K \times \{0,1\}) \to X^v(\mathfrak{M})$ of $\gamma^q(\mathfrak{P}) \times 0$ and $\gamma^q(\mathfrak{P}) \times 1$, respectively. For each $a \in \mathbf{Vert}(K) \subset X$, define ψ as $\psi((a,0)) = \varphi^b_{\mathfrak{M}} \circ \zeta^b_{\mathfrak{N}}(a)$ and $\psi((a,1)) = a$. For any simplex $\langle (a_0,0), \ldots, (a_i,0), (a_i,1), \ldots, (a_k,1) \rangle$ in $K \times \{0,1\}$, we have a simplex $a_0 \cdots a_k$ of $K = K(\gamma^q(\mathfrak{P}))$, so that there is a member P of $\mathfrak{P}$ such that $a_0, \ldots, a_k \in P$. We have to show that $\langle \varphi^b_{\mathfrak{M}} \circ \zeta^b_{\mathfrak{N}}(a_0), \ldots, \varphi^b_{\mathfrak{M}} \circ \zeta^b_{\mathfrak{N}}(a_i), a_i, \ldots, a_k \rangle$ forms a simplex in $X^v(\mathfrak{M})$. Since $\mathfrak{P} \preccurlyeq^* \mathfrak{N} \preccurlyeq^* \mathfrak{M}$, there are $N \in \mathfrak{N}$ and $M \in \mathfrak{M}$ such that $St(P, \mathfrak{P}) \subset N$ and $St(N, \mathfrak{N}) \subset M$. Hence, by definitions of $\varphi^b_{\mathfrak{M}}$ and $\zeta^b_{\mathfrak{N}}$, M includes all vertices of $\langle \varphi^b_{\mathfrak{M}} \circ \zeta^b_{\mathfrak{N}}(a_0), \ldots, \varphi^b_{\mathfrak{M}} \circ \zeta^b_{\mathfrak{N}}(a_i), a_i, \ldots, a_k \rangle$.

(2-3) For each $\mathfrak{M}$, $\mathfrak{N} = {}^*\mathfrak{M}$, $\mathfrak{P} = {}^*\mathfrak{N}$, and $z^q \in C^c(X)$, we see $\zeta^b_{\mathfrak{M} q} \circ \varphi^b_{\mathfrak{N} q}(z^q(\mathfrak{P})) \sim z^q(\mathfrak{P})$. This exactly shows $\zeta^b_{*q} \circ \varphi^b_{*q}(z^q) = z^q$. Let $K = K(z^q(\mathfrak{P}))$ be the subcomplex of $X^c(\mathfrak{P})$ formed by simplexes of $z^q(\mathfrak{P})$

and their faces. By Lemma 6.1.2, it is sufficient for our purpose to show the existence of simplicial map ψ on $K \times \{0,1\}$ to $L = X^c(\mathfrak{M})$ such that $\zeta^b_{\mathfrak{M}q} \circ \varphi^b_{\mathfrak{M}q}(z^q(\mathfrak{P}))$ and $z^q(\mathfrak{P})$ are images under the induced map $\psi_{q+1} : C_{q+1}(K \times \{0,1\}) \to X^c(\mathfrak{M})$ of $z^q(\mathfrak{P}) \times 0$ and $z^q(\mathfrak{P}) \times 1$, respectively. For each $a \in \mathbf{Vert}(K) \subset \mathfrak{P}$, define ψ as $\psi((a,0)) = \zeta^b_{\mathfrak{M}} \circ \varphi^b_{\mathfrak{M}}(a)$ and $\psi((a,1)) = a$. For any simplex $\langle (a_0, 0), \ldots, (a_i, 0), (a_i, 1), \ldots, (a_k, 1) \rangle$ in $K \times \{0,1\}$, we have a simplex $a_0 \cdots a_k$ of $K = K(z^q(\mathfrak{P}))$, so that sets $a_0, \ldots, a_k \in \mathfrak{P}$ satisfy $a_0 \cap \cdots \cap a_k \neq \emptyset$. We have to show that $\langle \zeta^b_{\mathfrak{M}} \circ \varphi^b_{\mathfrak{M}}(a_0), \ldots, \zeta^b_{\mathfrak{M}} \circ \varphi^b_{\mathfrak{M}}(a_i), a_i, \ldots, a_k \rangle$ forms a simplex in $X^c(\mathfrak{M})$. By definitions of $\varphi^b_{\mathfrak{M}}$ and $\zeta^b_{\mathfrak{M}}$, vertex $\zeta^b_{\mathfrak{M}} \circ \varphi^b_{\mathfrak{M}}(a_j)$ ($0 \leq j \leq i$) is a set in $M_j \in \mathfrak{M}$ such that for a certain $x_j \in a_j$ and its neighborhood $N_j \in \mathfrak{N}$, relation $M_j \supset St(N_j; \mathfrak{N})$ holds. Since $a_0 \cap \cdots \cap a_k \neq \emptyset$, there is a set $N \in \mathfrak{N}$ such that $a_0 \cup \cdots \cup a_k \subset St(a_0; \mathfrak{P}) \subset N$. Since $(N_j; \mathfrak{N})$ includes N for each $j = 0, \ldots, i$, M_j includes N for each $j = 0, \ldots, i$. Hence $M_1 \cap \cdots \cap M_i \cap a_i \cap \cdots a_k \supset a_0 \cap \cdots \cap a_k \neq \emptyset$, so that $\langle \zeta^b_{\mathfrak{M}} \circ \varphi^b_{\mathfrak{M}}(a_0), \ldots, \zeta^b_{\mathfrak{M}} \circ \varphi^b_{\mathfrak{M}}(a_i), a_i, \ldots, a_k \rangle$ is a simplex in $X^c(\mathfrak{M})$. ∎

6.2 Vietoris–Begle Mapping and Local Connectedness

6.2.1 *Vietoris–Begle mapping*

It is sometimes convenient to use the notion of a *reduced* set of 0-cycles and *reduced* 0-*th homology groups*. A reduced 0-th homology group is obtained by considering only 0-cycles (= 0-chains) in which the sum of the coefficients is 0. Instead of 0-th homology group $H_0(X) = Z_0(X)/B_0(X)$, the reduced homology group will be denoted by $\tilde{H}_0(X) = \tilde{Z}_0(X)/B_0(X)$, where $\tilde{Z}_0(X) = \{z \in Z_0(X) | z = \sum \alpha_i \sigma_i^0$ and $\sum \alpha^i = 0\}$. Topological space X is called *acyclic* under a certain homology theory, if (1) X is non-empty, (2) homology groups $H_q(X)$ are 0 for all $q > 0$, and (3) 0-th homology group $H_0(X)$ equals the coefficient group F (or equivalently 0-th reduced homology group $\tilde{H}_0(X)$ equals 0).[12]

---

[12] Note that the basis of $Z_0(X) = C_0(X)$ is $\{\sigma^0 | \sigma^0 = \langle x \rangle, x \in X\}$ and $Z_0(X)/\tilde{Z}_0(X)$ may be identified with F. One can also check that all 0-boundaries belong to $\tilde{Z}_0(X)$. If every 0-cycle (= 0-chain) in which the sum of coefficients is 0 is a 0-boundary (i.e., if $\tilde{H}_0(X) = 0$), $Z_0(X)/B_0(X)$ may therefore be identified with F. On the other hand, if there is a 0-cycle such that the sum of its coefficients is 0 and it is not a 0-boundary, then the set of all such chains forms a subspace of $\tilde{Z}_0(X)$ and the dimension of $Z_0(X)/B_0(X)$ that equals the dimension of $(Z_0(X)/\tilde{Z}_0(X)) \times (\tilde{Z}_0(X)/B_0(X))$ is greater than two; hence it cannot be identified with F.

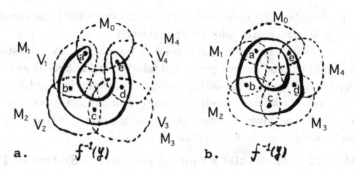

Figure 34: Vietoris–Begle mapping of order n

Let X and Y be compact Hausdorff spaces. For Vietoris $\mathfrak{M}$-complex $X^v(\mathfrak{M})$ and subset W of X, the set of all Vietoris $\mathfrak{M}$-simplexes whose vertices are points in W forms a subcomplex of $X^v(\mathfrak{M})$ and is denoted by $X^v(\mathfrak{M}) \cap W$. Then continuous function f of X onto Y is called a *Vietoris–Begle mapping of order n* if for each covering $\mathfrak{M}$ of X and for each $y \in Y$, there is covering $\mathfrak{P} = \mathfrak{P}(\mathfrak{M}, y)$ of X with $\mathfrak{P} \preccurlyeq \mathfrak{M}$ such that each q-dimensional ($0 \leq q \leq n$) Vietoris $\mathfrak{P}$-cycle $z^q(\mathfrak{P}) \in X^v(\mathfrak{P}) \cap f^{-1}(y)$ bounds (can be identified with an image of the boundary operator of) a $q+1$-dimensional Vietoris $\mathfrak{M}$-chain $c^{q+1}(\mathfrak{M}) \in X^v(\mathfrak{M}) \cap f^{-1}(y)$, where all 0-dimensional cycles are chosen in the reduced sense. (See Figure 34a, for $\mathfrak{M} = \{M_0, M_1, M_2, M_3, M_4\}$, refinement $\mathfrak{V} = \{V_1, V_2, V_3, V_4\}$ satisfies the condition. In Figure 34b, however, for any refinement of $\mathfrak{M} = \{M_0, M_1, M_2, M_3, M_4\}$, we cannot expect 1-dimensional cycles of type $ab+bc+cd+de+ea$ bound a 2-dimensional $\mathfrak{M}$-chain in $f^{-1}(y)$.) Continuous onto function $f : X \to Y$ is said to be a *Vietoris mapping* if compact set $f^{-1}(y)$ is acyclic for all $y \in Y$, i.e., $H_n^v(f^{-1}(y)) = 0$ for all $n > 0$ and $\tilde{H}_0^v(f^{-1}(y)) = 0$. If f is a Vietoris–Begle mapping of order n for all n, by definition of the inverse limit, f is a Vietoris mapping.[13] The converse is also true in our special settings. In this chapter, we see the following two important theorems: (i) if coefficient group F is a field, Vietoris mapping is a Vietoris–Begle mapping of order n for all n (Theorem 6.2.2), and (ii) if $f : X \to Y$ is a Vietoris–Begle mapping of order n, there are isomorphisms between $H_q^v(X)$ and $H_q^v(Y)$ ($0 \leq q \leq n$) (Theorem 6.3.3). In this section,

---

[13] Under induced inverse system $(H_n^v(\mathfrak{M}) = Z_n^v(\mathfrak{M})/B_n^v(\mathfrak{M}), h_{*n}^{\mathfrak{M}\mathfrak{N}})_{\mathfrak{M},\mathfrak{N} \in Cover(f^{-1}(y))}$, for each n and $\mathfrak{M}$, by taking $\mathfrak{V}$ in the condition for Vietoris–Begle mapping of order n, images of n-th $\mathfrak{N}$-cycle, $h_{*n}^{\mathfrak{M}\mathfrak{N}}(z^n(\mathfrak{N}))$, belong to $B_n^v(\mathfrak{M}) = 0 \in H_n^v(\mathfrak{M})$ for all $\mathfrak{N} \preccurlyeq \mathfrak{V}$. This means that every element of inverse limit $H_n^v(f^{-1}(y))$ must be 0.

we see (i). Assertion (ii) is treated in the next section after the concept of Vietoris–Begle barycentric subdivision is defined.

Since coefficient group F is supposed to be a field, inverse systems of Vietoris and Čech type chains, cycles, boundaries, and homology groups are systems of vector spaces. All n-dimensional chains, cycles, and boundary groups of nerves (defining Čech homology groups) are especially finite dimensional. For an inverse system of finite dimensional vector spaces, we know the following result on *essential elements*.[14]

LEMMA 6.2.1: (Essential Elements for Inverse System of Finite Dimensional Vector Spaces) *Let* $(E_i, \pi_{ij})_{i,j \in I, j \geq i}$ *over directed set* $(I, \geq)$ *be an inverse system of finite dimensional vector spaces. Then for every i there is element $j_0 \geq i$ such that for all $j \geq j_0$, every element x_i of $\pi_{ij}(E_j) \subset E_i$ is an essential element of E_i, i.e., $x_i \in \pi_{ik}(E_k)$ for all $k \geq i$.*

PROOF: The set of essential elements of E_i is subspace $H_i = \bigcap_{j \geq i} \pi_{ij}(E_j)$. Since E_i is finite dimensional, the dimension of H_i is also finite: n. Then there are finite elements $k_1, \ldots, k_n$ of I such that $H_i = \bigcap_{j=1}^{n} \pi_{ik_j}(E_{k_j})$. Let j_0 be an element of I such that $j_0 \geq j_k$ for each $k = 1, \ldots, n$. Then for all $j \geq j_0$, we have $\pi_{ij}(E_j) = \pi_{ij_0}(\pi_{j_0 j}(E_j)) \subset \pi_{ij_0}(E_{j_0}) = \pi_{ij_k}(\pi_{j_k j_0}(E_{j_0})) \subset \pi_{ij_k}(e_{j_k})$ for each $k = 1, \ldots, n$. Hence, for each $j \geq j_0$, $\pi_{ij}(E_j) \subset H_i = \bigcap_{j=1}^{n} \pi_{ik_j}(E_{k_j})$. ∎

Since the inverse system for the Čech homology group (for compact Hausdorff space X) is a system of finite dimensional vector spaces, it follows from Lemma 6.2.1 that for each covering $\mathfrak{M}$ of X, there is a refinement $\mathfrak{N} \prec \mathfrak{M}_0 = {}^*\mathfrak{M}$ such that if $z^q(\mathfrak{N}) \in Z_k^c(\mathfrak{N})$ is a q-dimensional $\mathfrak{N}$-cycle of X, then $p_q^{\mathfrak{M}_0 \mathfrak{N}}(z^q(\mathfrak{N}))$ is the $\mathfrak{M}_0$-coordinate of a Čech cycle. By taking the finest $\mathfrak{N}$ for $q = 0, 1, \ldots, k$ and taking $\mathfrak{P} = {}^*\mathfrak{N}$, we have the following theorem:

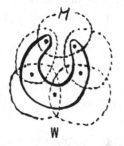
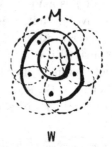

---

[14] This concept of importance in the homology theory of system of groups is due to Čech (1932). See also Lefschetz (1942, p. 79) and Steenrod (1936) for elementary compact coefficient groups.

THEOREM 6.2.2: (Vietoris Mapping and Vietoris–Begle Mapping of Order n) *Let $\mathfrak{M}$ be a covering of compact Hausdorff space X and W be a compact subset of X such that every q-dimensional Čech reduced cycle in W ($0 \le q \le k$) bounds a $q+1$-dimensional Čech chain in W. ($\tilde{H}_q^c(W) = 0$).*[15] *Then refinement $\mathfrak{P}$ of $\mathfrak{M}$ exists such that every q-dimensional Vietoris $\mathfrak{P}$-cycle on W ($0 \le q \le k$) bounds a $q+1$-dimensional Vietoris $\mathfrak{M}$-chain on W. Hence, Vietoris mapping is a Vietoris–Begle mapping of order n for all n.*

PROOF: Take refinements $\mathfrak{P} = {}^*\mathfrak{N}$ and $\mathfrak{N}$ of $\mathfrak{M}_0 = {}^*\mathfrak{M}$ as stated in the previous paragraph. Let $\gamma_\mathfrak{P}^q$ be a q-dimensional Vietoris $\mathfrak{P}$-cycle on W ($0 \le q \le k$). Denote by $\zeta_\mathfrak{N}^b : X^v(\mathfrak{P}) \to X^c(\mathfrak{N})$ the simplicial mapping defined in the proof of Theorem 6.1.3. Then $\zeta_{\mathfrak{N}q}^b(\gamma_\mathfrak{P}^q)$ is a q-dimensional Čech $\mathfrak{N}$-cycle ($0 \le q \le k$). By definition of $\mathfrak{N}$, $p_q^{\mathfrak{M}_0\mathfrak{N}}\zeta_\mathfrak{N}^b(\gamma_\mathfrak{P}^q)$ is the $\mathfrak{M}_0$-coordinate of Čech cycle z^q on W. Since $\tilde{H}_q^c(W) = 0$, this Čech cycle bounds a $q+1$ chain so that $p_q^{\mathfrak{M}_0\mathfrak{N}}\zeta_\mathfrak{N}^b(\gamma_\mathfrak{P}^q) \sim 0$ on $C_q^c(\mathfrak{M}_0)$. It follows that $\varphi_\mathfrak{M}^b p_q^{\mathfrak{M}_0\mathfrak{N}}\zeta_\mathfrak{N}^b(\gamma_\mathfrak{P}^q) \sim 0$ on $W^v(\mathfrak{M}) = X^v(\mathfrak{M}) \cap W$, where $\varphi_\mathfrak{M}^b$ is the simplicial mapping defined in the proof of Theorem 6.1.3 and $X^v(\mathfrak{M}) \cap W$ denotes the subcomplex of Vietoris $\mathfrak{M}$-simplexes on W. Hence, the first assertion of this theorem follows if we see $\varphi_\mathfrak{M}^b p_q^{\mathfrak{M}_0\mathfrak{N}}\zeta_\mathfrak{N}^b(\gamma_\mathfrak{P}^q) \sim \gamma_\mathfrak{P}^q$ on $X^v(\mathfrak{M}) \cap W$. We can see it by completely repeating the same argument with (2-2) in the proof of Theorem 6.1.3. The second assertion follows immediately from the first if we set $W = f^{-1}(y)$ for Vietoris mapping $f : X \to Y$ and point $y \in Y$. ∎

6.2.2 Locally connected spaces

Besides Vietoris–Begle mapping, there is another important concept for fixed-point arguments under Čech-type homology: *local connectedness*. In Čech-type homology theory, the family of open coverings, **Cover**(X), on compact Hausdorff space X is used to describe the two fundamental features of topological arguments: (i) the measure of connectivity (represented by the intersection property among open sets), and (ii) the measure of convergence or approximation (as a net of refinements of coverings). All analytic concepts are changed into algebraic ones through the above two

---

[15]For notational convenience, let us define here $\tilde{H}_q^c(W)$ for each $q > 0$ as $\tilde{H}_q^c(W) = H_q^v(W)$. Of course, the condition says that W is an acyclic set in the sense of Čech homology (equivalently, in the sense of Vietoris homology by considering the Čech–Vietoris equivalence arguments in Theorem 6.1.3).

channels. In the following, it is especially important to notice the second feature, so that each covering $\mathfrak{M} \in \mathbf{Cover}(X)$ is used as a sort of metric or a norm, and $\mathbf{Cover}(X)$ is used as if it were uniformity in describing the total convergence properties for space X. To emphasize that we are choosing a covering or a refinement for the second purpose, we call it a *norm covering* or a *norm refinement* instead of simply saying covering or refinement.

Local connectedness is defined as a purely homological notion to generalize the concept of *absolute neighborhood retracts* frequently used under the framework of metrizable spaces.[16] Consider compact Hausdorff space Y and $\mathfrak{M} \in \mathbf{Cover}(Y)$. A *realization* of simplicial complex K in $Y^v(\mathfrak{M})$ is a chain map τ of K into $Y^v(\mathfrak{M})$. *Partial realization* τ' of K is a chain map defined on subcomplex L of K such that $\mathbf{Vert}(L) = \mathbf{Vert}(K)$. For norm covering $\mathfrak{N} \in \mathbf{Cover}(X)$ and realization τ of K, write norm$(\tau) \leq \mathfrak{N}$ if for each simplex σ of K, there is a set $N \in \mathfrak{N}$ that contains underlying space $|\tau\sigma|$ of chain $\tau\sigma$.[17] For partial realization τ' of K (a chain mapping on subcomplex L of K), we write norm$(\tau') \leq \mathfrak{N}$ only when for each simplex σ in K (not in L) there is a set $N \in \mathfrak{N}$ that contains underlying space $|\tau'\sigma'|$ of chain $\tau'\sigma'$ for all face σ' of σ in L.

DEFINITION 6.2.3: (Locally Connected Space) Topological space X is said to be *locally connected* (abbreviated by lc) if for each norm covering $\mathfrak{E} \in \mathbf{Cover}(X)$, norm refinement $\mathfrak{J}(\mathfrak{E}) \preccurlyeq \mathfrak{E}$ satisfies the following condition: for each covering $\mathfrak{M}$, there is a refinement $\mathfrak{N}(\mathfrak{M}, \mathfrak{E})$ such that every partial realization τ' of finite complex K into $X^v(\mathfrak{N}(\mathfrak{M}, \mathfrak{E}))$ with norm$(\tau') \leq \mathfrak{J}(\mathfrak{E})$ may be extended to realization τ into $X^v(\mathfrak{M})$ with norm$(\tau) \leq \mathfrak{E}$.

Intuitively, the definition says that for X we can take covering $\mathfrak{J}(\mathfrak{E})$ to roughly grasp the shape of X by identifying members of $\mathfrak{J}(\mathfrak{E})$ as connected

---

[16] Topological space X is an *absolute neighborhood retract* (*ANR*) if for every normal space Y, for every closed subset $B \subset Y$, and for every continuous function $f : B \to X$, neighborhood U of B and continuous function $f' : U \to X$ exist such that the restriction of f' on B equals f. Generally, for subset $A \subset Z$ of topological space Z, A is a *retract* of Z if there is a continuous function $g : Z \to A$ such that the restriction of g on A is identity. A is a *neighborhood retract* (*NR*) of Z if A is a retract of a certain neighborhood V in Z. It is clear by the definition that if X is ANR, its open subsets, finite products, and retracts are also ANRs.

[17] For a value under a homomorphism, parenthesis is abbreviated as $\tau\sigma = \tau(\sigma)$. Note also that the underlying space of chain $\tau\sigma$ is the underlying space of the corresponding complex defined by all simplexes of $\tau\sigma$ (that appeared with non-zero coordinates in the formal summation). The situation where set $N \in \mathfrak{N}$ contains $|\tau\sigma|$ is denoted by diam $|\tau\sigma| \leq \mathfrak{N}$.

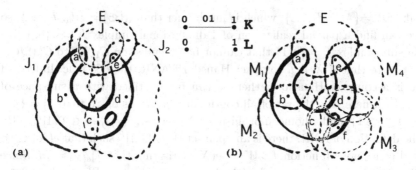

Figure 35: Covering $\mathfrak{J}$ in definition of lc space

components without holes of X. See Figure 35. If $\mathfrak{E} = \{E, J_1, J_2\}$, $\mathfrak{J}(\mathfrak{E}) = \{J, J_1, J_2\}$, and $\mathfrak{M} = \{M_1, M_2, M_3, M_4\}$, then for any refinement $\mathfrak{N}(\mathfrak{M},\mathfrak{E})$ of $\mathfrak{E}$, partial realization τ' of K on L such that $\tau'_0(0) \mapsto \langle a \rangle$ and $\tau'_0(1) \mapsto \langle b \rangle$ fails to be extended in a way such that norm(τ) $\leq \mathfrak{E}$. If we take $\mathfrak{J}(\mathfrak{E})$ as $\{J_1, J_2\}$, we can take $\mathfrak{N}(\mathfrak{M},\mathfrak{E})$ as $\mathfrak{N}(\mathfrak{M},\mathfrak{E}) = \mathfrak{M}$. In this case, however, $\mathfrak{J}(\mathfrak{E})$ still fails to grasp the hole in J_2. If we refine $\mathfrak{M}$ in such a way that M_3 is further refined into three open sets, $M_{3-1}, M_{3-2}, M_{3-3}$, surrounding the hole with $\bigcap_{i=1}^{3} M_{3-i} = \emptyset$, then using c, d, f in the picture, we obtain a partial realization of 2-dimensional simplex $K = \{0, 1, 2, 01, 12, 20, 123\}$ on subcomplex $L = \{0, 1, 2\}$ that cannot be extended on $C_2(K)$ to $C_2^v(\mathfrak{M})$.

A converging sequence with a limit point may provide a simple example for a compact Hausdorff space that is not lc. Even a connected compact Hausdorff space may fail to be lc, as the example (*topologist's sine curve*) in Figure 36 shows. In the picture (including segment $[v_0 v_1]$ of limit points), any sufficiently small ϵ-ball near $y \in [v_0 v_1]$ will crop part of the sine curve as a set of infinitely many segments. As in the case with a converging sequence, we cannot find an appropriate $\mathfrak{J}(\mathfrak{E})$ representing the shape of X, i.e., if we

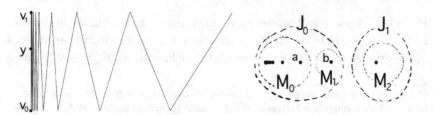

Figure 36: Topologist's sine curve with limit points

take $\mathfrak{M} = \{M_0, M_1, \ldots\}$, which is much finer than $\mathfrak{J}(\mathfrak{E}) = \{J_0, J_1, \ldots\}$, so we can find a partial realization of 1-dimensional simplex $K = \{0, 1, 01\}$ on subcomplex $L = \{0, 1\}$ that cannot be extended on $C_1(K)$ to $C_1^v(\mathfrak{M})$.

Note that if X is a compact Hausdorff ANR, then X is lc. Indeed, if X is a compact Hausdorff, then X can be identified with a subspace of $[0,1]^F$, where F is the set of all continuous functions on X to $[0, 1] = \{y \in R | 0 \leq y \leq 1\}$, under homeomorphism $e : X \ni x \to (f(x))_{f \in F} \in [0,1]^F$.[18] By the definition of ANR, there is an open set $W \subset [0,1]^F$ such that $e(X) \subset W$ and continuous function $f : W \to e(X)$ satisfying that $f|_{e(X)} = id$. Since $[0,1]^F$ can also be identified with the convex subset of R^F, we have base U for the topology of $[0,1]^F$ such that each $U \in U$ can be identified with a convex subset of R^F. That is, each set of $n + 1$ points in U is associated with a continuous image in U of an n-dimensional simplex in R^F whose $n + 1$ vertices are mapped to those $n + 1$ points. Hence, if for any $\mathfrak{E}$ given, we construct $\mathfrak{J}(\mathfrak{E})$ through elements of $\{U | U \in U, U \subset W\}$, then for each set of $n + 1$ points in $J \in \mathfrak{J}(\mathfrak{E})$, we have continuous function g on an n-dimensional simplex in R^E to X such that g maps the vertices of the simplex to those $n + 1$ points. For $\mathfrak{M} \in Cover(X)$, take $\mathfrak{N}(\mathfrak{M}, \mathfrak{E})$ as a refinement of $\mathfrak{M}$ and $\mathfrak{J}(\mathfrak{E})$.

It is clear from the definition that if X is lc, then $X \times X$ is also lc. Moreover, compact Hausdorff lc space has the following strong properties.

THEOREM 6.2.4: (Begle 1950b) *If X is a compact Hausdorff lc space, the next three assertions hold: (a), (b), and (c).*

(a) *There is covering $\mathfrak{N}_0$ of X such that if z is a Vietoris cycle such that $z(\mathfrak{N}) \sim 0$ on $X^v(\mathfrak{N})$ for some $\mathfrak{N} \preccurlyeq \mathfrak{N}_0$, then $z \sim 0$.*

(b) *There is finite complex K such that the homology groups of X are isomorphic to the corresponding groups of K.*

(c) *Each covering $\mathfrak{M}$ of X has a normal refinement $\mathfrak{M}'$, i.e., a refinement such that for each cycle $z_{\mathfrak{M}'}$ on $X^v(\mathfrak{M}') \subset X^v(\mathfrak{M})$, there is a Vietoris cycle z such that $z(\mathfrak{M}) = z_{\mathfrak{M}'}$.*

PROOF: These results correspond to Lemmas 1, 2, and 3 in Begle (1950b). I provide here a brief sketch of the proof by roughly following the arguments in the original paper. To see assertion (a), define $\mathfrak{N}_0$ as $\mathfrak{N}_0 = \mathfrak{J}(\mathfrak{E})$, where

---

[18] Let X be a normal space and F be the set of all continuous functions on X to $[0, 1]$. By Urysohn's Lemma (see Mathematical Appendix I), one can probe that X is identified with a subspace of $[0,1]^F$ under homeomorphism $e : X \ni x \to (f(x))_{f \in F} \in [0,1]^F$. (See also Kelley (1955, p. 117).)

6: The Čech Type Homology Theory and Fixed Points

$\mathfrak{E}$ is an arbitrary covering of X. (For example, suppose that $\mathfrak{E} = \{X\}$.) Since the definition of lc is intended that $\mathfrak{J}(\mathfrak{E})$ represents the shape of X by identifying each $J \in \mathfrak{J}(\mathfrak{E})$ with connected components having no holes, we can check that given any Vietoris cycle (of a certain order) $\{z(\mathfrak{M})| \mathfrak{M} \in \mathbf{Cover}(X)\}$, $z(\mathfrak{N}) \sim 0$ on $X^v(\mathfrak{N})$ for some $\mathfrak{N} \preccurlyeq \mathfrak{N}_0$ means $z(\mathfrak{N}') \sim 0$ on $X^v(\mathfrak{N}')$ for all $\mathfrak{N}' \preccurlyeq \mathfrak{N}$ using the definition of lc (use $\mathfrak{N}'' = \mathfrak{N}(\mathfrak{N}',\mathfrak{E})$, chain C such that $\partial C = z(\mathfrak{N}'')$, and partial realization τ' of $|C|$ in $X^v(\mathfrak{N}'')$).

For assertion (b), consider $\mathfrak{N}_0$ in assertion (a), a star refinement $\mathfrak{N}_1 \preccurlyeq^* \mathfrak{N}_0$, and the nerve of covering $\mathfrak{N}_1$, $X^c(\mathfrak{N}_1)$. For each $M \in \mathfrak{N}_1$, define element $x^M \in M$ arbitrarily. Then we can define K as the Vietoris subcomplex of $X^v(\mathfrak{N}_0)$ such that $\mathbf{Vert}(K) = \{x^M| M \in \mathfrak{N}_1\}$ and $\langle x^{M_0} \cdots x^{M_n}\rangle \in K \subset X^v(\mathfrak{N}_0)$ if and only if $\langle M_0 \cdots M_n\rangle \in X^c(\mathfrak{N}_1)$. Mapping $X^c(\mathfrak{N}_1) \ni \langle M_0 \cdots M_n\rangle \mapsto \langle x^{M_0} \cdots x^{M_n}\rangle \in K \subset X^v(\mathfrak{N}_0)$ may be identified with $\varphi^b_{\mathfrak{N}_0 n}$ defined from (6.5). Take a further star refinement $\mathfrak{N}_2 \preccurlyeq^* \mathfrak{N}_1$ and consider mapping $\zeta^b_{\mathfrak{N}_2} : X \ni x \mapsto M_x \in \mathfrak{N}_2$ in (6.4). Then, as in the proof of Theorem 6.1.3, we can check that $\varphi^b_{\mathfrak{N}_0 n} \circ \zeta^b_{\mathfrak{N}_2 n}$ induces an isomorphism between $H^v_n(X)$ and $H_n(K)$.

For the last assertion (c), use $\mathfrak{N}_1$ in (b), and let $\mathfrak{M}'$ be $\mathfrak{N}(\mathfrak{M},\mathfrak{N}_1)$ for each $\mathfrak{M}$. To show the result, there is no loss in generality to assume that $\mathfrak{M} \preccurlyeq \mathfrak{J}(\mathfrak{N}_1)$. Let z be a cycle on $X^v(\mathfrak{M}')$. For each $\mathfrak{M}_1 \preccurlyeq \mathfrak{M}'$, by considering $\mathfrak{M}_2 = \mathfrak{N}(\mathfrak{M}_1, \mathfrak{N}_1)$, we may define a cycle $z(\mathfrak{M}_1)$ on $X^v(\mathfrak{M}_1)$ as follows. Let τ' be the partial realization of $|z|$ in $X^v(\mathfrak{M}_2)$ defined as $\tau'\sigma^0 = \sigma^0$ for each 0-dimensional simplex in $|z|$. Since norm$(\tau') \leq \mathfrak{J}(\mathfrak{N}_1)$, there is a realization τ of $|z|$ in $X^v(\mathfrak{N}_1)$ with norm$(\tau) \leq \mathfrak{M}_1$. Define $z'(\mathfrak{M}_1)$ as $z'(\mathfrak{M}_1) = \tau z$. For the special case $\mathfrak{M}_1 = \mathfrak{M}'$, we may suppose τ to be equal to the identity chain map. Though family $\{z'(\mathfrak{N})| \mathfrak{N} \preccurlyeq \mathfrak{M}'\}$ may not form a Vietoris cycle, we can find a cofinal (with respect to directed set $\mathbf{Cover}(X)$) subfamily that may be utilized to define a Vietoris cycle. Consider cofinal subset $\{\mathfrak{P}| \mathfrak{N}(\mathfrak{P},\mathfrak{N}_1) \preccurlyeq \mathfrak{M}'\}$ of $\mathbf{Cover}(X)$, and for each $\mathfrak{P} \in \{\mathfrak{P}| \mathfrak{N}(\mathfrak{P},\mathfrak{N}_1) \preccurlyeq \mathfrak{M}'\}$, define $z(\mathfrak{P})$ as $z(\mathfrak{P}) = z'(\mathfrak{N}(\mathfrak{P},\mathfrak{N}_1))$. Note that $z(\mathfrak{M}) = z'(\mathfrak{N}(\mathfrak{M},\mathfrak{N}_1)) = z'(\mathfrak{M}') = z$. To show that we may construct from $\{z(\mathfrak{P})| \mathfrak{N}(\mathfrak{P},\mathfrak{N}_1) \preccurlyeq \mathfrak{M}'\}$ a Vietoris cycle (i.e., for each $\mathfrak{P}' \preccurlyeq \mathfrak{P}''$, we have $z(\mathfrak{P}') \sim z(\mathfrak{P}'')$ on $X^v(\mathfrak{P}'')$), exploit that for all refinements $\mathfrak{N}'$ and $\mathfrak{N}''$ of a covering of the form $\mathfrak{N}(\mathfrak{P},\mathfrak{N}_1)$ that refines $\mathfrak{M}'$, we have $z(\mathfrak{N}') \sim z(\mathfrak{N}'')$ on $X^v(\mathfrak{P})$ (to see this, use partial realization ρ' of $K = |z| \times \{0,1\}$ in $X^v(\mathfrak{N}(\mathfrak{P},\mathfrak{N}_1))$ such that on the base of K, ρ' is the chain map on $|z|$ to $X^v(\mathfrak{N}')$ and on the top of K, ρ' is the chain map on $|z|$ to $X^v(\mathfrak{N}'')$). ∎

6.3 Nikaido's Analogue of Sperner's Lemma

In this section we see the important second half of the Vietoris–Begle mapping theorem; (ii) if $f : X \to Y$ is a Vietoris–Begle mapping of order n, there are isomorphisms between $H_q^v(X)$ and $H_q^v(Y)$ ($0 \leq q \leq n$). For this proof, we need the concept of barycentric subdivision under the framework of Vietoris complexes. After the proof of Vietoris–Begle mapping theorem, we also see an extension of Sperner's lemma originally given by Nikaido (1959) as the first application.

6.3.1 *Vietoris–Begle barycentric subdivision*

Let Y be a compact Hausdorff space. Consider coverings $\mathfrak{N} \in \mathbf{Cover}(Y)$ and $\mathfrak{R} \in \mathbf{Cover}(Y)$ of Y. In the following, for Vietoris $\mathfrak{M}$-chain $c(\mathfrak{M}) \in C_q^v(\mathfrak{M})$, denote by $K(c(\mathfrak{M}))$ the complex of all simplexes that appear with positive coefficients in $c(\mathfrak{M})$ and by $\text{diam}\,|c(\mathfrak{M})| \leq \mathfrak{N}$ that there is element $N \in \mathfrak{N}$ in which all vertices of $K(c(\mathfrak{M}))$ belong. Moreover, for each q-dimensional chain $c^q \in C_q^v(\mathfrak{M})$ and $y \in Y$, we denote by $y * c$ the $(q+1)$-dimensional $\{Y\}$-chain defined as the extension of the operation $y * \langle a_0 \cdots a_k \rangle = \langle y a_0 \cdots a_k \rangle$ for each oriented k-dimensional simplex $\langle a_0 \cdots a_k \rangle$.[19] $\mathfrak{R}\mathfrak{M}$-*barycentric subdivision* of k-dimensional Vietoris $\mathfrak{R}$-simplex $\sigma^k \in Y^v(\mathfrak{R})$ is chain map $Sd_q : C_q^v(\mathfrak{R}) \to C_q^v(\mathfrak{M})$, satisfying the following conditions:

(SD1) For each 0-dimensional simplex y_0 of $K(\sigma^k)$, $Sd_0(y_0) = y_0$.

(SD2) For each q-dimensional simplex $\langle y_0 \cdots y_q \rangle$ ($0 < q \leq k$) in $K(\sigma^k)$, $y \in Y$ exists such that $y * Sd_{q-1}(\langle y_0 \cdots \hat{y}_i \cdots y_q \rangle) \in C_q^v(\mathfrak{M})$ for each i and $Sd_q(\langle y_0 \cdots y_q \rangle) = \sum_{i=0}^q (-1)^i y * Sd_{q-1}(\langle y_0 \cdots \hat{y}_i \cdots y_q \rangle)$.

(SD3) $\text{diam}\,|Sd_k \sigma^k| \leq \mathfrak{N}$.

Note that as long as the existence of y for each q-dimensional $\mathfrak{R}$-simplex $\langle y_0 \cdots y_q \rangle$ stated in (SD2) is assured, conditions (SD1) and (SD2) may be considered a process to construct Sd_q, $q = 0, 1, \cdots$. By mathematical induction, we can verify for each $q > 0$ that $\partial_q Sd_q(\langle y_0 \cdots y_q \rangle) = Sd_{q-1}\partial_q(\langle y_0 \cdots y_q \rangle)$, so that the Sd_q constructed is indeed a chain map.

Consider n-*skeleton* $Y_n^v(\mathfrak{R}) \subset Y^v(\mathfrak{R})$ of $Y^v(\mathfrak{R})$, the subcomplex formed by all k-dimensional ($0 \leq k \leq n$) Vietoris $\mathfrak{R}$-simplexes on Y.

---

[19] Note that above, $\{Y\} \in \mathbf{Cover}(Y)$ is taken as a covering of Y.

An n-dimensional $\mathfrak{RN}$-barycentric subdivision of Y is a chain map $\{Sd_q^{\mathfrak{RN}} : C_q^v(Y_n^v(\mathfrak{R})) \to C_q^v(\mathfrak{N})\}$ such that for each k-dimensional simplex σ^k ($0 \le k \le n$), the restriction of $\{Sd_q^{\mathfrak{RN}}\}$ on the chain of subcomplex of $Y_n^v(\mathfrak{R})$ defined by σ^k is an $\mathfrak{RN}$-barycentric subdivision of σ^k.

Next, assume a continuous onto map f on compact Hausdorff space X to Y. For each pair of coverings $\mathfrak{M} \in \mathbf{Cover}(X)$ and $\mathfrak{N} \in \mathbf{Cover}(Y)$ such that $\mathfrak{M} \preccurlyeq \{f^{-1}(N) \mid N \in \mathfrak{N}\}$, f induces simplicial map $X^v(\mathfrak{M}) \ni a_0 \cdots a_k \mapsto f(a_0) \cdots f(a_k) \in Y^v(\mathfrak{N})$ so that chain map $\{f_q : C_q^v(\mathfrak{M}) \to C_q^v(\mathfrak{N})\}$. Then as we can see in the next theorem, if f is a Vietoris–Begle mapping of order n, there is a chain map $\tau = \{\tau_q\}$ on $(n+1)$-skeleton of $Y^v(\mathfrak{R})$ to $X(\mathfrak{M})$ such that $\{f_q \circ \tau_q\}$ is an $(n+1)$-dimensional $(\mathfrak{RN})$-barycentric subdivision of Y. Moreover, given $\mathfrak{M}$, such refinement $\mathfrak{R}$ may be taken arbitrarily small and corresponding τ's may be defined as (Vietoris homologically) unique.

THEOREM 6.3.1: Let X and Y be compact Hausdorff spaces and let $f : X \to Y$ be a Vietoris–Begle mapping of order n. For each $\mathfrak{M} \in \mathbf{Cover}(X)$ and $\mathfrak{N} \in \mathbf{Cover}(Y)$ such that $\mathfrak{M} \preccurlyeq \{f^{-1}(N) \mid N \in \mathfrak{N}\}$, cover $\mathfrak{R} = \mathfrak{R}(\mathfrak{M}, \mathfrak{N}) \in \mathbf{Cover}(Y)$ and chain map $\tau = \{\tau_q\}$ on $(n+1)$-skeleton of $Y^v(\mathfrak{R})$ to $X^v(\mathfrak{M})$ exist such that chain map $\{f_q \circ \tau_q\}$ is an n-dimensional $(\mathfrak{RN})$-barycentric subdivision of Y. Moreover, for any $\mathfrak{S} \in \mathbf{Cover}(Y)$, there exist $\mathfrak{R}'$ and τ' satisfying the same condition with $\mathfrak{R}$ and τ such that $\mathfrak{R}' \preccurlyeq \mathfrak{S}$ and $\tau_q'(z^q) \sim \tau_q(z^q)$ in $C_q^v(\mathfrak{M})$ for all $z^q \in Z_q^v(\mathfrak{R}')$.

The above theorem shows an essential feature of Vietoris–Begle mapping and plays crucial roles in the proof of the Vietoris–Begle mapping theorem. Before proving it, I introduce one technical lemma. In Lemma 6.1.2, we have seen one of the simplest kinds of *prismatical relations* that may be utilized to show equivalence between two cycles. Another convenient (though slightly more complicated) method exists in forming prisms. Denote by $\{0, 1, I\}$ the 1-dimensional abstract complex formed by 2 0-dimensional simplices 0 and 1 with 1-dimensional simplex I whose boundaries are 0 and 1 under relation $\partial_1(I) = 1 - 0$. For simplicial complex K, the *product complex* of K and $\{0, 1, I\}$ denoted by $K \times \{0, 1, I\}$ is the family of simplexes of form $\sigma \times 0$, $\sigma \times 1$, and $\sigma \times I$, where σ runs through all simplexes in K. Boundary relations on $K \times \{0, 1, I\}$ are defined as $\partial(\sigma \times 0) = (\partial \sigma) \times 0$, $\partial(\sigma \times 1) = (\partial \sigma) \times 1$, and $\partial(\sigma \times I) = (\partial \sigma) \times I + (\sigma \times 1) - (\sigma \times 0)$. (See Figure 37.) Note that $K \times \{0, 1, I\}$ is no longer a simplicial complex. The subcomplex of $K \times \{0, 1, I\}$ constructed by all simplexes of form $\sigma \times 0$ may clearly be identified with K and is called the *base* of $K \times \{0, 1, I\}$. An

Figure 37: Prism $K \times \{0,1,I\}$

isomorphism also exists between K and the subcomplex of all simplexes of form $\sigma \times 1$ called the *top* of $K \times \{0,1,I\}$. Then for each cycle z on K, we immediately have $\partial(z \times I) = (z \times 1) - (z \times 0)$ so that $z \times 1 \sim z \times 0$ in $K \times \{0,1,I\}$. Therefore, as before (Lemma 6.1.2), if chain mapping θ exists on $K \times \{0,1,I\}$ to a certain simplicial complex L, we have the following:

LEMMA 6.3.2: *Assume that chain mapping θ exists on $K \times \{0,1,I\}$ to simplicial complex L. For two images, $\theta_{q+1}(z^q \times 0)$ and $\theta_{q+1}(z^q \times 1)$, in q-th chain group $C_q(L)$ of q-cycle $z^q \in C_q(K)$ (through induced homomorphism $\theta_{q+1} : C_{q+1}(K \times \{0,1,I\}) \to C_q(L)$), we have $\theta_{q+1}(z^q \times 0) \sim \theta_{q+1}(z^q \times 1)$ on L.*

PROOF OF THEOREM 6.3.1: We use four steps. Step 1 is devoted to preparing the basic tools. In Step 2, we construct $\mathfrak{R}$. Step 3 is used to define τ. Step 4 is assigned for the constructions of $\mathfrak{R}'$ and τ'.

(Step 1) By the definition of Vietoris–Begle mapping, there is covering $\mathfrak{P}(\mathfrak{M},y)$ for each $y \in Y$ and $\mathfrak{M}$. Consider closed (compact) subset $X \backslash St(f^{-1}(y); {}^*\mathfrak{P}(\mathfrak{M},y))$. Then the image under f of $X \backslash St(f^{-1}(y); {}^*\mathfrak{P}(\mathfrak{M},y))$ is also a closed (compact) subset of normal space Y disjointed from $\{y\}$. Given $\mathfrak{N} \in \mathbf{Cover}(Y)$, choose $Q(\mathfrak{M},\mathfrak{N},y) \ni y$ as an element of $^*\mathfrak{N}$ and $\mathfrak{Q}(\mathfrak{M},\mathfrak{N})$ as a finite subcovering of the covering $\{Q(\mathfrak{M},\mathfrak{N},y)|y \in Y\}$. Then covering $\mathfrak{Q}(\mathfrak{M},\mathfrak{N})$ satisfies that if B is a subset of Y such that $B \subset Q$ for some $Q \in \mathfrak{Q}(\mathfrak{M},\mathfrak{N})$, there is point $y \in Y$ such that $St(y; {}^*\mathfrak{N}) \supset B$ and $St(f^{-1}(y); {}^*\mathfrak{P}(\mathfrak{M},y)) \supset f^{-1}(B)$. In this proof we

6: The Čech Type Homology Theory and Fixed Points

call this y the corresponding point of Y to B and use it as if it were the barycenter of points in B.

(Step 2) Hence, for each $\mathfrak{M} \in \mathbf{Cover}(X)$ and $\mathfrak{N} \in \mathbf{Cover}(Y)$, $\mathfrak{Q}(\mathfrak{M},\mathfrak{N}) \in \mathbf{Cover}(Y)$ satisfies that for every q-dimensional $\mathfrak{Q}(\mathfrak{M},\mathfrak{N})$-simplex $\langle y_0 \cdots y_q \rangle$, $(0 \leq q \leq n)$, there is point $y \in Y$ such that $y * \langle y_0 \cdots y_q \rangle$ is a *$\mathfrak{N}$-simplex and $St(f^{-1}(y); {}^*\mathfrak{P}(\mathfrak{M}, y)) \supset f^{-1}(\{y_0, \ldots, y_q\})$. This suggests the possibility to obtain a sequence of refinements $\mathfrak{M}_1 \preccurlyeq \cdots \preccurlyeq \mathfrak{M}_{n+1} = \mathfrak{M}$ with refinements $\mathfrak{N}_0 \preccurlyeq \cdots \preccurlyeq \mathfrak{N}_{n+1} = \mathfrak{N}$ such that $\mathfrak{M}_k \preccurlyeq \{f^{-1}(N) | N \in \mathfrak{N}_k\}$ for each $k = 1, \ldots, n+1$, and for each q-dimensional $\mathfrak{N}_q$-simplex $(q = 0, \ldots, n)$ $\langle y_0 \cdots y_q \rangle$, $y \in Y$ exists such that $y * \langle y_0 \cdots y_q \rangle$ is a *$\mathfrak{N}_{q+1}$-simplex and $St(f^{-1}(y); {}^*\mathfrak{P}(\mathfrak{M}_{q+1}, y)) \supset f^{-1}(\{y_0, \ldots, y_q\})$. (As we see in the next step, under the definition of barycentric subdivision (SD1)–(SD3), this property shows that for each $n+1$-dimensional $\mathfrak{N}_0$-simplex we can define an $\mathfrak{N}_0 \mathfrak{N}_{n+1}$-barycentric subdivision.) Indeed, given $\mathfrak{N}_{n+1} = \mathfrak{N}$ and $\mathfrak{M}_{n+1} = \mathfrak{M}$, set $\mathfrak{N}_n = \mathfrak{Q}(\mathfrak{M}_{n+1}, {}^*\mathfrak{N}_{n+1}) \preccurlyeq {}^{**}\mathfrak{N}_{n+1}$. Note that with $\mathfrak{Q}(\mathfrak{M}_{n+1}, {}^*\mathfrak{N}_{n+1})$ associates finite $y_{n+1,i}$'s such that $\mathfrak{Q}(\mathfrak{M}_{n+1}, {}^*\mathfrak{N}_{n+1})$ consists of $Q(\mathfrak{M}_{n+1}, {}^*\mathfrak{N}_{n+1}, y_{n+1,i})$'s. Let $\mathfrak{M}_n$ be a common refinement of coverings ${}^*\mathfrak{P}(\mathfrak{M}_{n+1}, y_{n+1,i})$'s and $\{f^{-1}(N) | N \in \mathfrak{N}_n\}$. Set $\mathfrak{N}_{n-1} = \mathfrak{Q}(\mathfrak{M}_n, {}^*\mathfrak{N}_n)$. Repeat the process until we obtain $\mathfrak{N}_0$. Define $\mathfrak{R}$ as $\mathfrak{R} = \mathfrak{R}(\mathfrak{M}, \mathfrak{N}) = \mathfrak{N}_0$.

(Step 3) Let us define τ_q $(0 \leq q \leq n)$ on chains of $Y^v(\mathfrak{R}) = Y^v(\mathfrak{N}_0)$ to $X^v(\mathfrak{M})$. Consider a 0-dimensional Vietoris $\mathfrak{R}$-simplex, σ^0, of $Y^v(\mathfrak{R})$. σ^0 may be identified with point y_0 in Y. Define $\tau(\sigma^0)$ as 0-dimensional Vietoris $\mathfrak{M}_0$-simplex ξ^0 of $X^v(\mathfrak{M}_0)$ that may be identified with arbitrary point $x_0 \in f^{-1}(y_0) \subset X$. Then we have $f_0 \circ \tau_0(\sigma^0) = \sigma^0 = Sd_0(\sigma^0)$, so that we obtain τ_0 by linearly extending it. Next, consider k-dimensional Vietoris $\mathfrak{R}$-simplex, σ^k, of $Y^v(\mathfrak{R})$ $(0 < k \leq n+1)$. Suppose that for each $(k-1)$-dimensional $\mathfrak{R}$-simplex σ^{k-1}, $\tau_{k-1}(\sigma^{k-1})$ is already defined and satisfies that $f_{k-1} \circ \tau_{k-1}(\sigma^{k-1})$ is a $\mathfrak{R} {}^*\mathfrak{N}_{k-1}$-barycentric subdivision of σ^{k-1} with the relation of chain map $\partial_{k-2} \circ \tau_{k-1} = \tau_{k-2} \circ \partial_{k-1}$, where τ_{k-2} for $k = 1$ is defined to be a 0-map. In the following, we see that we may define $\tau_k(\sigma^k)$ to satisfy that $\partial_{k-1} \circ \tau_k = \tau_{k-1} \circ \partial_k$ and $f_k \tau_k \sigma^k$ is a $\mathfrak{R} {}^*\mathfrak{N}_k$-barycentric subdivision of σ^k for each k-dimensional Vietoris $\mathfrak{R}$-simplex σ^k. Then by mathematical induction, we may extend the definition of τ_k until it is finally defined on all elements of the $(n+1)$-skeleton of $Y(\mathfrak{R})$. Since $\partial_k \sigma^k$ is an $\mathfrak{R}$-chain, $\tau_{k-1} \partial_k \sigma^k$ is already defined and is a $\mathfrak{M}_k$-cycle since $\partial_{k-1} \tau_{k-1} \partial_k \sigma^k = \tau_{k-2} \partial_{k-1} \partial_k \sigma^k = 0$. By assumption $f_{k-1} \tau_{k-1} \partial_k \sigma^k = f_{k-1} \tau_{k-1} \sum_{i=0}^{k}(-1)^i \sigma_i^{k-1} = \sum_{i=0}^{k}(-1)^i f_{k-1} \tau_{k-1} \sigma_i^{k-1}$

belongs to $C_{k-1}^v(^{**}\mathfrak{N}_{k-1})$, where σ_i^{k-1}'s are $k+1$ $(k-1)$-dimensional face of σ^k, and $f_{k-1}\tau_{k-1}\sigma_i^{k-1}$ is a $\mathfrak{R}\,^*\mathfrak{N}_{k-1}$-barycentric subdivision of σ_i^{k-1} for each i. It follows that all vertices of the $^{**}\mathfrak{N}_{k-1}$-chain, $f_{k-1}\tau_{k-1}\partial_k\sigma^k = f_{k-1}\tau_{k-1}\sum_{i=0}^k(-1)^i\sigma_i^k = \sum_{i=0}^k(-1)^k f_{k-1}\tau_{k-1}\sigma_i^k$, belong to $St(R_0; ^{**}\mathfrak{N}_{k-1}) \subset St(^{**}N_{k-1}; ^{**}\mathfrak{N}_{k-1})$ for an $R_0 \in \mathfrak{R}$ having all vertices of σ^k as its elements and $^{**}N_{k-1} \in \,^{**}\mathfrak{N}_{k-1}$ such that $R_0 \subset \,^{**}N_{k-1}$. Since there exists $^*N_{k-1} \in \,^*\mathfrak{N}_{k-1}$ such that $St(^{**}N_{k-1}; ^{**}\mathfrak{N}_{k-1}) \subset \,^*N_{k-1}$, we have diam$|f_{k-1}\tau_{k-1}\partial_k\sigma^k| \leq \,^*\mathfrak{N}_{k-1}$. Then $\mathfrak{N}_{k-1} = \mathfrak{Q}(\mathfrak{M}_k, ^*\mathfrak{N}_k)$ implies that there is corresponding point $y = y_{k,i} \in Y$, $Q(\mathfrak{M}_k, ^*\mathfrak{N}_k, y_{k,i}) \in \mathfrak{Q}(\mathfrak{M}_k, ^*\mathfrak{N}_k)$, to $|f_{k-1}\tau_{k-1}\partial_k\sigma^k|$ satisfying the following two relations.[20]

$$St(y; ^{**}\mathfrak{N}_k) \supset |f_{k-1}\tau_{k-1}\partial_k\sigma^k| \tag{6.6}$$

$$St(f^{-1}(y); ^*\mathfrak{P}(\mathfrak{M}_k, y)) \supset f^{-1}(|f_{k-1}\tau_{k-1}\partial_k\sigma^k|) \supset |\tau_{k-1}\partial_k\sigma^k|. \tag{6.7}$$

Denote by z^{k-1} the cycle $\tau_{k-1}\partial_k\sigma^k \in Z_{k-1}^v(\mathfrak{M}_{k-1})$ and let $x_1, \ldots, x_\ell$ be the vertices of $K(z^{k-1})$. Note that by (6.7), there are finite $x_1', \ldots, x_\ell' \in f^{-1}(y)$ and $^*P_1, \ldots, ^*P_\ell \in \,^*\mathfrak{P}(\mathfrak{M}_k, y)$ such that $x_1' \in \,^*P_1, \ldots, x_\ell' \in \,^*P_\ell$ and $x_1 \in \,^*P_1, \ldots, x_\ell \in \,^*P_\ell$. By defining mapping μ on **Vert**$(K(z^{k-1}) \times \{0,1\})$ to X as $\mu(x_i, 0) = x_i$ for each vertex $(x_i, 0)$ in the base of $K(z^{k-1}) \times \{0,1\}$ and $\mu(x_i, 1) = x_i'$ for each vertex $(x_i, 1)$ in the top of $K(z^{k-1}) \times \{0,1\}$. It is easy to check that μ is a simplicial map. Indeed, if $((a_0, 0), \ldots, (a_i, 0), (a_i, 1), \ldots, (a_m, 1))$ is a simplex in $K(z^{k-1}) \times \{0,1\}$, then $((a_0, 0), \ldots, (a_m, 1))$ is a simplex in $K(z^{k-1})$, so that element $M_{k-1} \in \mathfrak{M}_{k-1}$ exists such that $a_0, \ldots, a_m \in M_{k-1}$. Since a_i equals some x_j, and both $(x_j, 0)$ and $(x_j, 1)$ are in $^*P_j$, all vertices in $(a_0, \ldots, a_i, \mu(a_i, 1), \ldots, \mu(a_m, 1))$ belong to $St(M_{k-1}, ^*\mathfrak{P}(\mathfrak{M}_k, y))$. By considering that $\mathfrak{M}_{k-1} \preccurlyeq \,^*\mathfrak{P}(\mathfrak{M}_k, y)$, they belong to an element of $\mathfrak{P}(\mathfrak{M}_k, y)$, so that μ maps $K(z^{k-1})$ simplicially to $X^v(\mathfrak{P}(\mathfrak{M}_k, y))$. Let us use μ to define $\tau_k(\sigma^k)$ as follows: Set $\xi_1^k = \mu(\Phi_k(z^{k-1}))$, where Φ_k is the prismatic chain homotopy defined in Equations (6.1)–(6.3). By (6.3), we have $\partial_k(\mu\Phi_k z^{k-1}) = \mu(z^{k-1} \times 1) - \mu(z^{k-1} \times 0) = \mu(z^{k-1} \times 1) - z^{k-1}$. Since $\mu(z^{k-1} \times 1)$ is a cycle on $X^v(\mathfrak{P}(\mathfrak{M}_k, y)) \cap f^{-1}(y)$, there is chain ξ_2^k on $X^v(\mathfrak{P}(\mathfrak{M}_k, y)) \cap f^{-1}(y)$ such that $\partial_k\xi_2^k = \mu(z^{k-1} \times 1)$. Then if we set $\tau_k(\sigma^k) = \xi_2^k - \xi_1^k$, we have $\partial_k\tau_k\sigma^k = z^{k-1} = \tau_{k-1}\partial_{k-1}\sigma^k$, so that τ_k satisfies the condition for the chain map. Moreover, since $f_k(\tau_k\sigma^k) = f_k(\xi_2^k - \xi_1^k) = f_k(\xi_2^k) - f_k(\mu(\Phi_k(z^{k-1})))$, we may also rewrite it

---

[20] For Vietoris $\mathfrak{P}$-chain c, $|c|$ denotes the set of all vertices of simplexes that appeared in c with positive coefficients.

6: The Čech Type Homology Theory and Fixed Points   155

as $f_k(\xi_2^k) - \hat{\mu}(\hat{\Phi}_k(f_{k-1}z^{k-1})) = f_k(\xi_2^k) - \hat{\mu}(\hat{\Phi}_k(f_{k-1}\tau_{k-1}\partial_k\sigma^k)) = f_k(\xi_2^k) - \hat{\mu}(\hat{\Phi}_k(\mathrm{Sd}_{k-1}\partial_k\sigma^k))$, where $\hat{\Phi}$ is the prismatic chain homotopy on complex $K(f_{k-1}(z^{k-1}))$ to $K(f_{k-1}(z^{k-1})) \times \{0,1\}$ and $\hat{\mu}$ is defined on $K(f_{k-1}(z^{k-1}))$ in exactly the same way as μ, i.e., $\hat{\mu}(f(x_i),0) = f(x_i)$ and $\hat{\mu}(f(x_i),1) = f(x_i') = y$. Since $St(y; {}^{**}\mathfrak{N}_k) \supset |f_{k-1}\tau_{k-1}\partial_k\sigma^k|$, μ is a simplicial map on $K(f_{k-1}(z^{k-1})) \times \{0,1\}$ to $Y^v({}^*\mathfrak{N}_k)$. Moreover, $f_k(\tau_k\sigma^k)$ is clearly the connection of y with $\mathrm{Sd}_{k-1}\partial_k\sigma^k$ with diam $|\mathrm{Sd}_k\sigma^k| \leq {}^*\mathfrak{N}_k$.

(Step 4) Take $\mathfrak{M}_1' \preccurlyeq \cdots \preccurlyeq \mathfrak{M}_{n+1}'$ and $\mathfrak{N}_0' \preccurlyeq \cdots \preccurlyeq \mathfrak{N}_{n+1}'$ in the same way as $\mathfrak{M}_1 \preccurlyeq \cdots \preccurlyeq \mathfrak{M}_{n+1}$ and $\mathfrak{N}_0 \preccurlyeq \cdots \preccurlyeq \mathfrak{N}_{n+1}$ except for the process to define $\mathfrak{N}_k$ ($k \leq n$). Let us define $\mathfrak{N}_k'$ as a common refinement of $\mathfrak{Q}(\mathfrak{M}_{k+1}', {}^*\mathfrak{N}_{k+1}')$, ${}^*\mathfrak{N}_k$, and $\mathfrak{S}$ for each $k \leq n$. Define $\mathfrak{R}'$ as $\mathfrak{N}_0'$ and τ_k' ($0 \leq k \leq n+1$) in exactly the same way as τ_k. We now check for each $\mathfrak{R}'$-cycle z^n, $\tau_n(z^n) = \tau_n'(z^n)$. For this purpose, it is sufficient by Lemma 6.3.2 to show mapping θ to $X^v(\mathfrak{M})$ such that for each $\sigma^k \times 0$, $\theta(\sigma^k \times 0) = \tau_k(\sigma^k)$, and for each $\sigma^k \times 1$, $\theta(\sigma^k \times 1) = \tau_k'(\sigma^k)$, ($0 \leq k \leq n$), may be extended as a chain mapping on $K(z^n) \times \{0,1,I\}$. On the base and top of $K(z^n) \times \{0,1,I\}$, θ clearly defines chain maps since we have $\partial_k(\theta_k\sigma^k \times 0) = \partial_k(\tau_k(\sigma^k)) = \tau_{k-1}(\partial_k\sigma^k) = \theta_{k-1}(\partial_k\sigma^k \times 0)$ and $\partial_k(\theta_k\sigma^k \times 1) = \partial_k(\tau_k'(\sigma^k)) = \tau_{k-1}'(\partial_k\sigma^k) = \theta_{k-1}(\partial_k\sigma^k \times 1)$.

Let us consider a 0-dimensional simplex σ^0 in $K(z^n)$ and $\sigma^0 \times I \in K(z^n) \times \{0,1,I\}$. By definition (in Step 3) $f_0\tau_0\sigma^0 = f_0\tau_0'\sigma^0 = \sigma^0$ and both $\tau_0(\sigma^0)$ and $\tau_0'(\sigma^0)$ are points in $f^{-1}(\sigma^0) = f^{-1}(|f_0\tau_0\sigma^0|) = f^{-1}(|f_0\tau_0'\sigma^0|) \supset |\tau_0\sigma^0| \cup |\tau_0'\sigma^0|$. Then it is automatically satisfied that y ($y = \sigma^0$) exists such that

$$St(y; \mathfrak{N}_1) \supset |\sigma^0| \text{ and}$$

$$St(f^{-1}(y); {}^*\mathfrak{P}(\mathfrak{M}_1, y)) \supset f^{-1}(\sigma^0).$$

Note that $\theta\partial(\sigma^0 \times I) = \tau(\sigma^0) - \tau'(\sigma^0)$. Hence, we have $St(f^{-1}(y); {}^*\mathfrak{P}) \supset |\theta\partial(\sigma^0 \times I)|$ (Figure 38). Let us consider simplicial complex $K = K(\tau(\sigma^0) - \tau'(\sigma^0))$ and mapping $\omega : \mathbf{Vert}(K \times \{0,1\})$ to X such that $\omega(a,0) = a$ and $\omega(a,1) = y^a$, where y^a is an element of $f^{-1}(y)$ satisfying $\{a, y^a\} \subset {}^*P$ for some ${}^*P \in {}^*\mathfrak{P}$. Such y^a exists since $St(f^{-1}(y); {}^*\mathfrak{P}) \supset |\theta\partial(\sigma^0 \times I)|$. Then ω is a simplicial map on $K \times \{0,1\}$ to $X^v(\mathfrak{P})$. As before, let us define ξ_1^1 as $\xi_1^1 = \omega(\Phi(\tau_0\sigma^0 - \tau_0'\sigma^0))$, where Φ denotes the prismatic chain homotopy. Note that $\partial\xi_1^1 = \omega((\tau_0\sigma^0 - \tau_0'\sigma^0) \times 1) - (\tau_0\sigma^0 - \tau_0'\sigma^0)$. Now $\omega((\tau_0\sigma^0 - \tau_0'\sigma^0) \times 1)$ is a 0-cycle (by the previous equation) on $X^v(\mathfrak{P}) \cap f^{-1}(y)$, there is a 1-chain ξ_2^1 on $X^v(\mathfrak{M}_1) \cap f^{-1}(y)$ such that $\partial\xi_2^1 = \omega((\tau_0\sigma^0 - \tau_0'\sigma^0) \times 1)$. Define $\theta(\sigma^0 \times I)$ to be $\xi_2^1 - \xi_1^1$. Then θ satisfies the condition of chain map $\partial\theta = \theta\partial$

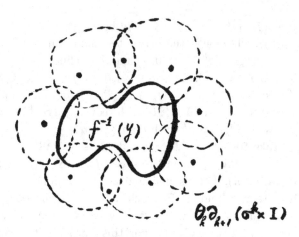

Figure 38: y and $\theta\partial(\sigma^k \times I)$

for $\sigma^0 \times I$ for each 0-dimensional σ^0. Clearly, $f|\xi_2^1 - \xi_1^1|$ is the connection of y and $\sigma^0 = y$, so that diam $f|\xi_2^1 - \xi_1^1| \leq {}^*\mathfrak{N}_1$.

Next assume that $\theta(\sigma^m \times I)$ is defined for each $m \leq k$ in such a way that $\partial\theta = \partial\theta$, $\theta(\sigma^m \times I) \in \mathfrak{M}_{m+1}$, and diam $f|\theta(\sigma^m \times I)| \leq {}^*\mathfrak{N}_{m+1}$. Let σ^k be a k-dimensional simplex of $K(z^n)$. Then $\theta(\partial(\sigma^k \times I))$ is already defined. Since $\theta(\partial(\sigma^k \times I)) = \theta((\partial\sigma^k) \times I) + \theta(\sigma^k \times 1) - \theta(\sigma^k \times 0)$, we have $f|\theta(\partial(\sigma^k \times I))| \subset f|\theta(\partial\sigma^k)| \cup f|\tau_k(\sigma^k)| \cup f|\tau'_k\sigma^k|$. By considering facts diam $f|\tau_k(\sigma^k)| \leq {}^*\mathfrak{N}_k$ and diam $f|\tau'_k(\sigma^k)| \leq {}^*\mathfrak{N}'_k \preccurlyeq {}^*\mathfrak{N}$, we have $St(R'; \mathfrak{N}_k) \supset f|\tau_k(\sigma^k)| \cup f|\tau'_k(\sigma^k)|$, where R' denotes an element of $\mathfrak{R}'$ to which all vertices of σ^k belong. It is also true by assumption that for each $(k-1)$-dimensional face σ^{k-1} of σ^k, diam $f|\theta(\sigma^{k-1} \times I)| \leq {}^*\mathfrak{N}_k$, so we have diam $f|\theta\partial(\sigma^k \times I)| \leq \mathfrak{N}_k = \mathfrak{Q}(\mathfrak{M}_{k+1}, {}^*\mathfrak{N}_{k+1})$. Hence, we have point y such that $Q(\mathfrak{M}_{k+1}, {}^*\mathfrak{N}_{k+1}, y) \in \mathfrak{Q}(\mathfrak{M}_{k+1}, {}^*\mathfrak{N}_{k+1})$,

$$St(y; {}^*\mathfrak{N}_{k+1}) \supset f|\theta\partial(\sigma^k \times I)| \quad \text{and}$$
$$St(f^{-1}(y); {}^*\mathfrak{P}(\mathfrak{M}_{k+1}, y)) \supset f^{-1}f|\theta\partial(\sigma^k \times I)|.$$

It follows that we have $St(f^{-1}(y); {}^*\mathfrak{P}(\mathfrak{M}_{k+1}, y)) \supset |\theta\partial(\sigma^k \times I)|$. (See Figure 38.) Consider again simplicial complex $K = K(\theta\partial(\sigma^k \times I))$ and mapping $\omega : \mathbf{Vert}(K \times I)$ to X. Then we may define $\theta(\sigma^k \times I)$ in exactly the same way as before until $k = n$ in such a way that $\partial\theta(\sigma^k \times I) = \theta\partial(\sigma^k \times I)$, $\theta(\sigma^k \times I) \in \mathfrak{M}_{k+1}$, and diam $f|\theta(\sigma^k \times I)| \leq {}^*\mathfrak{N}_{k+1}$. ∎

6.3.2 Vietoris–Begle mapping theorem

Let X and Y be two compact Hausdorff spaces and $f: X \to Y$ a continuous mapping. For each covering $N \in \mathbf{Cover}(Y)$, $\mathfrak{M}(\mathfrak{N}) = \{f^{-1}(N)|N \in \mathfrak{N}\}$ is a covering of X. It is clear that f maps each $\mathfrak{M}(\mathfrak{N})$-simplex to $\mathfrak{N}$-simplex to induce a simplicial mapping on $X^v(\mathfrak{M}(\mathfrak{N}))$ to $Y^v(\mathfrak{N})$ and chain mapping $\{f_q^{\mathfrak{N}}\}$. Given q-dimensional Vietoris cycle $\gamma^q = \{\gamma^q(\mathfrak{M})|\mathfrak{M} \in \mathbf{Cover}(X)\}$ of X, define $f_q(\gamma^q)$ as the q-dimensional Vietoris cycle of Y, $\{f_q^{\mathfrak{N}}(\gamma^q(\mathfrak{M}(\mathfrak{N})))|\mathfrak{N} \in \mathbf{Cover}(Y)\}$. The mapping of γ^q to $f_q(\gamma^q)$ clearly induces a homomorphism. The next theorem shows that f_q indeed induces an isomorphism (Figure 39).

THEOREM 6.3.3: (Vietoris–Begle Mapping Theorem: Begle 1950a) *Let X and Y be compact Hausdorff spaces. If $f: X \to Y$ is a Vietoris–Begle mapping of order n, an isomorphism exists between $H_q^v(X)$ and $H_q^v(Y)$ for each $q = 0, 1, \ldots, n$.*

PROOF: We use three steps to prove the assertion. In Step 1, we construct n-dimensional Vietoris cycle $\{\gamma^n(\mathfrak{M})\}$ of X from $\{z^n(\mathfrak{N})\}$ of Y. With it, we can see in Step 2 that the homomorphism induced by f between $H_q^v(X)$ and $H_q^v(Y)$ for each $q = 0, 1, \ldots, n$ is onto. The homomorphism is seen to be one to one in Step 3.

(Step 1) With each $\mathfrak{M} \in \mathbf{Cover}(X)$ is associated covering $\mathfrak{N}(\mathfrak{M}) \in \mathbf{Cover}(Y)$ such that $\mathfrak{M} \preccurlyeq \{f^{-1}(N)|N \in \mathfrak{N}(\mathfrak{M})\}$. If $\mathfrak{M} = \{f^{-1}(N)|N \in \mathfrak{N}\}$ for some $\mathfrak{N}$, it is always assumed that $\mathfrak{N}(\mathfrak{M})$ equals one such $\mathfrak{N}$. Let $z^n = \{z^n(\mathfrak{N})|\mathfrak{N} \in \mathbf{Cover}(X)\}$ (or simply $\{z^n(\mathfrak{N})\}$) be an n-dimensional

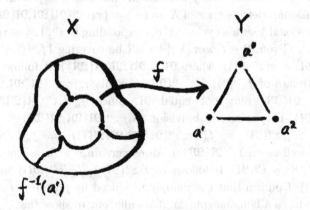

Figure 39: Isomorphism under Vietoris–Begle mapping of order n

Vietoris cycle of Y. For each covering $\mathfrak{M} \in \mathbf{Cover}(X)$, define $\gamma^n(\mathfrak{M})$ as $\gamma^n(\mathfrak{M}) = \tau_n(z^n(\mathfrak{R}(\mathfrak{M},\mathfrak{N}(\mathfrak{M}))))$, where $\tau = \{\tau_n\}$ and $\mathfrak{R}(\mathfrak{M},\mathfrak{N})$ are the chain mapping and the covering defined in Theorem 6.3.1.

We see that $\gamma^n = \{\gamma^n(\mathfrak{M})\}$ is an n-dimensional Vietoris cycle. Since every $\gamma^n(\mathfrak{M}) = \tau_n(z^n(\mathfrak{R}(\mathfrak{M},\mathfrak{N}(\mathfrak{M}))))$, an image of cycle, is obviously an n-dimensional Vietoris $\mathfrak{M}$-cycle, all we have to show is $\gamma^n(\mathfrak{M}) \sim h_n^{\mathfrak{M}\mathfrak{M}''}(\gamma^n(\mathfrak{M}''))$ for each pair $\mathfrak{M}'' \preccurlyeq \mathfrak{M}$. That is, $\tau_n(z^n(\mathfrak{R}(\mathfrak{M},\mathfrak{N}(\mathfrak{M})))) \sim h_n^{\mathfrak{M}\mathfrak{M}''}(\tau_n''(z^n(\mathfrak{R}(\mathfrak{M}'',\mathfrak{N}(\mathfrak{M}'')))))$ for each $\mathfrak{M}'' \preccurlyeq \mathfrak{M}$, where τ'' is the chain mapping associated with $\mathfrak{R}(\mathfrak{M}'',\mathfrak{N}(\mathfrak{M}''))$. Denote $\mathfrak{R}(\mathfrak{M}'',\mathfrak{N}(\mathfrak{M}''))$ by $\mathfrak{R}''$ and $\mathfrak{R}(\mathfrak{M},\mathfrak{N}(\mathfrak{M}))$ by $\mathfrak{R}$. If we omit inclusion map h_n, we have to show $\tau_n(z^n(\mathfrak{R})) \sim \tau_n''(z^n(\mathfrak{R}''))$.

In Step 4 of the proof of the second assertion in Theorem 6.3.1, we may choose $\mathfrak{M}_1' \preccurlyeq \cdots \preccurlyeq \mathfrak{M}_{n+1}'$ and $\mathfrak{N}_0' \preccurlyeq \cdots \preccurlyeq \mathfrak{N}_{n+1}'$ as common refinements not only of series $\{\mathfrak{M}_k\}$ and $\{\mathfrak{N}_k\}$ constructing τ (in Step 3) for $\mathfrak{M}$ and $\mathfrak{N}$ but also of other streams $\{\mathfrak{M}_k''\}$ and $\{\mathfrak{N}_k''\}$ combined with chain map τ'' for $\mathfrak{M}''$ and $\mathfrak{N}''$ that satisfy the same condition with $\mathfrak{M}$ and $\mathfrak{N}$. Since the construction of τ' is independent of τ and τ'', by repeating the same argument (to construct θ' instead of θ), we can see $\tau_n'(z^n) \sim \tau_n(z^n)$ and $\tau_n'(z^n) \sim \tau_n''(z^n)$ in $C_n^v(\mathfrak{M})$ for all $z^n \in Z_n^v(\mathfrak{R}')$.

That is, refinement $\mathfrak{R}'$ that is common to $\mathfrak{R} = \mathfrak{R}(\mathfrak{M},\mathfrak{N}(\mathfrak{M}))$ and $\mathfrak{R}'' = \mathfrak{R}(\mathfrak{M}'',\mathfrak{N}(\mathfrak{M}''))$ exists with chain map τ' such that $\tau'(z^n(\mathfrak{R}')) \sim \tau(z^n(\mathfrak{R}'))$ and $\tau'(z^n(\mathfrak{R}')) \sim \tau''(z^n(\mathfrak{R}'))$, where τ and τ'' are the chain maps associated respectively with $\mathfrak{R}$ and $\mathfrak{R}''$. Hence we have $\tau(z^n(\mathfrak{R}')) \sim \tau''(z^n(\mathfrak{R}'))$. Since z^n is a Vietoris cycle, we know $h_n^{\mathfrak{R}\mathfrak{R}'}(z^n(\mathfrak{R}')) \sim z^n(\mathfrak{R})$ and $h_n^{\mathfrak{R}''\mathfrak{R}'}(z^n(\mathfrak{R}')) \sim z^n(\mathfrak{R}'')$, so we have $\tau(z^n(\mathfrak{R})) \sim \tau''(z^n(\mathfrak{R}''))$.

(Step 2) Now we see that f induces an onto homomorphism. Let z^n be an n-dimensional Vietoris cycle of X and $\gamma^n = \{\tau_n(z^n(\mathfrak{R}(\mathfrak{M},\mathfrak{N}(\mathfrak{M}))))\}$ be the n-dimensional Vietoris cycle of Y corresponding to z^n. Let us verify that $f_q(\gamma^n) \sim z^n$. Given $\mathfrak{N} \in \mathbf{Cover}(Y)$, let $\mathfrak{M}$ be covering $\{f^{-1}(N)|N \in \mathfrak{N}\}$. Then $\gamma^n(\mathfrak{M}) = \tau(z^n(\mathfrak{R}))$, where $\mathfrak{R} = \mathfrak{R}(\mathfrak{M},\mathfrak{N}(\mathfrak{M}))$. It follows that the $\mathfrak{N}$-th coordinate of $f_n(\gamma^n)$, $f_n^{\mathfrak{N}}(\gamma^n(\mathfrak{M}))$, equals $f_n^{\mathfrak{N}}\tau_n z^n(\mathfrak{R}(\mathfrak{M},\mathfrak{N}(\mathfrak{M})))$. Note that $\mathfrak{N}(\mathfrak{M})$ may not equal $\mathfrak{N}$. Since $f_n^{\mathfrak{N}}\tau_n z^n(\mathfrak{R}(\mathfrak{M},\mathfrak{N}(\mathfrak{M})))$ is $(\mathfrak{R}\mathfrak{N}(\mathfrak{M}))$-barycentric subdivision of $z^n(\mathfrak{R}(\mathfrak{M},\mathfrak{N}(\mathfrak{M})))$, we have $z^n(\mathfrak{R}) \sim \mathrm{Sd}_n z^n(\mathfrak{R}) = f_n^{\mathfrak{N}}(\tau_n z^n(\mathfrak{R}(\mathfrak{M},\mathfrak{N}(\mathfrak{M})))) = f_n^{\mathfrak{N}}(\gamma^n(\mathfrak{M}))$ on $Y^v(\mathfrak{N})$ (as well as on $Y^v(\mathfrak{N}(\mathfrak{M}))$). Moreover, since z^n is a Vietoris cycle, we have $z^n(\mathfrak{R}) \sim z^n(\mathfrak{N})$. It follows that $z^n(\mathfrak{N}) \sim f_n^{\mathfrak{N}}(\gamma^n(\mathfrak{M}))$ on $Y^v(\mathfrak{N})$.

(Step 3) Confirm that the mapping induced by f is one to one. Since f clearly induces a homomorphism, it is sufficient to show that $f_n(\gamma^n) \sim 0$ means $\gamma^n \sim 0$ for each n-dimensional Vietoris cycle γ^n of X. Given

6: The Čech Type Homology Theory and Fixed Points   159

$\mathfrak{M} \in \mathit{Cover}(X)$, choose $\mathfrak{N} = \mathfrak{N}(\mathfrak{M})$ and $\mathfrak{R} = \mathfrak{R}(\mathfrak{M}, \mathfrak{N}(\mathfrak{M}))$ as before, and let $\mathfrak{U}$ be covering $\mathfrak{U} = \{f^{-1}(R) | R \in \mathfrak{R}\}$. Moreover let us recall sequence $\{\mathfrak{M}_k\}$ of refinements of $\mathfrak{M}$ defined in the proof of Theorem 6.3.1 and $\mathfrak{V}$ a common refinement of $\mathfrak{U}$ and all $\mathfrak{M}_k$'s.

Since γ^n is an n-dimensional Vietoris cycle, $\gamma^n(\mathfrak{V}) \sim \gamma^n(\mathfrak{U})$ on $X^v(\mathfrak{U})$. Then we have $f_n^{\mathfrak{R}} \gamma^n(\mathfrak{V}) \sim f_n^{\mathfrak{R}} \gamma^n(\mathfrak{U})$ on $Y^v(\mathfrak{R})$. But if $f_n(\gamma^n) \sim 0$, $\mathfrak{R}$-th coordinate of $f_n(\gamma^n)$, $f_n^{\mathfrak{R}} \gamma^n(\mathfrak{M}(\mathfrak{R})) = f_n^{\mathfrak{R}} \gamma^n(\mathfrak{U})$ satisfies $f_n^{\mathfrak{R}} \gamma^n(\mathfrak{U}) \sim 0$ on $Y^v(\mathfrak{R})$. Hence, we have $f_n^{\mathfrak{R}}(\gamma^n(\mathfrak{V})) \sim 0$, so $\tau_n(f_n^{\mathfrak{R}}(\gamma^n(\mathfrak{V}))) \sim 0$, where $\tau = \{\tau_n\}$ is the chain map associated with $\mathfrak{R} = \mathfrak{R}(\mathfrak{M}, \mathfrak{N})$. Now it is possible to show $\tau_n(f_n^{\mathfrak{R}}(\gamma^n(\mathfrak{V}))) \sim \gamma^n(\mathfrak{V})$ on $X^v(\mathfrak{M})$. Indeed, consider $K = K(\gamma^n(\mathfrak{V}))$ and the product cell-complex $K \times \{0, 1, I\}$ with chain map θ defined on the base and top of $K \times \{0, 1, I\}$ to $X^v(\mathfrak{M})$ as $\theta(\sigma^k \times 0) = \sigma^k$ and $\theta(\sigma^k \times 1) = \tau_k f_k \sigma^k$ for each simplex σ^k of K. We may extend θ as a chain map on $K \times \{0, 1, I\}$ in exactly the same way with the process stated in the proof of Theorem 6.3.1. (In Step 4, substitute $\tau_k f_k \sigma^k$ for $\tau_k \sigma^k$ and σ^k for $\tau'_k(\sigma^k)$.) Then we have $\tau_n(f_n^{\mathfrak{R}}(\gamma^n(\mathfrak{V}))) \sim \gamma^n(\mathfrak{V})$ on $X^v(\mathfrak{M})$, so that $\gamma^n(\mathfrak{V}) \sim 0$ since $\tau_n f_n \gamma^n(\mathfrak{V}) \sim 0$ on $X^v(\mathfrak{M})$. Since γ^n is a Vietoris cycle, $\gamma^n(\mathfrak{V}) \sim \gamma^n(\mathfrak{M})$. Thus $\gamma^n(\mathfrak{M}) \sim 0$ on $X^v(\mathfrak{M})$, so $\gamma^n \sim 0$. ∎

6.3.3 Analogue of Sperner's lemma

Nikaido (1959) treats a theorem that may be considered an extension of Sperner's lemma based on the Vietoris–Begle mapping theorem. Let X and Y be compact Hausdorff spaces. Suppose that Y may be identified (under homeomorphism) with n-dimensional simplex $\langle a^0 a^1 \cdots a^n \rangle$ in Euclidean $(n+1)$-space R^{n+1}. Moreover, assume that continuous onto function $f : X \to Y$ exists. For each k-dimensional face $a^{i_0} \cdots a^{i_k}$ of $a^0 \cdots a^n$, denote by $[a^{i_0} \cdots a^{i_k}]$ the set of all convex combination of points of $\{a^0, \ldots, a^k\}$. In this section, we call $f^{-1}([a^{i_0} \cdots a^{i_k}])$ a k-face of X. For point x of X, smallest dimensional face $a^{i_0} \cdots a^{i_k}$ exists such that $f(x) \in [a^{i_0} \cdots a^{i_k}]$, the *carrier* of $f(x)$. We also call such an $f^{-1}([a^{i_0} \cdots a^{i_k}])$ the carrier of x (Figure 40).

Consider covering $\mathfrak{M} \in \mathit{Cover}(X)$ of X and Vietoris $\mathfrak{M}$-complex $X^v(\mathfrak{M})$. Denote by $K(Y)$ simplicial complex $K(\langle a^0 a^1 \cdots a^n \rangle)$. Suppose chain map $\tau = \{\tau_q\}$ on chains of $K(Y)$ exists to chains of $X^v(\mathfrak{M})$, $\tau_q : C_q(K(Y)) \to C_q^v(\mathfrak{M})$, satisfying the following two conditions:

(T1) $|\tau_k(\langle a^{i_0} \cdots a^{i_k} \rangle)| \subset f^{-1}([a^{i_0} \cdots a^{i_k}])$ for any k-face $a^{i_0} \cdots a^{i_k}$ of Y.
(T2) $\tau_0(a^i)$ is a single point for each vertex a^i of Y.

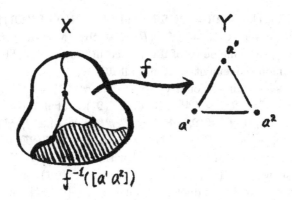

Figure 40: Faces and carriers

We can always construct such a τ when f is a Vietoris mapping. Indeed, if f is a Vietoris mapping, each compact k-face of X, $f^{-1}([a^{i_0} \cdots a^{i_k}])$, is acyclic by Theorems 6.2.2 and 6.3.3. Note that the number of all k-faces of X is finite, and they are also compact as inverse images of closed sets. By applying Theorem 6.2.2 repeatedly, for each covering $\mathfrak{M}$ of X, a sequence of coverings of X, $\mathfrak{M}_0 \preccurlyeq \mathfrak{M}_1 \preccurlyeq \cdots \mathfrak{M}_{n-1} \preccurlyeq \mathfrak{M}_n = \mathfrak{M}$, may be obtained that satisfy that for each k-face $f^{-1}([a^{i_0} \cdots a^{i_k}])$ of X and $\ell \leq k$, each $(\ell - 1)$-dimensional Vietoris $\mathfrak{M}_{\ell-1}$-cycle in $f^{-1}([a^{i_0} \cdots a^{i_k}])$ bounds an ℓ-dimensional Vietoris $\mathfrak{M}_\ell$-chain in $f^{-1}([a^{i_0} \cdots a^{i_k}])$. Define for each 0-dimensional face (vertex) a^i of Y, $\tau_0(a^i)$ as $\tau_0(a^i)$ equals an arbitrary point of $f^{-1}(a^i)$, and extend τ^0 linearly on $C_0(K(Y))$. Now assume that up to k ($0 \leq k \leq n-1$) the construction of τ_m ($m \leq k$) is finished and that for each m-dimensional face $\sigma^m = \langle a^{i_0} \cdots a^{i_m} \rangle$ of Y, $\tau_m(\sigma^m)$ is a Vietoris $\mathfrak{M}_m$-chain in $f^{-1}([a^{i_0} \cdots a^{i_m}])$. Take an arbitrary $(k+1)$-dimensional face $\sigma^{k+1} = \langle a^{i_0} \cdots a^{i_{k+1}} \rangle$ of Y. Then $\partial \sigma^{k+1}$ is a k-dimensional chain in Y so that $\tau_k \partial \sigma^{k+1}$ is well-defined as a Vietoris $\mathfrak{M}_k$-chain in $a^{i_0} \cdots a^{i_{k+1}}$. Note however, that $\partial \tau_k \partial \sigma^{k+1} = \partial \partial \tau_k \sigma^{k+1} = 0$, so $\tau_k \partial \sigma^{k+1}$ is indeed a k-dimensional Vietoris $\mathfrak{M}_k$-cycle in $a^{i_0} \cdots a^{i_{k+1}}$. Hence, by considering the relation between $\mathfrak{M}_k$ and $\mathfrak{M}_{k+1}$, $\tau_k \partial \sigma^{k+1}$ bounds a $(k+1)$-dimensional Vietoris $\mathfrak{M}_{k+1}$-chain in $f^{-1}([a^{i_0} \cdots a^{i_{k+1}}])$, δ^{k+1}. Therefore, by defining $\tau_{k+1} \sigma^{k+1}$ as δ^{k+1} and extending it linearly on $C_{k+1}(K(Y))$, we obtain τ_{k+1}. It follows that by mathematical induction, we obtain τ.[21] (See Figure 41.)

---

[21] One can find such arguments to construct a chain map based on acyclicity in many places like the construction of the Vietoris–Begle barycentric subdivision in Theorem 6.3.1 and the definition of the extended Lefschetz number for the mappings of class D in Chapter 7, Section 7.1. The construction method is axiomatized in homology theory and called *the method of acyclic models*.

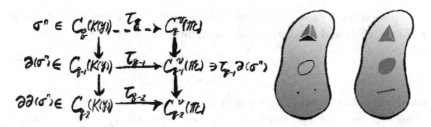

Figure 41: Method of acyclic models

Operator τ may be considered a generalization process of barycentric subdivision. If $X = Y$ and f is identity mapping, and if y in (SD2) can be taken as $y = (y_0 + \cdots + y_q)/(q+1)$ under the linear structure, one may verify that chain map Sd satisfies conditions (T1) and (T2).

Vertex assignment v is a mapping on $X = \mathbf{Vert}(X^v(\mathfrak{M}))$ to $\{a^0, a^1, \ldots, a^n\} = \mathbf{Vert}(K(Y))$ such that for each $x \in X$, $v(x)$ is a vertex of the carrier of $f(x)$. Obviously, v is a simplicial mapping on $X^v(\mathfrak{M})$ to $K(Y)$, so it induces a chain homomorphism that we also denoted by v or $\{v_q\}$, $v_q : C_q^v(\mathfrak{M}) \to C_q(K(Y))$. Given vertex assignment v, we call n-dimensional simplex σ^n in $X^v(\mathfrak{M})$ *regular* if $v_n(\sigma^n) = \langle a^0 a^1 \cdots a^n \rangle$ or $v_n(\sigma^n) = -\langle a^0 a^1 \cdots a^n \rangle$. It is also convenient to define *sign* $\epsilon(\sigma^m)$ of m-simplex of $X^v(\mathfrak{M})$ for each $m = 0, 1, \ldots, n$ as $\epsilon(\sigma^m) = 1$, if $v_m(\sigma^m) = \langle a^0 a^1 \cdots a^m \rangle$, $\epsilon(\sigma^m) = -1$ if $v_m(\sigma^m) = -\langle a^0 a^1 \cdots a^m \rangle$, and $\epsilon(\sigma^m) = 0$ otherwise. In the next lemma, we use J as an index set for all n-dimensional simplexes in $X^v(\mathfrak{M})$.[22]

LEMMA 6.3.4: (Nikaido 1959: Sperner's Lemma) *Let $\tau_n(\langle a^0 a^1 \cdots a^n \rangle) = \sum_{j \in J} \alpha_j \sigma_j^n$, where τ denotes the chain map defined above. Then $\sum_{j \in J} \alpha_j \epsilon(\sigma_j^n) \neq 0$. At least one regular simplex exists for an arbitrary vertex assignment.*

PROOF: Note that in the above expression, $\tau_n(\langle a^0 a^1 \cdots a^n \rangle) = \sum_{j \in J} \alpha_j \sigma_j^n$, the value of τ_n, $\sum_{j \in J} \alpha_j \sigma_j^n$, is a finite sum by definition of the chain map, so $\alpha_j = 0$ except for finitely many $j \in J$. By condition (T2), the lemma is clearly true for $n = 0$. In the following we show the lemma using mathematical induction over n. Let K be an index set for all $(n-1)$-dimensional simplexes in $X^v(\mathfrak{M})$. We call $(n-1)$-dimensional simplex σ^{n-1} in $X^v(\mathfrak{M})$ *regular* if $v_q(\sigma^{n-1}) = \langle a^1 \cdots a^n \rangle$ or $v_q(\sigma^{n-1}) = -\langle a^1 \cdots a^n \rangle$. Assume that the lemma is true for $n-1$, i.e., for

---

[22] Recall that we only treat finite chains, so in the formal summation all but a finite number of coefficients are 0.

f restricted on $f^{-1}([a^1 \cdots a^n])$ to $K(\langle a^1 \cdots a^n \rangle)$, τ restricted on chains of $K(\langle a^1 \cdots a^n \rangle)$, and arbitrary vertex assignment v on X to $\{a^1 \cdots a^n\}$,

$$\tau_{n-1}(\langle a^1 \cdots a^n \rangle) = \sum_{k \in K} \beta_k \epsilon(\sigma_k^{n-1}) \neq 0,$$

where the summation is taken over all $k \in K$ for notational simplicity. (There is no problem since $\epsilon(\sigma_k^{n-1}) = 0$ for all $\sigma_k^{n-1} \notin X^v(\mathfrak{M}) \cap f^{-1}([a^1 \cdots a^n])$ by the definition of ϵ.) For our purpose, it is sufficient to show that

$$\sum_{j \in J} \alpha_j \epsilon(\sigma_j^n) = \sum_{k \in K} \beta_k \epsilon(\sigma_k^{n-1}).$$

(Step 1) First, notice that

$$\sum_{j \in J} \alpha_j \epsilon(\sigma_j^n) = \sum_{j \in J} \alpha_j \sum_{k \in K} [\langle \sigma_k^{n-1} \rangle : \langle \sigma_j^n \rangle] \epsilon(\sigma_k^{n-1}),$$

where $[\cdot : \cdot]$ denotes the incidence number. Indeed, when σ_j^n is regular, there is one and only one regular $(n-1)$-face σ_k^{n-1} of σ_j^n. Let $\langle \sigma_k^{n-1} \rangle = \langle u_1 \cdots u_n \rangle$. If $[\langle \sigma_k^{n-1} \rangle : \langle \sigma_j^n \rangle] = 1$, then by using a certain point $u_0 \in X$, we may write $\langle \sigma_j^n \rangle = \langle u_0 u_1 \cdots u_n \rangle$. Hence, $v_n(\sigma_j^n) = \langle v(u_0) v(u_1) \cdots v(u_n) \rangle = \pm \langle a^0 a^1 \cdots a^n \rangle$ if and only if $v_{n-1}(\sigma_k^{n-1}) = \langle v(u_1) \cdots v(u_n) \rangle = \pm \langle a^1 \cdots a^n \rangle$. Therefore, $\epsilon(\sigma_j^n) = \epsilon(\sigma_k^{n-1})$. If $[\langle \sigma_k^{n-1} \rangle : \langle \sigma_j^n \rangle] = -1$, then we may write $\langle \sigma_j^n \rangle = -\langle u_0 u_1 \cdots u_n \rangle$. Hence, $v_n(\sigma_j^n) = -\langle v(u_0) v(u_1) \cdots v(u_n) \rangle = \pm \langle a^0 a^1 \cdots a^n \rangle$ if and only if $v_{n-1}(\sigma_k^{n-1}) = \langle v(u_1) \cdots v(u_n) \rangle = \mp \langle a^1 \cdots a^n \rangle$. Therefore, $\epsilon(\sigma_j^n) = -\epsilon(\sigma_k^{n-1})$. In each case, we have $\epsilon(\sigma_j^n) = \sum_{k \in K} [\langle \sigma_k^{n-1} \rangle : \langle \sigma_j^n \rangle] \epsilon(\sigma_k^{n-1})$. When σ_j^n is not regular, we must show that $\sum_{k \in K} [\langle \sigma_k^{n-1} \rangle : \langle \sigma_j^n \rangle] \epsilon(\sigma_k^{n-1}) = 0$ even if σ_j^n has regular faces. Suppose that σ_i^{n-1} is a regular face of σ_j^n and let $\langle \sigma_i^{n-1} \rangle = \langle u_1 \cdots u_n \rangle$. There is a point u_0 of X such that $\mathbf{Vert}(\sigma_j^n) = \{u_0, u_1, \ldots, u_n\}$. Since σ_j^n is not regular, there is an m such that $v(u_0) = v(u_m)$. Let σ_k^{n-1} be the face of σ_j^n whose vertices are $\{u_0, u_1, \ldots, u_n\} \setminus \{u_m\}$. Let $\langle \sigma_k^{n-1} \rangle = \langle w_1 \cdots w_n \rangle$. Clearly, σ_j^n has exactly two regular faces, σ_i^{n-1} and σ_k^{n-1}. Then, if $[\langle \sigma_i^{n-1} \rangle : \langle \sigma_j^n \rangle] = 1$ and $[\langle \sigma_k^{n-1} \rangle : \langle \sigma_j^n \rangle] = \pm 1$, we have $\langle \sigma_j^n \rangle = \langle u_0 u_1 \cdots u_n \rangle$ and $\langle \sigma_j^n \rangle = \pm \langle u_m w_1 \cdots w_n \rangle$. Since $\langle u_0 u_1 \cdots u_n \rangle = -\langle u_m u_1 \cdots u_{m-1} u_0 u_{m+1} \cdots u_n \rangle$, we have $\langle u_m w_1 \cdots w_n \rangle = \pm \langle u_0 u_1 \cdots u_n \rangle = \mp \langle u_m u_1 \cdots u_{m-1} u_0 u_{m+1} \cdots u_n \rangle$, so that $\langle v(w_1) v(w_2) \cdots v(w_n) \rangle = \mp \langle v(u_1) \cdots v(u_{m-1}) v(u_0) v(u_{m+1}) \cdots v(u_n) \rangle = \mp \langle v(u_1) v(u_2) \cdots v(u_n) \rangle$. It follows that $\epsilon(\sigma_k^{n-1}) = \mp \epsilon(\sigma_i^{n-1})$. In exactly the same way, if $[\langle \sigma_i^{n-1} \rangle : \langle \sigma_j^n \rangle] = -1$ and $[\langle \sigma_k^{n-1} \rangle : \langle \sigma_j^n \rangle] = \pm 1$, we obtain that $\epsilon(\sigma_k^{n-1}) = \pm \epsilon(\sigma_i^{n-1})$. Therefore, we have $[\langle \sigma_i^{n-1} \rangle :$

6: The Čech Type Homology Theory and Fixed Points

$\langle\sigma_j^n\rangle]\epsilon(\sigma_i^{n-1}) + [\langle\sigma_k^{n-1}\rangle : \langle\sigma_j^n\rangle]\epsilon(\sigma_k^{n-1}) = 0$ in all cases, so $\sum_{k\in K}[\langle\sigma_k^{n-1}\rangle : \langle\sigma_j^n\rangle]\epsilon(\sigma_k^{n-1}) = 0$.

(Step 2) Next, we see that

$$\sum_{j\in J}\alpha_j\sum_{k\in K}[\langle\sigma_k^{n-1}\rangle : \langle\sigma_j^n\rangle]\epsilon(\sigma_k^{n-1}) = \sum_{k\in K}\beta_k\epsilon(\sigma_k^{n-1}).$$

Note that since $\tau_n(\langle a^0\cdots a^n\rangle) = \sum_{j\in J}\alpha_j\sigma_j^n$, we have

$$\partial_n(\tau_n(\langle a^0\cdots a^n\rangle)) = \partial_n\left(\sum_{j\in J}\alpha_j\sigma_j^n\right) = \sum_{j\in J}\alpha_j\partial_n(\sigma_j^n)$$

$$= \sum_{j\in J}\alpha_j\sum_{k\in K}[\langle\sigma_k^{n-1}\rangle : \langle\sigma_j^n\rangle]\sigma_k^{n-1}.$$

Moreover, since $\partial\tau = \tau\partial$, we also have

$$\partial_n(\tau_n(\langle a^0\cdots a^n\rangle)) = \tau_{n-1}\partial_n(\langle a^0\cdots a^n\rangle) = \sum_{i=0}^n(-1)^i\tau_{n-1}(\langle a^0\cdots\hat{a}^i\cdots a^n\rangle),$$

where the circumflex accent denotes the omission of vertex a^i. It follows that

$$\sum_{j\in J}\alpha_j\sum_{k\in K}[\langle\sigma_k^{n-1}\rangle : \langle\sigma_j^n\rangle]\sigma_k^{n-1} = \sum_{i=0}^n(-1)^i\tau_{n-1}(\langle a^0\cdots\hat{a}^i\cdots a^n\rangle).$$

Since $\tau_{n-1}(\langle a^0\cdots\hat{a}^i\cdots a^n\rangle) \subset f^{-1}([a^0,\cdots,\hat{a}^i,\cdots,a^n])$ (Condition (T1)), by considering that each σ_k^{n-1} appearing in formal summation $\tau_{n-1}(\langle a^0\cdots\hat{a}^i\cdots a^n\rangle)$ except for $i = 0$ cannot be regular, the coefficient of each regular σ_k^{n-1} ($k \in K$) must equal its coefficient in $\tau_{n-1}(\langle a^1\cdots a^n\rangle)$, so we must have

$$\sum_{j\in J}\alpha_j[\langle\sigma_k^{n-1}\rangle : \langle\sigma_j^n\rangle] = \beta_k$$

for each regular σ_k^{n-1} ($k \in K$). Since $\epsilon(\sigma_k^{n-1}) = 0$ for each σ_k^{n-1} that is not regular, we have

$$\sum_{j\in J}\alpha_j\sum_{k\in K}[\langle\sigma_k^{n-1}\rangle : \langle\sigma_j^n\rangle]\epsilon(\sigma_k^{n-1}) = \sum_{k\in K}\beta_k\epsilon(\sigma_k^{n-1}). \blacksquare$$

6.4 Eilenberg–Montgomery's Theorem

By combining Lemma 6.3.4 with the Vietoris–Begle mapping theorem, we obtain the following coincidence theorem. Although the result may be considered a special case of Eilenberg–Montgomery–Begle's fixed-point theorem, we prove it directly and use it to show a simple version of Eilenberg–Montgomery's theorem.

THEOREM 6.4.1: (Nikaido 1959) *Let X be a compact Hausdorff space and Y a set homeomorphic to finite dimensional simplex $a^0 a^1 \cdots a^n$. Suppose two continuous mappings f and θ exist on X to Y, one of which, f, is a Vietoris mapping. Then there is point $x \in X$ such that $f(x) = \theta(x)$.*

PROOF: Let us identify Y with $[a^0 a^1 \cdots a^n]$. Then every point $y \in Y$ may be uniquely represented as $y = \sum_{i=0}^n y_i a^i$, where $y_i \geq 0$ for all i and $\sum_{i=0}^n y_i = 1$. In the same way, we may represent $f(x)$ and $\theta(x)$ as $(f_0(x), \ldots, f_n(x))$ and $(\theta_0(x), \ldots, \theta_n(x))$, respectively. Denote by F_i set $\{x \in X | f_i(x) \geq \theta_i(x)\}$. It is easy to check that for each k-face $a^{i_0} \cdots a^{i_k}$ of Y, $f^{-1}([a^{i_0} \cdots a^{i_k}]) \subset \bigcup_{j=0}^k F_{i_j}$. Then we may define vertex assignment v as $v(x) = a^i$ for vertex a^i of the carrier of x $(f^{-1}([a^{i_0} \cdots a^{i_k}]))$ such that $x \in F_i$. Since for Vietoris mapping we may construct chain map τ in Lemma 6.3.4, we may obtain regular n-simplex σ^n in $X^v(\mathfrak{M})$. Therefore, there is at least one $M \in \mathfrak{M}$ such that $M \cap F_i \neq \emptyset$ for all $i = 0, \ldots, n$. Now, assume that $\bigcap_{i=0}^n F_i = \emptyset$. Then family $\{F_i^c = X \backslash F_i | i = 0, \ldots, n\}$ may be considered a covering of X. If we apply the same argument for $\mathfrak{M}$ to $\{F_i^c = X \backslash F_i | i = 0, \ldots, n\}$, we obtain an element of $\{F_i^c = X \backslash F_i | i = 0, \ldots, n\}$ that intersects with all F_i's, which is impossible since $F_i^c \cap F_i = \emptyset$ for all i. Hence, we have $\bigcap_{i=0}^n F_i \neq \emptyset$. Now, checking that any element $x \in \bigcap_{i=0}^n F_i$ satisfies $f(x) = \theta(x)$ is easy. ∎

By using Theorem 6.4.1, we can easily obtain the following simple version of Eilenberg–Montgomery's fixed-point theorem.

THEOREM 6.4.2: (Eilenberg–Montgomery's Fixed-Point Theorem: Euclidean Space) *Let Y be a set homeomorphic to finite dimensional simplex $a^0 a^1 \cdots a^n$. If $\varphi : Y \to Y$ is an acyclic valued correspondence having a closed graph, then φ has a fixed point.*

PROOF: Let X be the graph of φ, $G_\varphi \subset Y \times Y$. Since φ has a closed graph, G_φ is a compact Hausdorff space. Consider two projections: $f : X = G_\varphi \ni (x, y) \mapsto x \in Y$ and $\theta : X = G_\varphi \ni (x, y) \mapsto y \in Y$ (Figure 42). Since φ is acyclic valued, f is a Vietoris mapping. Therefore, by Theorem 6.4.1,

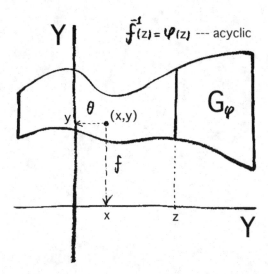

Figure 42: Eilenberg–Montgomery fixed-point theorem

there is point $x^* \in X = G_\varphi \subset Y \times Y$ such that $f(x^*) = \theta(x^*)$. This means, however, the first and the second coordinates of x^* are identical, i.e., x^* may be represented as (x, x). Hence, we have $(x, x) \in G_\varphi$, so that $x \in \varphi(x)$. ∎

Note that both the above two theorems include Brouwer's fixed-point theorem we used in Chapter 2 without a proof.

Bibliographic Notes

The theorems and arguments in this chapter are based on papers by Begle (1950a), Begle (1950b), and Nikaido (1959). For the basic concepts in algebraic topology, I also use notions from books by Lefschetz (1942), Eilenberg and Steenrod (1952), Hocking and Young (1961), and Granas and Dugundji (2003). With respect to the Vietoris and Čech homology concepts, however, except for brief introductory arguments, we have to prepare tools and ideas directly from such papers like Begle (1942), Begle (1950a, 1950b), Eilenberg and Montgomery (1946), Spanier (1948), etc.

Chapter 7

Convex Structure and Fixed-Point Index

7.1 Lefschetz's Fixed-Point Theorem and Its Extensions

This chapter treats theorems that relate the fixed-point arguments in Chapters 2 and 3 to homological arguments in Chapter 6. Our general treatment of fixed-point arguments in compact Hausdorff spaces without vector-space structures, concepts like Browder- and Kakutani-type mappings, are directly related in this section to the homological fixed-point formula of Lefschetz (1937). Moreover, it is also possible to use these concepts to extend the notion of the Lefschetz number and to obtain an extension of Lefschetz's fixed-point theorem. We can also develop our arguments into a cohomology and an index theory in the later sections.

In this section, we treat compact Hausdorff lc space X. As stated before (see p. 148), this class of compact Hausdorff spaces includes absolute neighborhood retracts (ANRs). The homology groups of X are isomorphic to the corresponding groups of a finite complex (Theorem 6.2.4), and the classical results of Lefschetz (1937) and Eilenberg and Montgomery (1946) can be shown as extended (Begle 1950b) in this case.

The *Lefschetz number* of continuous mapping $f : X \to X$ is the summation of a trace of homomorphisms, $\text{trace}\,(f_i) : H_i^v(X) \to H_i^v(X)$,

$$\sum_{i=0}^{\infty}(-1)^i \text{trace}\,(f_i), \tag{7.1}$$

which is well-defined since all $H_i^v(X)$ are finite dimensional and $H_i^v(X) = 0$ for all sufficiently large i. Intuitively, for every dimension i, the basis of $C_i^v(\mathfrak{M})$ (hence, of $H_i^v(\mathfrak{M})$) is given by i-dimensional simplexes in $X^v(\mathfrak{M})$, so that if f completely maps all points in a certain simplex to other simplexes, the trace of linear mapping f_i should necessarily be 0

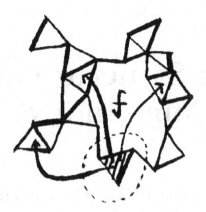

Figure 43: Lefschetz number 0

(Figure 43). Lefschetz's fixed-point theorem is nothing but a restatement of this intuitive observation; i.e., if there is no fixed point, the trace of all such linear functions should equal 0.

The purpose of this section is to relate such profound algebraic features of fixed-point arguments with our fixed-point theorems and methods for general Kakutani-type mappings.

7.1.1 Convex structures and mappings of Browder type (class $\mathscr{B}$)

Before relating Kakutani-type mappings with arguments for Lefschetz's fixed-point theorem, we see how methods for Browder-type mappings may be recaptured through the framework of Čech-type homology theory.

Let E be a Hausdorff space on which a convex structure (a concept of combination among finite points with real coefficients) is defined, and let X be a non-empty compact subset that may not necessarily be convex. Mapping $\varphi : X \to 2^X$ is of *class $\mathscr{B}$* if φ has a fixed-point-free convex extension having local intersection property on $X \backslash \mathcal{F}ix(\varphi)$. Figure 44 represents a typical situation for mapping $\varphi : X \to 2^X$ of type $\mathscr{B}$, where x and x' are not in $\mathcal{F}ix(\varphi)$. Under the convex structure defined in Chapter 2, Section 2.2, if X is convex, then class $\mathscr{B}$ mapping is nothing but a mapping satisfying condition (K*) in Chapter 2, Section 2.3.

The local intersection property on $X \backslash \mathcal{F}ix(\varphi)$ for a convex extension of mapping φ of class $\mathscr{B}$ enables us to replace the relation among the open coverings of $X \backslash \mathcal{F}ix(\varphi)$ with convex combination of points. See Figure 45,

7: Convex Structure and Fixed-Point Index   169

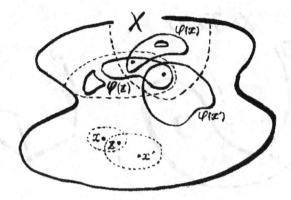

Figure 44:   Mapping of class $\mathcal{B}$

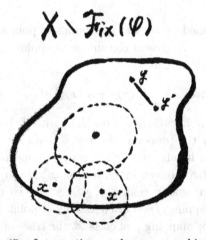

Figure 45:   Intersections and convex combinations

where y and y' are points in the convex extensions of $\varphi(x)$ and $\varphi(x')$, respectively, satisfying the local intersection property near x and x'. If the neighborhoods of x and x' have an intersection point in $X\backslash\boldsymbol{Fix}(\varphi)$, then the convex combination of y and y' belongs to X since point $z \in X\backslash\boldsymbol{Fix}(\varphi)$ exists such that both y and y' belong to a convex extension of $\varphi(z)$.

For mapping φ such that $\boldsymbol{Fix}(\varphi) = \emptyset$, such neighborhoods form a covering of X, and a convex combination of points (y, y', etc.) constructs a complex that may be considered an approximation of X (see Figure 46). Clearly, the complex may also be characterized as the nerves of the covering formed by the neighborhoods of x, x', etc. Note that the partition of unity

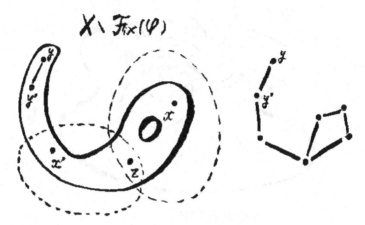

Figure 46: Realization of Čech complex

for the covering formed by the neighborhoods of points, $x, x', \ldots, \alpha : X \to [0,1], \alpha' : X \to [0,1], \ldots$ gives a continuous mapping on X to complex K formed by points $y, y', \ldots$, as

$$f^\varphi : X \ni x \mapsto \alpha(x)y + \alpha'(x)y' + \cdots \in |K|.$$

The continuous mapping restricted on $|K|$ to itself, however, never has a fixed point since by the property of class $\mathscr{B}$ mapping $\varphi, x^* \in U(x)$, $x^* \in U(x'), \ldots$, (neighborhoods of $x, x', \ldots$, resp.), means $y, y', \ldots$, belong to the fixed-point-free convex extension of $\varphi(x^*)$, so x^* cannot be any convex combination among points $y, y', \ldots$. As we can see below, for such continuous mapping f^φ, Lefschetz's fixed-point arguments may be applicable; hence, for mapping φ of class $\mathscr{B}$, the trace of homology mapping $f_q^\varphi : H_q^v(|K|) \to H_q^v(|K|)$ for each $q = 0, 1, 2, \ldots$, of f^φ (a certain kind of linear approximation of φ) is 0 for sufficiently fine K as long as φ has no fixed point.

7.1.2 Convex structures and mappings of Kakutani type (class $\mathscr{K}$)

In the last part of Chapter 2, we treated a wide class of mappings, the Kakutani type, and we saw that (1) the fixed-point property holds (Theorem 2.3.5), and (2) a directional structure on which the dual space representation of φ has local intersection property as long as φ has no fixed points (p. 53, Figure 23) may be definable.

7: Convex Structure and Fixed-Point Index

Assume that X is a compact subset of Hausdorff topological space having convex structure $(\{f_A | A \in \mathscr{F}(X)\}, C)$. Mapping $\varphi : X \to 2^X \setminus \{\emptyset\}$ is of *class $\mathscr{K}$* if for each $x \in X$, there is closed convex set K_x such that (1) $(x \notin \varphi(x)) \Rightarrow (x \notin K_x)$, and (2) there is an open neighborhood U_x of x satisfying that $\forall z \in U_x$, $\varphi(z) \subset K_x$.[1] Note that for mapping φ of class $\mathscr{K}$, each neighborhood U_x of x may be chosen arbitrarily small. Of course, class $\mathscr{K}$ mapping is nothing but the Kakutani-type mappings in Chapter 2, Section 2.3 if X is convex.

For mapping $\varphi : X \to X$ of class $\mathscr{K}$, let us define the *Lefschetz number* of φ in a generalized sense. Since X is compact and Hausdorff, for each mapping $\varphi : X \to 2^X$ of class $\mathscr{K}$, there is at least one covering $\mathfrak{M} = \{M_1, \ldots, M_m\}$ of X such that for each $i = 1, \ldots, m$, there is a convex set K_i satisfying that $(z \in M_i) \Rightarrow \varphi(z) \subset K_i$. As stated above, $\mathfrak{M}$ may be chosen arbitrarily small, so we may suppose that $\mathfrak{M} \preccurlyeq {}^*\mathfrak{N}_0$, where $\mathfrak{N}_0 \in \mathbf{Cover}(X)$ is the covering for lc space X stated in Theorem 6.2.4-(a). The nerve of any covering $\mathfrak{N} \preccurlyeq {}^*\mathfrak{M} \preccurlyeq {}^*\mathfrak{N}_0$ gives a finite dimensional (ordinary simplicial) homology group that is isomorphic to $H_n^v(X)$ for any dimension n. The isomorphism is induced by a composite of mappings $\varphi_{\mathfrak{N}_0 n}^b : C_n^c({}^*\mathfrak{N}_0) \to C_n^v(\mathfrak{N}_0)$, projection $p_n^{*\mathfrak{N}_0 \mathfrak{M}}$, $\zeta_\mathfrak{N}^b : C_n^v({}^*\mathfrak{M}) \to C_n^c(\mathfrak{M})$, and inclusion $h_n^{*\mathfrak{M}\mathfrak{N}}$ to define the mapping between cycles as $\theta_n(z) = \varphi_n^b \circ p_n \circ \zeta_n^b \circ h_n(z(\mathfrak{N}))$. (See the proof of Theorem 6.2.4-(b) (Lemma 2 in Begle (1950b)).)

Let $\mathfrak{N} = \{N_1, \ldots, N_n\}$ be ${}^*\mathfrak{M}$. Take $a_1 \in N_1, \ldots, a_n \in N_n$ and $b_1 \in \varphi(a_1), \ldots, b_n \in \varphi(a_n)$ arbitrarily and respectively denote by A and B sets $\{a_1, \ldots, a_n\}$ and $\{b_1, \ldots, b_n\}$. Denote by $K(A)$ the complex with vertices in A such that $a_{i_0} \cdots a_{i_\ell} \in K(A)$ iff $\bigcap_{j=1}^\ell N_{i_j} \neq \emptyset$. Clearly, $K(A)$ is isomorphic to the nerve of covering $\mathfrak{N}$, so for an arbitrarily small refinement $\mathfrak{P}$ of ${}^*\mathfrak{N}$, homomorphism θ_n exists between cycles that define the isomorphism between homology groups:

$$\theta_n : Z_n^v(X) \to Z_n(K(A)) \tag{7.2}$$

for any dimension n, where $Z_n^v(X)$ denotes the set of all n-dimensional Vietoris cycles on X and $\theta_n(z) = \varphi_n^b \circ p_n \circ \zeta_n^b \circ h_n^{*\mathfrak{N}\mathfrak{P}}(z(\mathfrak{P}))$.

Since $\mathfrak{N}$ is a star refinement of $\mathfrak{M}$, complex $K(A)$ may be considered a subcomplex of $X^v(\mathfrak{M})$. Define abstract complex $K(B)$ with set of vertices B as $b_{i_0} \cdots b_{i_\ell} \in K(B)$ if and only if $\mathrm{co}\{b_{i_j} : j = 0, \ldots, \ell\} \subset X$. Then, we may obtain simplicial mapping $\tau : K(A) \to K(B)$ such that $\tau(a_i) = b_i$

---

[1] Since K_x is closed, we may suppose $U_x \cap K_x = \emptyset$ without loss of generality as long as $x \notin K_x$.

for each $i = 1, \ldots, n$. Moreover, under convex structure on X, by taking $B' \supset B$ sufficiently large, the restriction of $f_{B'}$ on $K(B)$, we may obtain continuous mapping r on a standard realization of $K(B)$ into X. Hence, we have homomorphism $r_n \circ \tau_n \circ \theta_n : H_n^v(X) \to H_n^v(X)$ whose trace is well-defined for each dimension n. Note that these mappings depend on how we choose $\mathfrak{M}, \mathfrak{P}, A, B$. For mapping φ of class $\mathscr{K}$, define the Lefschetz number $\Lambda(\varphi)$ as the minimum of the natural numbers given by such traces as

$$\Lambda(\varphi) = \min_{\mathfrak{M},\mathfrak{P},A,B} \sum_{i=0}^{\infty} (-1)^i \operatorname{trace}(r_i \circ \tau_i \circ \theta_i). \qquad (7.3)$$

We can verify that this number also characterizes the existence of fixed points in exactly the same way as ordinary Lefschetz numbers even for wide class of mappings $\mathscr{K}$. All we have to show is that if φ of class $\mathscr{K}$ has no fixed point, there is at least one set of $\mathfrak{M}$, A, and B under which $\operatorname{trace}(r_i \circ \tau_i \circ \theta_i) = 0$ for any dimension i. Verifying it is a routine task, however, if we recall the definition of θ_n (i.e., all we have to consider is $\mathfrak{P}$-simplexes that may be taken as small as possible).

7.1.3 Acyclic valued directional structures and mappings of class $\mathscr{D}$

The arguments in the previous subsection for a generalization of Lefschetz's fixed-point theorem may also be applicable to such cases where each K_x characterizing the mapping of class $\mathscr{K}$ is not convex but acyclic.

Let X be a compact Hausdorff lc space. Mapping $\varphi : X \to 2^X \setminus \{\emptyset\}$ is of *class $\mathscr{D}$* if for each $x \in X$, closed acyclic set K_x exists such that (1) $(x \notin \varphi(x)) \Rightarrow (x \notin K_x)$, and (2) open neighborhood U_x of x satisfies that $\forall z \in U_x, \varphi(z) \subset K_x$. As before, since K_x is closed, we may suppose $U_x \cap K_x = \emptyset$ without loss of generality as long as $x \notin K_x$. Note also that for mapping φ of class $\mathscr{K}$, each neighborhood U_x of x may be chosen arbitrarily small. In standard cases, non-empty convex sets are acyclic, so the discussion for class $\mathscr{D}$ mapping below may also be considered a generalization of the previous argument for class $\mathscr{K}$ mappings (Figure 47).

Since X is compact and Hausdorff, for mapping $\varphi : X \to 2^X$ of class $\mathscr{D}$, at least one covering $\mathfrak{M} = \{M_1, \ldots, M_m\}$ of X exists such that for each $i = 1, \ldots, m$, acyclic set K_i exists that satisfies that $(z \in M_i) \Rightarrow (\varphi(z) \subset K_i)$. Since $\mathfrak{M}$ may be chosen arbitrarily small, we may suppose that $\mathfrak{M} \preccurlyeq \mathfrak{N}_0$, where $\mathfrak{N}_0 \in \mathbf{Cover}(X)$ is the covering for lc space X stated

7: Convex Structure and Fixed-Point Index 173

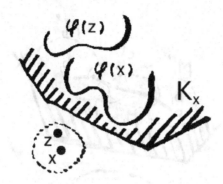

Figure 47:   Mappings of classes $\mathscr{K}$ and $\mathscr{D}$

in Theorem 6.2.4-(a) as before. The nerve of any covering $\mathfrak{N} \preccurlyeq {^{**}}\mathfrak{M} \preccurlyeq \mathfrak{N}_0$ provides a finite dimensional simplicial homology group that is isomorphic to $H_n^v(X)$ for each dimension n. The isomorphism is induced by a composite of mappings $\varphi_{\mathfrak{M}n}^b : C_n^c({^*}\mathfrak{M}) \to C_n^v(\mathfrak{M})$, projection $p_n^{\mathfrak{M}\, {^*}\mathfrak{M}}$, $\zeta_{{^*}\mathfrak{M}}^b : C_n^v({^*}\mathfrak{M}) \to C_n^c({^*}\mathfrak{M})$, and inclusion $h_n^{{^{**}}\mathfrak{M}\,\mathfrak{N}}$ as $\theta_n(z) = \varphi_n^b \circ p_n \circ \zeta_n^b \circ h_n(z(\mathfrak{N}))$.

Let k be the dimension of the nerve of $\mathfrak{N}$. We shall define a sequence of the refinements of $\mathfrak{N}$,

$$\mathfrak{M}_0 \preccurlyeq \mathfrak{M}_1 \preccurlyeq \cdots \preccurlyeq \mathfrak{M}_{k-1} \preccurlyeq \mathfrak{M}_k \preccurlyeq \mathfrak{M}_{k+1} \preccurlyeq \mathfrak{N} \preccurlyeq {^{**}}\mathfrak{M} \preccurlyeq {^*}\mathfrak{M} \preccurlyeq \mathfrak{M} \preccurlyeq \mathfrak{N}_0, \tag{7.4}$$

as follows: Let $\mathfrak{M}_{k+1} = \mathfrak{N}$. For ℓ such that $0 \leq \ell \leq k$, define $\mathfrak{M}_\ell$ as a refinement of ${^{**}}\mathfrak{M}_{\ell+1}$ such that for each compact acyclic $K_i \in \{K_1, \ldots, K_m\}$, any ℓ-dimensional Vietoris $\mathfrak{M}_\ell$-cycle of K_i bounds a chain in $\mathfrak{M}_{\ell+1}$ of K_i. (This is always possible by Theorem 6.2.2.) Note that for each pair of $\mathfrak{M}_\ell$ and $\mathfrak{M}_{\ell+1}$ and dimension n, homomorphism $\theta_n^{\ell+1\ell} = \varphi_n^b \circ p_n \circ \zeta_n^b \circ h_n$ between $C_n^v(\mathfrak{M}_{\ell+1})$ and $C_n^v(\mathfrak{M}_\ell)$, which induces the isomorphism among homology groups, exists.

Let us define chain homomorphism $\tau = \{\tau_q\}$ on the k-skeleton of $X^v(\mathfrak{M}_0)$ to $X^v(\mathfrak{N})$. First, denote by $\mathfrak{L} = \{L_0, L_1, \ldots, L_s\}$ cover ${^*}\mathfrak{M}$. By the definition of $\varphi_{\mathfrak{M}n}^b$, $\varphi_{\mathfrak{M}n}^b(L_i) = x_{L_i} \in L_i$, and an $M_j \in \mathfrak{M}$ exists such that $St(L_i; \mathfrak{L}) \subset M_j$. Define a_i as $a_i = x_{L_i}$ and K_{a_i} as the corresponding K_j for each $i = 0, \ldots, s$. Then we have for each $x \in L_i$, $\varphi(x) \subset K_{a_i}$ for all i. With respect to a_i, fix a point $b_i \in \varphi(a_i) \subset K_{a_i}$ for each i.

For 0-dimensional simplex $\sigma^0 = \langle x^0 \rangle$ of $X^v(\mathfrak{M}_0)$, image $\theta_0 \circ \theta_0^{n+1\,n} \circ \cdots \circ \theta_0^{10}(x^0)$ is by definition one of points $a_0, \ldots, a_s : a_i$. Define $\tau_0(\sigma^0)$ as $\tau_0(\sigma^0) = b_i$ and extend it linearly on $C_0^v(\mathfrak{M}_0)$ to $C_0^v(\mathfrak{M}_0) \subset C_0^v(\mathfrak{N})$.

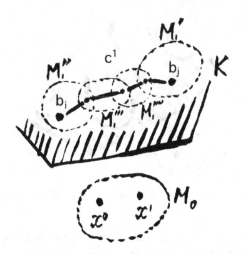

Figure 48: Class $\mathscr{D}$ mapping and $\mathfrak{M}_1$-chain

Next, for 1-dimensional simplex $\sigma^1 = \langle x^0 x^1 \rangle$ of $X^v(\mathfrak{M}_0)$, we may write $\tau_0 \partial(\sigma^1) = \tau_0(x^0 - x^1)$ as $b_i - b_j$, where $b_i = \tau_0(x^0)$ and $b_j = \tau_0(x^1)$. Of course, $b_i - b_j$ may also be considered an $\mathfrak{M}_0$-cycle (in the reduced sense).[2] Hence, by definition of $\mathfrak{M}_0$ relative to $\mathfrak{M}_1$, we have a $\mathfrak{M}_1$-chain c^1 such that $\partial(c^1) = b_i - b_j$ (Figure 48). Define $\tau_1(\sigma^1)$ as $\tau_1(\sigma^1) = c^1$ and extend it linearly on $C_1^v(\mathfrak{M}_0)$ to $C_1^v(\mathfrak{M}_1) \subset C_1^v(\mathfrak{M})$. Clearly, $\partial \tau_1 = \tau_0 \partial$ holds.

Now, assume that for all dimension $q < \ell$, $(2 \leq \ell \leq k)$, τ_q is defined on $C_q^v(\mathfrak{M}_0)$ to $C_q^v(\mathfrak{M}_q) \subset C_q^v(\mathfrak{M})$ and $\partial \tau_q = \tau_{q-1} \partial$ holds. Then for ℓ-dimensional simplex σ^ℓ of $X^v(\mathfrak{M}_0)$, chain $c = \tau_{\ell-1} \partial(\sigma^\ell)$ is well-defined. Since $\partial(c) = \partial \tau_{\ell-1} \partial(\sigma^\ell) = \tau_{\ell-2} \partial \partial(\sigma^\ell) = 0$, c is indeed $\mathfrak{M}_{\ell-1}$-cycle. Hence, by definition of $\mathfrak{M}_{\ell-1}$ relative to $\mathfrak{M}_\ell$, we have a $\mathfrak{M}_\ell$-chain c^ℓ such that $\partial(c^\ell) = c$. Define $\tau_\ell(\sigma^\ell)$ as $\tau_\ell(\sigma^\ell) = c^\ell$ and extend it linearly on $C_\ell^v(\mathfrak{M}_0)$ to $C_\ell^v(\mathfrak{M}_\ell) \subset C_1^v(\mathfrak{M})$. Clearly, $\partial \tau_\ell = \tau_{\ell-1} \partial$ holds.

Hence, by induction, we have successfully obtained chain map $\tau = \{\tau_q\}$ on the k-skeleton of $X^v(\mathfrak{M}_0)$ to $X^v(\mathfrak{M}_{k+1}) = X^v(\mathfrak{N}) \subset X^v(\mathfrak{M})$; i.e., we have

$$\tau_q : C_q^v(\mathfrak{M}_0) \to C_q^v(\mathfrak{N}) \subset C_q^v(\mathfrak{M}) \tag{7.5}$$

for all $q = 0, 1, \ldots, k$. The homology groups of $X^v(\mathfrak{M}_0)$ and $X^v(\mathfrak{N})$ are isomorphic under the isomorphism induced by $\theta^{n+1n} \circ \cdots \circ \theta^{10}$. Since both are isomorphic to the corresponding group of a finite complex, trace(τ_q)

---

[2]Every point of X may be considered a 0-dimensional $\mathfrak{M}_0$-simplex. Note also that in Theorem 6.2.2, 0-dimensional cycles should be taken in the reduced sense.

7: Convex Structure and Fixed-Point Index 175

is well-defined for all q and $\sum_{i=0}^{\infty}(-1)^i \operatorname{trace}(\tau_i)$ is finite. Although the definition of τ depends on $\mathfrak{M}$, $\mathfrak{N}$, and especially sets A of all a_i's and B of all b_i's, we may define as before the minimum of such values

$$\Lambda(\varphi) = \min_{\mathfrak{M},\mathfrak{N},A,B} \sum_{i=0}^{\infty}(-1)^i \operatorname{trace}(\tau_i) \qquad (7.6)$$

as an extended Lefschetz number for mapping φ of class $\mathscr{D}$. By considering the definition of $\theta_n^{\ell+1\ell}$'s, we obtain the following extension of Lefschetz's fixed-point theorem.

THEOREM 7.1.1: (Extension of Lefschetz's Fixed-Point Theorem)
Let X be a compact Hausdorff lc space. Mapping φ of class $\mathscr{D}$ has a fixed point if $\Lambda(\varphi) \neq 0$.

7.2 Cohomology Theory for General Spaces

In this section, we see a cohomology theory that may be considered the dual of Vietoris homology theory for compact Hausdorff space X. The concepts of *homology* and *cohomology groups for a pair of sets* will also be introduced here.

As before, we consider the set of covering $\boldsymbol{Cover}(X)$ of X, and denote for each $\mathfrak{M} \in \boldsymbol{Cover}(X)$, the set of all Vietoris $\mathfrak{M}$-simplexes by $X^v(\mathfrak{M})$. Let A be a subset of X and consider subcomplex $A^v(\mathfrak{M})$ of $X^v(\mathfrak{M})$. It is clear that all the values of boundary operators ∂_n's and inclusion mappings $h_n^{\mathfrak{N}\mathfrak{M}}$'s, restricted on $A^v(\mathfrak{M})$, are closed in subcomplex $A^v(\mathfrak{M})$ of $X^v(\mathfrak{M})$. Hence, for pair (X, A), we can define the *q-th homology group of $X^v(\mathfrak{M})$ modulo $A^v(\mathfrak{M})$*, $H_q^v(X, A; \mathfrak{M})$, as the q-th homology group based on chain group $C_q^v(X, A; \mathfrak{M})$ such that

$$C_q^v(X, A; \mathfrak{M}) = \frac{C_q^v(\mathfrak{M})}{C_q^v(\mathfrak{M}_A)}, \qquad (7.7)$$

where $C_q^v(\mathfrak{M}_A)$ denotes the chain group defined on $A^v(\mathfrak{M})$. If $A = \emptyset$, $C_q^v(\mathfrak{M}_A) = 0$, so we may identify $H_q^v(X, A; \mathfrak{M})$ with $H_q^v(\mathfrak{M})$. The inverse limit

$$H_q^v(X, A) = \varprojlim_{\mathfrak{M}} H_q^v(X, A; \mathfrak{M}) \qquad (7.8)$$

is the *q-th Vietoris homology group for a pair of sets (X, A)*. We may alternatively define $H_q^v(X, A)$ as $H_q^v(X, A) = Z_q^v(X, A)/B_q^v(X, A)$, where

$Z_q^v(X, A)$ is the inverse limit, $\varprojlim_{\mathfrak{M}} Z_q^v(X, A; \mathfrak{M})$, the group of Vietoris cycles, and $B_q^v(X, A)$ is the inverse limit $\varprojlim_{\mathfrak{M}} B_q^v(X, A; \mathfrak{M})$, the group of Vietoris boundaries. They can also all be defined from $C_q^v(X, A) = \varprojlim_{\mathfrak{M}} C_q^v(X, A; \mathfrak{M})$. All of these definitions are equivalent.

The q-th *cochain group of* $X^v(\mathfrak{M})$ *modulo* $A^v(\mathfrak{M})$, $C_v^q(X, A, \mathfrak{M})$, is the set of all functions on all q-simplexes of $X^v(\mathfrak{M})$ into coefficient field R extended linearly on $C_q^v(X, A; \mathfrak{M})$. (By definition of $C_q^v(X, A; \mathfrak{M})$, the differences for values on simplexes in $A^v(\mathfrak{M})$ are ignored.) $C_v^q(X, A; \mathfrak{M})$ is nothing but the dual of vector space $C_q^v(X, A; \mathfrak{M})$, so we may also consider coboundary operators $\{\delta_q : C_v^q(X, A; \mathfrak{M}) \to C_v^{q+1}(X, A; \mathfrak{M})\}$ as the dual of boundary operators $\{\partial_{q+1} : C_{q+1}^v(X, A; \mathfrak{M}) \to C_{q+1}^v(X, A; \mathfrak{M})\}$. That is, for each $f \in C_v^q(X, A; \mathfrak{M})$, we assign (equivalence class modulo $C_q^v(\mathfrak{M}_A)$ of) element

$$\delta_q(f) : \sigma^{q+1} \mapsto f(\partial_{q+1}\sigma^{q+1}) \tag{7.9}$$

of $C_v^{q+1}(X, A; \mathfrak{M})$. By definition it is clear that $\delta_{q+1} \circ \delta_q = 0$, so the set of images of δ^{q+1}, $B^q(X, A; \mathfrak{M})$, is a subset of the kernel of δ_q, $Z^q(X, A; \mathfrak{M})$. Elements of $B^q(X, A; \mathfrak{M})$ and $Z^q(X, A; \mathfrak{M})$ are called the q-th $\mathfrak{M}$-coboundaries and q-th $\mathfrak{M}$-cocycles, respectively. Quotient $H_v^q(X, A; \mathfrak{M}) = Z^q(X, A; \mathfrak{M})/B^q(X, A; \mathfrak{M})$ is the q-th $\mathfrak{M}$-cohomology group for pair (X, A). Direct limit (as the dual of inverse limit in Equation (7.8))

$$H_v^q(X, A) = \varinjlim_{\mathfrak{M}} H_v^q(X, A; \mathfrak{M}) \tag{7.10}$$

is the *q-th Vietoris cohomology group for a pair of sets* (X, A). When $A = \emptyset$, we write $H_v^q(X)$ instead of $H_v^q(X, A)$.

Element f of $C_v^q(X, A; \mathfrak{M})$, q-dimensional $\mathfrak{M}$-*cochain* of $X^v(\mathfrak{M})$ modulo $A^v(\mathfrak{M})$, may also be identified with mapping $\bar{f}$ on a subset of $(q+1)$-product X^{q+1} to R. For simplicity, let $\mathfrak{M} = \{X\}$, and then $\bar{f}$ is defined on the entire X^{q+1} as follows:

$$\bar{f} : X^{q+1} \ni (a_0, a_1, \ldots, a_q) \mapsto f(\langle a_0, a_1, \ldots, a_q\rangle) \in R. \tag{7.11}$$

With each q-simplex $\sigma^q = \langle a_0 a_1 \cdots a_q \rangle$ in $X^v(\mathfrak{M})$, f assigns a value $f(\sigma^q) \in R$ that may be considered the value for $(a_0, a_1, \ldots, a_q) \in X^{q+1}$ under $\bar{f}$. Since f is linearly extended on $C_q^v(X, A; \mathfrak{M})$, the value under $\bar{f}$ is unique up to even permutations (changing signs for an odd permutation) among coordinates:

$$\bar{f}(\ldots, a_i, \ldots, a_j, \ldots) = -\bar{f}(\ldots, a_j, \ldots, a_i, \ldots). \tag{7.12}$$

If X is identified with a vector space over R, we may also consider $\bar{f}$ to be multi-linear on X^{q+1}. In such cases, $\bar{f}$ is called an *alternating tensor* of order $q+1$ on X.

7.3 Dual-System Structure and Differentiability

Suppose that as in the previous section, we have cohomology groups $H_v^q(X)$, $q = 0, 1, \ldots$, for topological space X. For each $k = 0, 1, 2, \ldots$, let us define a *directional form* of order $k+1$ on X as a mapping ω that assigns for each $x \in X$ a k-cochain $\omega(x) \in C_v^k(X)$.

For $k = 0$, directional form ω of order 1 means that at each point x of X, we have function $\omega(x)$ that gives a value in R for each 0-simplex of X. By identifying 0-simplex with a point of X, we may recognize $\omega(x)$ as a function on X to R. A typical example for order 1 directional form may be obtained when X has directional structure $(X, W, V : W \to X)$. We can specify ω as a function on X to W since $w \in W$ may be considered a function on X to R through the characteristic function $\chi_{V(x,w)}$ of $V(x,w)$. The directional form ω of order 1 may be characterized as a tool (based on the directional structure, etc.) to give a first-order approximation for a given function on X to R. (If X has a vector structure, to give first-order derivatives at each point $x \in X$ of a function $f : X \to R$ is nothing but to offer one first-order approximation for f. Such an ω, without referring to the existence of original f, is called a *differential form* of order 1.)

Directional form ω of order $k+1$ for $k \geq 1$ gives at each $x \in X$ function $\omega(x)$ that assigns a value in R for each k-simplex $\langle x_0 x_1 \cdots x_k \rangle$ of X satisfying that $\omega(x)(\langle \cdots x_i \cdots x_j \cdots \rangle) = -\omega(x)(\langle \cdots x_j \cdots x_i \cdots \rangle)$. As in Equation (7.11), $\omega(x)$ may be identified with a function on a subset of X^{k+1} to R. The order $k+1$ directional form ω may be considered a tool, say, for $k+1$-th order approximation of function $f : X \to R$. (Of course, a typical example is the differential form of order $k+1$.)

7.4 Linear Approximation for Isolated Fixed Points

In this section we restrict our attention merely to simplexes in Euclidean space R^n. A *finite simplicial complex* in this section is, therefore, a finite collection of simplexes in some Euclidean space such that: (i) if $\sigma \in K$, then every face of σ also belongs to K, and (ii) if $\sigma, \sigma' \in K$, then $\sigma \cap \sigma'$ is a face of σ and σ' as long as $\sigma \cap \sigma' \neq \emptyset$. For a simplicial complex K, we denote by **Vert**(K) the set of all vertices of K as before.

Let X be a polyhedron, i.e., there is a finite simplicial complex, K, such that X is homeomorphic to the *underlying space* $|K| = \bigcup_{\sigma \in K} \sigma$ of K. Simplicial complex K is called a *simplicial decomposition* of X. Complex K with the homeomorphism is said a *triangulation* of X. Clearly, the same underlying space may be taken with different simplicial decompositions (e.g., K and the barycentric subdivision of K have the same underlying space), so that topological space X may have different triangulations.[3]

Given two simplicial complexes, K and L, a *simplicial map* is a map φ on **Vert**(K) to **Vert**(L) such that if $v_0 v_1 \cdots v_k \in K$, then $\varphi(v_0)\varphi(v_1)\cdots\varphi(v_k) \in L$. A simplicial map $\varphi : \mathbf{Vert}(K) \to \mathbf{Vert}(L)$ may naturally be extended to their underlying spaces as $\bar\varphi : |K| \ni \sum_{i=0}^k \alpha_i v_i \mapsto \sum_{i=0}^k \varphi(v_i) \in |L|$, an *affine extension*. A map on X to X is said to be *piecewise linear* if it is an affine extension of a simplicial map. Denote by $\mathscr{G}(X)$ the set of all single valued mappings $g : X \to X$ satisfying the following conditions:

(1) There is a simplicial decomposition K of X such that each $x \in \boldsymbol{Fix}(g)$ is a unique fixed point belonging to the interior of a certain highest dimensional simplex, σ_x, in K.
(2) For each $x \in \boldsymbol{Fix}(g)$, g restricted on σ^x is a piecewise linear isomorphism onto itself.
(3) g is continuous.

Let $f : X \to X$ be a continuous function whose fixed points are all isolated points in the relative interior of X. (Since $|K|$ is compact, $\boldsymbol{Fix}(f)$ is necessarily finite.) Moreover, assume that f is differentiable at each fixed point and the determinant of the derivative at each fixed point, $\det(Df(x))$, does not equal 0. Denote by $\mathscr{C}(X)$ the set of all such functions f. Clearly,

---

[3] The highest dimension of simplex $\sigma \in K$ is said to be the *dimension* of K. One can prove that the dimension of K depends merely on the underlying space $|K|$. (To see this, an important theorem of Brouwer known as the *invariance of domain* is necessary. See, e.g., Hurewicz and Wallman (1948), Rotman (1988), etc.)

7: Convex Structure and Fixed-Point Index

$\mathscr{C}(X)$ includes class $\mathscr{G}(X)$. The *Fixed-point index* for mapping $f \in \mathscr{C}(X)$ is the number

$$\sum_{x \in \boldsymbol{\mathcal{F}ix}(f)} \text{sign} \det(Df(x)), \qquad (7.13)$$

where the sign is 1 if $\det(Df(x)) > 0$ and -1 if $\det(Df(x)) < 0$.[4]

Let us extend the above definition for a fixed-point index for mappings that may not necessarily be continuous. Consider the ordinary (linear) dual-system structure on R^n (precisely, on $R^n \times R^n$), i.e., $(R^n, R^n, \langle \cdot, \cdot \rangle : R^n \times R^n \to R)$, where $\langle \cdot, \cdot \rangle$ denotes the inner product. The dual-system (or directional) structure is restricted on $X \subset R^n$ (precisely, on $X \times R^n \subset R^n \times R^n$) if we define V through $\langle \cdot, \cdot \rangle$ as $V(x, w) = \{y \in X | \langle y - x, w \rangle > 0\}$. Of course, "dual-system (directional) structure restricted on X" may be an abuse of language since the convexity of $V(x,w)$ is not assured. In the same way, let us consider the ordinary (linear) convex structure on R^n and restrict it on $X \subset R^n$. Note in this case, X itself may not be convex so that the usage of "convex structure restricted on X" may also be an abuse of language.

Suppose that $h : X \to X$ is a function of type $\mathscr{K}$ whose fixed points are all isolated in the relative interior of X. Moreover, assume that function f in $\mathscr{C}(X)$ exists such that for all $x \in X \backslash \boldsymbol{\mathcal{F}ix}(h)$, there exists open neighborhood U^x of x such that for all $z \in U^x$, we have $h(z) \in K_x$ and $f(z) \in K_x$, where K_x denotes the convex set depending on x in the definition of class $\mathscr{K}$ mappings. Such f for h may not be unique. If h has no fixed point, however, each f and f' satisfying the above condition for h are homotopic. Indeed, $F : X \times [0,1] \ni (x,t) \mapsto (1-t)f(x) + tf'(x) \in K_x \subset X$ defines the homotopy bridge between f and f'. Therefore, for such an h ($\boldsymbol{\mathcal{F}ix}(h) = \emptyset$), the index for h is uniquely determined by the index of f as 0.[5]

Remark: We may also extend the above arguments for compact Hausdorff lc space X as in the case with the Lefschetz number. In such cases, we

---

[4] Of course, this is not different from defining the fixed-point index through the linear operator associated with each $g \in \mathscr{G}$ under Condition (2). Granas and Dugundji (2003) treat details on this type of fixed-point indices.
[5] We may consider this type of extension for the index of the mapping for mappings of type $\mathscr{D}$, mappings having locally fixed directions, and so on. Especially if we repeat the construction under the method of acyclic models as in the case with the Lefschetz number, we may extend the definition to all $\mathscr{D}$ or $\mathscr{K}$ mappings. In such cases, a certain kind of approximation under linear functions near fixed points will be sufficient to characterize all types mappings through index arguments.

may utilize the finite subcomplex of $X^v(\mathfrak{N}_0)$ whose homology group is isomorphic to the Vietoris homology group of X used in the previous chapter. We have to assume, however, for isolated (finite) fixed point of h, we may appropriately choose the finite subcomplex so that fixed points belong to its relative interior for the piecewise linear or differential approximations. In the next section, I introduce another approach for index arguments in general lc spaces.

7.5 Indices for Compact Set of Fixed Points

In this section, we see another method for index arguments that do not depend on linear approximation or differentiability for mappings. (Instead, we need stronger requirements for the entire space.)

Let X be a compact Hausdorff lc space. By assumption, Vietoris homology groups of X are isomorphic to those of finite complex K. Suppose that the highest dimension of simplexes in K is n. Assume, moreover, that for function $f : X \to X$, there is an open neighborhood O_f of $\boldsymbol{\mathcal{F}ix}(f)$ such that the following isomorphism i_* exists:

$$i_* : H_n(X, X \backslash O_f) \to H_n(\Delta^n, \Delta^n \backslash \{b\}), \qquad (7.14)$$

where Δ^n represents the n-dimensional unit simplex in R^{n+1}, b denotes the barycenter of Δ^n, and $\Delta^n - b = \Delta^n \backslash \{b\}$. The supposition correspond to the interior fixed point assumption in the previous section, though this condition requests more.

We can verify that $H_n(\Delta^n, \Delta^n - b)$ is isomorphic to R.[6] Let $o \in H_n(\Delta^n, \Delta^n - 0)$ be the corresponding point of $1 \in R$ and $o_X \in H_n(X, X \backslash O_f)$ be the image under i_*^{-1} of o. When $f : X \to X$ is a function of class $\mathscr{D}$, by considering function $\bar{f} : X \ni x \mapsto (f(x), x) \in X \times X$ for f, we may obtain the following function between pairs of sets:

$$\bar{f} : (X, X \backslash \boldsymbol{\mathcal{F}ix}(f)) \to (X \times X, X \times X \backslash dX), \qquad (7.15)$$

where dX denotes the diagonal set of $X \times X$. Since X is lc and f is of class $\mathscr{D}$, we may construct chain map $\{\bar{f}_q\}$ on a certain finite subcomplex

---

[6] Δ^n and $\Delta^n - b$ may be identified (have the same homotopy type) with one point and $n - 1$ sphere, respectively. Technically, use the homology exact sequence of pairs of topological spaces (see, e.g., Rotman 1988, p. 96), with such identifications (homotopy equivalences). For this kind of isomorphisms used in this section (under the singular homology), see also Nakaoka (1977, pp. 60 & 77).

7: Convex Structure and Fixed-Point Index

of $X^v(\mathfrak{N})$ to itself for sufficiently fine $\mathfrak{N}$, in the same way with the method of acyclic model (as (7.5) in Section 7.1) for the argument of the Lefschetz number. By using it, we have the following,

$$\bar{f}_* : H_n(X, X\backslash\boldsymbol{Fix}(f)) \to H_n(X \times X, X \times X\backslash dX). \quad (7.16)$$

Furthermore, consider the composite mapping of the next homomorphism (based on the supposition (7.14) with limit arguments),

$$H_n(X \times X, X \times X\backslash dX) \to H_n(\Delta^n \times \Delta^n, \Delta^n \times \Delta^n\backslash d\Delta^n), \quad (7.17)$$

with the following two isomorphisms:

$$H_n(R^n \times R^n, R^n \times R^n\backslash dR^n) \simeq H_n(R^n \times R^n, (R^n - 0) \times R^n) \quad (7.18)$$

$$H_n(R^n \times R^n, (R^n - 0) \times R^n) \simeq H_n(R^n, R^n - 0). \quad (7.19)$$

The relation (7.18) comes from the homeomorphism $(x, y) \mapsto ((y+x)/2, y)$ and (7.19) may be derived from the identification of Δ^n with one point. Let $O_F \in H_n(X, X\backslash\boldsymbol{Fix}(f))$ be the image of $o_X \in H_n(X, X\backslash O_f)$ under the mapping induced from $X\backslash O_f \subset X\backslash\boldsymbol{Fix}(f)$. Then the image of $O_F \in H_n(X, X\backslash\boldsymbol{Fix}(f))$ under the above composite mapping, $I(f)o \in H_n(\Delta^n, \Delta^n\backslash 0)$, defines *index* $I(f) \in R$ of f.

In economic analysis, the meanings of index or the degree of mappings crucially depends on the homotopy invariance property. Hence, the following theorem is important. (The proof is essentially included in the argument for the construction of $\bar{f}_*$ under the method of acyclic models.)

THEOREM 7.5.1: *Let X be a compact Hausdorff lc space and let f be a mapping (function) of class $\mathscr{D}$. If there is a continuous function g such that for each $x \in X\backslash\boldsymbol{Fix}(f)$, there is an open set O_x such that for all $z \in O_x$, $g(z) \in K_x$, then $\bar{f}_* = \bar{g}_*$, so the index of f is uniquely determined.*

By the above assertion it is easy to obtain that if $\bar{f}_{0*} = \bar{g}_{0*}$, $\bar{f}_{1*} = \bar{g}_{1*}$, g_0 and g_1 are continuous and homotopic, and $\bigcup_{t=0}^{1}\boldsymbol{Fix}(f_t)$ is compact, then $I(f_0) = I(f_1)$.

Bibliographic Notes

The fixed-point indices treated in the previous two sections may easily be generalized to case with correspondences (of class $\mathscr{K}$ or $\mathscr{D}$) as for the Lefschetz number. The extension we have seen is different from the standard argument based on upper semicontinuity, such as Eilenberg and Montgomery (1946), Jaworowski (1958), etc. Arguments for class $\mathscr{K}$ or $\mathscr{D}$ are sufficiently novel even for single-valued mappings. Of course, the discussions are not covered by the concept of admissible maps, as in Górniewicz (1976). For cohomological arguments for general spaces in Section 7.2, see Spanier (1948, Appendix A).

The generalization of the index or the degree of mappings for correspondences are also treated by mathematical economists such as Mas-Colell (1974) and McLennan (1991). Their arguments are special treatments for upper semicontinuous contractible valued cases. For recent research see Giraud (2001), where some basic algebraic methods are also introduced.

Chapter 8
Applications to Related Topics

8.1 KKM, KKMS, and Core Existence

In Chapter 2, the basic assumptions for fixed-point existence, (K*), (K♯), and (K1), are conditions guaranteeing that we can find a certain locally common "direction" of $\varphi(x)$ at each non-fixed point x of φ. The concept and the method of the proof may also be utilized to obtain an extension of the Knaster–Kuratowski–Mazurkiewicz (KKM) theorem. In the next extension of the KKM theorem, to show the existence of point $x^* \in \bigcap_{i \in I} C_i$, condition (KKM1) requests that we find a locally common direction (C_j) at each x such that $x \notin \bigcap_{i \in I} C_i$. Condition (KKM1) is automatically satisfied when every C_i is closed, though it is not necessarily the case (note also that in the condition, j may not equal i). Furthermore, we do not restrict I to be a finite set.[1]

THEOREM 8.1.1: (Extension of KKM Theorem) *Let X be a non-empty compact convex subset of Hausdorff topological vector space E. Let $\{C_i\}_{i \in I}$ be a family of subsets of X, and let $\{x_i\}_{i \in I}$ be a family of points of X such that for each finite subfamily $\{x_{i_1}, \ldots, x_{i_n}\}$ of $\{x_i\}_{i \in I}$ and for each $x \in X$, $x \in \mathrm{co}\{x_{i_1}, \ldots, x_{i_n}\} \Rightarrow x \in \bigcup_{k=1}^{n} C_{i_k}$. Assume the following condition:*

(KKM1) *For each $x \in X$, if $\exists i \in I, x \notin C_i$, then open neighborhood $U(x)$ of x in X and index $j \in I$ exist such that $U(x) \cap C_j = \emptyset$.*

Then $\exists x^* \in X$, $x^* \in \bigcap_{i \in I} C_i$.

PROOF: Note that by the assumption, each $x_i \in \mathrm{co}\{x_i\}$ is an element of C_i, so every C_i is non-empty. Assume that for each $x \in X$, $x \notin \bigcap_{i \in I} C_i$.

---
[1] Of course, one may obtain the same results under abstract convexity settings without using vector space structures.

Then by (KKM1), for each $x \in X$, index $j(x) \in I$ and open neighborhood $U(x)$ of x exist such that $U(x) \cap C_{j(x)} = \emptyset$. Since X is compact, finite points $x^1, \ldots, x^m$ such that family $\{U(x^1), \ldots, U(x^m)\}$ covers X. Let $\beta^i : X \to [0,1], i = 1, \ldots, m$ be a partition of unity subordinate to $\{U(x^1), \ldots, U(x^m)\}$ and let $f(x) = \sum_{i=1}^{m} \beta^i(x) x_{j(x^i)}$ for each $x \in \Delta = \operatorname{co}\{x_{j(x^i)} \,|\, i = 1, \ldots, m\}$. Then, $f : \Delta \to \Delta$ has a fixed point by Brouwer's fixed-point theorem. On the other hand, since for every $x \in \Delta$, x cannot be spanned by those x_j's such that x does not belong to the union among C_j's (i.e., for each $\{x_{i_1}, \ldots, x_{i_n}\}$, $x \notin \bigcup_{k=1}^{n} C_{i_k} \Rightarrow x \notin \operatorname{co}\{x_{i_1}, \ldots, x_{i_n}\}$), and since $\beta^i(x) > 0$ iff $x \in U(x^i)$ (so $x \notin C_{j(x^i)}$), f has no fixed point: a contradiction. ∎

In the same way, we obtain an extension of Knaster–Kuratowski–Mazurkiewicz–Shapley's (KKMS) theorem. Let I be a finite set. Family $\mathscr{B}$ of non-empty subsets of I is said to be *balanced* if and only if there are non-negative coefficients $\lambda(B)$, $B \in \mathscr{B}$, such that for each $i \in I$, $\sum_{i \ni B} \lambda(B) = 1$. (If we can take a set of affinely independent points, $\{x_i \,|\, i \in I\}$, in a certain Euclidean space, and if for each non-empty $J \subset I$, we denote by x_J the barycenter of the simplex spanned by $\{x_i \,|\, i \in J\}$, then the condition may be restated as "$\mathscr{B}$ is balanced if and only if the barycenter x_I can be represented as a convex combination among barycenters, $x_B, B \in \mathscr{B}$.")

Given n points and n closed sets indexed by $I = \{1, \ldots, n\}$, the KKM theorem assures the existence of the intersection among n closed sets. It also assures that given n points and $2^n - 1$ sets indexed by $\mathscr{I} = \{J \,|\, J \subset \{1, \ldots, n\}, J \neq \emptyset\}$, the existence of the "balanced" family $\mathscr{B} \subset \mathscr{I}$ such that the intersection among the closed sets indexed by $\mathscr{B}$ is non-empty.

We extend the KKMS theorem by simply using a condition like (K1) (actually we use a coincidence theorem proved in Section 8.3 through a fixed-point theorem based on condition (K1)). The condition (KKMS1) in the next theorem says that (if we are allowed to specify $\hat{y}$ as the barycenter of the simplex spanned by $\{x_i \,|\, i \in I\}$) as long as the family of C_J's to which point x belongs is not balanced, we can find a locally common direction (p^x) of the family of C_J's to which point z belongs for every z near x. One can easily verify that correspondence G in the following lemma is closed convex valued and upper semi-continuous when I is finite and all C_J's are closed, so (KKMS1) is automatically satisfied when I is finite and all C_J's are closed (an ordinary case: c.f. proof of Theorem 5.3.1 of Ichiishi (1983) or Ichiishi (1981b)). Note also (as the previous theorem) that we do not restrict I to be a finite set. Since $\hat{y}$ in the theorem is arbitrary, when I is finite and $\hat{y}$

may be identified with the barycenter of $\{x_i \,|\, i \in I\}$, the theorem asserts that x^* exists such that $\mathscr{B} = \{J \,|\, x^* \in C_J\}$ is balanced.

THEOREM 8.1.2: (Extension of KKMS Theorem) *Let X be a nonempty compact convex subset of Hausdorff topological vector space E. Let $\{x_i \,|\, i \in I\}$ be a family of points of X such that $\operatorname{co}\{x_i \,|\, i \in I\} = X$, and let $\{C_J \,|\, J \subset I, \sharp J < \infty\}$ be a family of subsets of X such that for each finite subset $\{x_{i_1}, \ldots, x_{i_n}\}$ of $\{x_i \,|\, i \in I\}$, we have $\operatorname{co}\{x_{i_1}, \ldots, x_{i_n}\} \subset \bigcup_{J \subset \{i_1, \ldots, i_n\}} C_J$. Moreover, let $\hat{y}$ be an arbitrary element of X and denote by x_J point $(1/\sharp J) \sum_{i \in J} x_i$ for each finite subset J of I. Assume the following condition:*

(KKMS1) *For each $x \in X$, if $\hat{y} \notin G(x) = \operatorname{co}\{x_J \,|\, x \in C_J\}$, vector $p^x \in E'$ and open neighborhood $U(x)$ of x in X exist such that for all $z \in U(x)$ and for all $w \in \hat{y} - G(z)$, $p^x(w) > 0$.*

Then $\exists x^* \in X$, $\hat{y} \in G(x^*)$.

PROOF: Let $\varphi(x) = \hat{y}$ and $\psi(x) = G(x)$ for all $x \in X$. By (KKMS1), φ and ψ satisfy condition (i) of Theorem 8.3.1. Hence, it remains to be shown that φ and ψ satisfy condition (ii) of Theorem 8.3.1. For each $x \in X$ such that $0 \notin \varphi(x) - \psi(x) = \hat{y} - G(x)$, by considering that $X = \operatorname{co}\{x_i \,|\, i \in I\}$, let T be a finite subset of I such that $x \in \operatorname{co}\{x_i \,|\, i \in T\}$, and let $x = \sum_{i \in T} \alpha_i x_i$, where $\alpha_i > 0$ for all $i \in I$ and $\sum_{i \in T} \alpha_i = 1$. Since $\operatorname{co}\{x_i \,|\, i \in T\} \subset \bigcup_{J \subset T} C_J$, there is a subset J of T such that $x \in C_J$. Let $w = \hat{y} - x_J \in \hat{y} - G(x) = \varphi(x) - \psi(x)$ and let $\lambda(x) = \alpha_J = \min\{\alpha_i \,|\, i \in J\}$. Then $x + \lambda(x)w = \sum_{i \in T} \alpha_i x_i + \alpha_J(\hat{y} - (1/\sharp J) \sum_{i \in J} x_i)$ is clearly an element of $\operatorname{co}(\{\hat{y}\} \cup \{x_i \,|\, i \in T\}) \subset X$. Hence, φ and ψ also satisfy condition (ii) of Theorem 8.3.1, and point x^* exists such that $\varphi(x^*) \cap \psi(x^*) \neq \emptyset$. ∎

Ichiishi (1988) pointed out (and developed further in Ichiishi and Idzik (1990) and Ichiishi and Idzik (1991)) that the KKMS theorem can be reformulated as an extension of the following theorem of Sperner:

(Sperner Theorem) Let X be the convex hull of an affinely independent subset $\{x_1, \ldots, x_n\}$ in Hausdorff topological vector space E. Let $\{C_i\}_{i \in I = \{1, \ldots, n\}}$ be a closed covering of X. Assume that $x \in \operatorname{co}\{x_i \,|\, i \in I, i \neq j\}$ means $x \in C_j$. Then, $\bigcap_{i \in I} C_i \neq \emptyset$.

As an extension of Sperner's theorem, we have the subsequent alternative form of KKMS theorem:

(Alt KKMS: Ichiishi 1988) Let X be the convex hull of an affinely independent subset $\{x_1, \ldots, x_n\}$ in Hausdorff topological vector space E. Let $\{C_J\}_{J \subset I = \{1,\ldots,n\}, J \neq \emptyset}$ be a closed covering of X. Assume that for each $T \subset I$, $x \in \mathrm{co}\{x_i \mid i \in T\}$ means $x \in \bigcup_{J \supset (I \setminus T)} C_J$. Then $\bigcap_{J \in \mathscr{B}} C_J \neq \emptyset$ for a certain balanced family $\mathscr{B}$.

In the above formulation, condition "covering" may be reduced since under condition "$x \in \mathrm{co}\{x_i \mid i \in T\} \Rightarrow x \in \bigcup_{J \supset (I \setminus T)} C_J$," by taking $T = I$, we have "$x \in \mathrm{co}\{x_i \mid i \in I\} \Rightarrow x \in \bigcup_{J \supset \emptyset} C_J$."

We can extend the theorem (Alt KKMS) in the same way with Theorem 8.1.2. For this theorem, however, we must assume I to be finite and fix $\hat{y}$ as the barycenter of $\{x_i \mid i \in I\}$. As before, condition (KKMS2) in the next theorem is automatically satisfied when each C_J is closed. Theorem asserts the existence of x^* such that $\{J \mid x^* \in C_J\}$ is balanced.

THEOREM 8.1.3: (Extension of Alt KKMS Theorem) *Let X be a non-empty compact convex subset of Hausdorff topological vector space E. Let $\{x_i \mid i \in I\}$ be a finite affinely independent family of points of X such that $\mathrm{co}\{x_i \mid i \in I\} = X$, and let $\{C_J \mid J \subset I\}$ be a family of subsets of X such that for each subset $\{x_{i_1}, \ldots, x_{i_n}\}$ of $\{x_i \mid i \in I\}$, we have $\mathrm{co}\{x_{i_1}, \ldots, x_{i_n}\} \subset \bigcup_{J \supset (I \setminus \{i_1, \ldots, i_n\})} C_J$. Moreover, let $\hat{y}$ be the barycenter of X and denote by x_J point $(1/\sharp J) \sum_{i \in J} x_i$ for each subset J of I. Assume the following condition:*

(KKMS2) *For each $x \in X$, if $\hat{y} \notin G(x) = \mathrm{co}\{x_J \mid x \in C_J\}$, vector $p^x \in E'$ and open neighborhood $U(x)$ of x in X exist such that for all $z \in U(x)$ and for all $w \in G(z) - \hat{y}, p^x(w) > 0$.*

Then $\exists x^* \in X, \hat{y} = x_I \in G(x^*)$.

PROOF: As before, let $\psi(x) = \hat{y}$ and $\varphi(x) = G(x)$ for all $x \in X$. By (KKMS2), φ and ψ satisfy condition (i) of Theorem 8.3.1, so it remains to be shown that φ and ψ satisfies condition (ii) of Theorem 8.3.1. For each $x \in X$ such that $0 \notin \varphi(x) - \psi(x) = G(x) - \hat{y}$, by considering that $X = \mathrm{co}\{x_i \mid i \in I\}$, let T be a subset of I such that $x \in \mathrm{co}\{x_i \mid i \in T\}$, and let $x = \sum_{i \in T} \alpha_i x_i$, where $\alpha_i > 0$ for all $i \in T$ and $\sum_{i \in T} \alpha_i = 1$. Note that $T \neq I$ since $\hat{y} = x_I \notin G(x)$. Since $\mathrm{co}\{x_i \mid i \in T\} \subset \bigcup_{J \supset (I \setminus T)} C_J$, there is set $J \subset I$ including $I \setminus T$ such that $x \in C_J$. Let $w = x_J - \hat{y} \in G(x) - \hat{y} = \varphi(x) - \psi(x)$. If we represent w by linear combinations among points in $\{x_i \mid i \in I\}$ as $w = \sum_{i \in I} \beta_i x_i, \beta_i = (1/\sharp J) - (1/\sharp I) > 0$ for all $i \in J \supset (I \setminus T)$ and $\beta_i = -(1/\sharp I) < 0$ for all $i \in (I \setminus J) \subset T$. Hence, by taking $\lambda(x) > 0$ sufficiently small, we are assured that $x + \lambda(x)w = \sum_{i \in I}(\alpha_i + \beta_i)x_i$

($\alpha_i = 0$ for all $i \notin T$) is a convex combination, so it is an element of X. Hence, φ and ψ also satisfy condition (ii) of Theorem 8.3.1, and there is point x^* such that $\varphi(x^*) \cap \psi(x^*) \neq \emptyset$. ∎

The KKMS theorem is used in game theory to characterize the core existence problem (Shapley 1973). For set I of players, denote by $\mathscr{I} = \{J \mid J \subset I, J \neq \emptyset\}$ the family of *coalitions*. A *non-side-payment game* is a correspondence $V : \mathscr{I} \to R^I$ such that for all $J \in \mathscr{I}$, $V(J)$ satisfies $\text{pr}_J(x) = \text{pr}_J(y) \leftrightarrow ((x \in V(J)) \wedge (y \in V(J)))$, where each $x \in V(J)$ is interpreted as a possible (utility) allocation among coalition J. The *core* of non-side-payment game V is a set $\textbf{\textit{Core}}(V) \subset R^I$ such that for all $x \in \textbf{\textit{Core}}(V)$, (1) $x \in V(I)$ (socially feasible), and (2) there is no coalition J such that for some $y \in V(J)$, $\text{pr}_j(y) > \text{pr}_j(x)$ for all $j \in J$ (cannot be improved). Game V is said to be *balanced* if for utility allocation $x \in R^I$, feasibility under a certain balanced family, $x \in \bigcap_{J \in \mathscr{B}} V(J)$ for a certain balanced family $\mathscr{B} \subset \mathscr{I}$, guarantees social feasibility, $x \in V(I)$. For balanced games, we can guarantee the non-emptiness of the core through the KKMS theorem by defining each C_J to represent the non-improvable utility frontier feasible under J. This type of core non-emptiness theorem is called a Scarf Theorem (Scarf 1967). It is also possible to generalize core non-emptiness arguments for cases including non-ordered preferences (see, e.g., Border 1984, and Yannelis 1991). Ichiishi and Idzik also utilized their (Alt KKMS)-type covering arguments to market allocation problems and equilibrium existence with indivisible goods (Ichiishi and Idzik 1999a, Ichiishi and Idzik 1999b).

8.2 Eaves' Theorem

The following theorem is known as Eaves' theorem.

THEOREM 8.2.1: (Eaves 1974) *Let S be a simplex of full dimension in R^ℓ and v be a function on S to R^ℓ such that $x + v(x) \in \text{int}\, S$ for all $x \in S \backslash \text{int}\, S$. Then, there is point $x^0 \in S$ such that for each neighborhood U of x^0 in S, $0 \in \text{co}\, v[U]$.*

In the theorem, int denotes the interior in R^ℓ and co denotes the convex hull. As we can see in Nishimura and Friedman (1981), Eaves' theorem enables us to construct economic equilibrium arguments without referring to the convexity and/or the continuity of individual preferences or best reply correspondences. Here, it is shown that Eaves' theorem may easily be generalized through our fixed-point theorems in Chapter 2.

At first, we see the following lemma, which is an immediate consequence of case (K1) of Theorem 2.3.4.

LEMMA 8.2.2: *Let X be a non-empty compact convex subset of R^ℓ, and f be a function on X to X. Then, there is point $x^0 \in X$ such that for all neighborhood U of x^0 in X, $\varphi(x) = f(x) - x$ satisfies $0 \in \text{co}\,\varphi[U]$.*

PROOF: Suppose that for all x in X, there is neighborhood U^x of x such that $0 \notin \text{co}\,\varphi[U^x]$. Then, there is vector p^x in the topological dual of R^ℓ such that $p^x(\varphi(z)) = p^x(f(z) - z) > 0$ for all $z \in U^x$. Hence, f satisfies the condition (K1) of Theorem 2.3.4, so that f has fixed point x^0, which is contradictory since $0 \neq \varphi(x) = f(x) - x$ for all $x \in X$. ∎

In the above proof, the separation argument crucially depends on the fact that the dimension of the total space is finite. Now, we prove the main theorem.

THEOREM 8.2.3: (**Generalization of Eaves' Theorem**) *Let X be a non-empty compact convex subset of R^ℓ, and v be a function on X to R^ℓ such that $x + v(x) \in X$ for all $x \in X \setminus \text{int}\,X$. Then, there is point $x^0 \in X$ such that for all neighborhood U of x^0 in X, $0 \in \text{co}\,v[U]$.*

PROOF: For each $x \in \text{int}\,X$, let λ_x be a positive real number such that $x + \lambda_x v(x) \in X$ and for each $x \in X \setminus \text{int}\,X$, let $\lambda_x = 1$. Let us define a function $f : X \to X$ as

$$f(x) = x + \lambda_x v(x).$$

By Lemma 8.2.2, there is $x^0 \in X$ such that for each neighborhood U of x^0, $0 \in \text{co}\,\{f(x) - x \mid x \in U\}$. That is, for certain natural number n, there are $x^1, \ldots, x^n \in X$ and $\alpha^1, \ldots, \alpha^n \in R_+$, $\sum_{i=1}^n \alpha^i = 1$, such that $0 = \sum_{i=1}^n \alpha^i \lambda_{x^i} v(x^i)$. Hence, if we define λ_0 as $\min\{\lambda_{x^1}, \ldots, \lambda_{x^n}\}$ and λ_i as $\frac{\lambda_{x^i}}{\lambda_0}$ for each $i = 1, \ldots, n$, we have

$$0 \in \text{co}\,\{\lambda_1 v(x^1), \ldots, \lambda_n v(x^n)\},$$

$\lambda_i \geq 1$ for all $i = 1, \ldots, n$. On the other hand, if $0 \notin \text{co}\,\{v(x^1), \ldots, v(x^n)\}$, p in the topological dual of R^ℓ exists such that $p(v(x^i)) > 0$ for all $i = 1, \ldots, n$. Hence, we have $0 \notin \{x \in R^\ell \mid p(x) > 0\} \supset \text{co}\,\{\lambda_1 v(x^1), \ldots, \lambda_n v(x^n)\}$, a contradiction. Therefore, we have $0 \in \text{co}\,\{v(x^1), \ldots, v(x^n)\}$, and x^0 satisfies the condition stated in the theorem. ∎

8: Applications to Related Topics

Note that Theorem 8.2.3 generalizes Theorem 8.2.1 in three ways, i.e., in Theorem 8.2.3, (i) X may not be a simplex, (ii) X may not be full dimensional, and (iii) $x + v(x)$ may not be an element of $\text{int}\, X$.

8.3 Fan–Browder's Coincidence Theorem

Let E be a Hausdorff topological vector space over R and let X be a compact convex subset of E. Fixed-point arguments in Chapter 2 may also be applied to correspondences on X to $E \supset X$ and the generalization directly gives the following extension of the well-known coincidence theorem in Fan (1969). Instead of the upper demi-continuity in a locally convex space, we use a weaker condition (i) in a Hausdorff topological vector space that intuitively asserts that not the values $\varphi(x)$ and $\psi(x)$ but the direction $\varphi(x) - \psi(x)$ satisfies the condition like (K1).

THEOREM 8.3.1: (Extension of Fan (1969, Theorem 6)) *Let X be a non-empty compact convex set in Hausdorff topological vector space E over R. Let φ, ψ be two correspondences on X to E satisfying the following conditions:*

(i) *For each $x \in X$ such that $0 \notin \varphi(x) - \psi(x)$, there exist vector $p^x \in E'$ and open neighborhood $U(x)$ of x in X satisfying $\forall z \in U(x)$, $\forall w \in \varphi(z) - \psi(z)$, $p^x(w) > 0$.*
(ii) *For every $x \in X$ such that $0 \notin \varphi(x) - \psi(x)$, there exist point $y \in X$, element $w \in \varphi(x) - \psi(x)$, and real number $\lambda(x) > 0$ satisfying $y - x = \lambda(x)w$.*

Then x^* exists such that $\varphi(x^*) \cap \psi(x^*) \neq \emptyset$.

PROOF: Assume that for all $x \in X$, $\varphi(x) \cap \psi(x) = \emptyset$, i.e., $0 \notin \varphi(x) - \psi(x)$. Let f be a function on X to itself such that $f(x) = x + \lambda(x) w^x$, where w^x is an element of $\varphi(x) - \psi(x)$ stated in the condition (ii). Then, by condition (i), f satisfies condition (K1) in Theorem 2.3.4, so f has a fixed point: a contradiction. ∎

The coincidence theorem of Fan (1969, Theorem 6) (sometimes called Fan–Browder's theorem) is frequently used in mathematical economics, especially in the proof of core existence theorems (see, e.g., Ichiishi 1983, Border 1984). We shall see below that Theorem 8.3.1 enables us to generalize the KKMS theorem (see Lemma 8.1.2).

In Fan (1969), the coincidence theorem is derived from a more general result (Theorem 5 in Fan (1969)). Though the above theorem is not a direct extension of Theorem 5 in Fan (1969), it is also easy to generalize our theorem using the same condition for the direction of difference $\varphi(x)-\psi(x)$.

8.4 L-majorized Mappings

Let I be a non-empty index set, and let $X = \prod_{i \in I} X^i$ be the product of subsets of a topological vector space E. Moreover, let $\phi : X \to X^i$ be a correspondence on X to a certain X^i. At first, we shall give the following definitions:[2]

(1) ϕ is *of class $\mathscr{L}$* if $\forall x = (x_j)_{j \in I} \in X$, $x_i \notin \operatorname{co} \phi(x)$ and $\forall y \in X^i$, $\phi^{-1}(y)$ is open in X.
(2) Correspondence $\Phi_x : X \to X^i$ is an $\mathscr{L}$-majorant of ϕ at x if Φ_x is of class $\mathscr{L}$ and there is open neighborhood U_x of x in X such that $\phi(z) \subset \Phi_x(z)$ for all $z \in U_x$.
(3) ϕ is $\mathscr{L}$-majorized if for all $x \in X$ such that $\phi(x) \neq \emptyset$, there is an $\mathscr{L}$-majorant of ϕ at x.

For the special case $I = \{i\}$, the following result is known.

THEOREM 8.4.1: (Yannelis–Prabhakar (1983, Corollary 5.1))
Let X be a non-empty, compact, convex subset of a Hausdorff topological vector space and $P : X \to X$ be an $\mathscr{L}$-majorized correspondence. Then x^* exists such that $P(x^*) = \emptyset$.

As stated before, our Theorems 3.1.2 and 3.1.5 are intended to generalize or to give an alternatively general condition for a maximal element existence theorem. If X is a subset of pseudo-metrizable space, it is really the case for we can see that $\mathscr{L}$-majorized mappings satisfy condition (K*).

LEMMA 8.4.2: *Let X be a non-empty, compact, convex subset of a pseudo-metrizable topological vector space and $P : X \to X^i$ be an $\mathscr{L}$-majorized correspondence. Then, there is convex non-empty valued correspondence $\Phi : X \to X$ such that $\forall x \in K = \{z \in X \mid P(x) \neq \emptyset\}$, $\Phi(x) \neq \emptyset$, $P(x) \subset \Phi(x)$, $x \notin \Phi(x)$, and for all $x \in K$, there exist*

---
[2] More generally, see, e.g., Tan and Yuan (1994).

neighborhood $U(x)$ of x in X and point $y^x \in X^i$ such that for each $z \in U(x) \cap K$, $y^x \in \Phi(z)$. (*That is, for* Φ, *condition* (K^*) *in Theorem* 2.1.6 *is satisfied.*)

PROOF: Since P is $\mathscr{L}$-majorized, for each $x \in K$, there are $\mathscr{L}$-majorant Φ_x of P at x and open neighborhood U_x of x in X such that $\forall z \in U_x$, $\phi(z) \subset \Phi_x(z)$. Since X is a subset of pseudo-metrizable space, K is also pseudo-metrizable. Hence, K is paracompact and we may suppose that the open cover $\{U_x\}_{x \in K}$ has a locally finite refinement $\{\hat{U}_x\}_{x \in J}$, $J \subset K$, $\hat{U}_x \subset U_x$ for each $x \in J$. For each $z \in K$, let $\Phi(z) = \bigcap_{x \in J, z \in \hat{U}_x} \Phi_x(z)$. Moreover, for each $z \notin K$, let $\Phi(z) = X$. Then for each $z \in K$, by letting $U(z)$ be intersection $\bigcap_{x \in J, z \in \hat{U}_x} \hat{U}_x$ and y^z be an arbitrary element of $P(z)$, correspondence $\Phi : X \to X$ satisfies all the conditions stated above. ∎

The notion of "$\mathscr{L}$-majorized" is quite useful for equilibrium existence arguments, and it is worth generalizing the concept into general spaces without the vector-space structure. Let I be a set. For each $i \in I$, let E^i be a Hausdorff space having convex structure $(\{f_A^i \mid A \in \mathscr{F}(E^i)\}, C^i)$ and X^i be a non-empty compact subset of E^i. Moreover, let $\phi : X = \prod_{j \in I} X^j \to X^i$ be a correspondence on X to a certain X^i. ϕ is *of class* $\mathscr{L}$ if $\forall x = (x_j)_{j \in I} \in X$, $x_i \notin C(\phi(x))$ and $\forall y \in X^i$, $\phi^{-1}(y)$ is open in X. Correspondence $\Phi_x : X \to X^i$ is an $\mathscr{L}$-majorant of ϕ at x if Φ_x is of class $\mathscr{L}$ and there is open neighborhood U_x of x in X such that $\phi(z) \subset \Phi_x(z)$ for all $z \in U_x$. ϕ is $\mathscr{L}$-majorized if for all $x \in X$ such that $\phi(x) \neq \emptyset$, there is $\mathscr{L}$-majorant of ϕ at x. One may prove the existence of maximal elements, Nash equilibrium, and equilibrium for abstract economies based on this condition. (For maximal element existence, use Theorem 2.3.2.)

8.5 Variational Inequality Problem

The existence of market equilibrium (Gale–Nikaido–Debreu's theorem) may mathematically be classified into a solution to the system of inequalities known as the *variational inequality problem*. For locally convex space E over R, compact convex subset K of its topological dual $(E', \sigma(E', E))$, and function $f : K \to E$, Element p^* of K satisfying the next inequality is known as a solution to the *variational inequality problem*.

$$\langle f(p^*), p^* \rangle \geq \langle f(p^*), q \rangle \quad \text{for all } q \in K. \tag{8.1}$$

One can easily see that the situation is nothing but asserting that "price p^* gives the highest evaluation on the excess demand under itself among all prices." (Hence, with the Walras' law, p^* is an equilibrium price.)

As we generalize the market equilibrium theorem, it is also possible to extend conditions for the existence of solutions to the variational inequality problems in our fashion. Condition (LDV*) in the following theorem asserts that "if p is not a solution to the variational inequality problem, then point q' exists that blocks all points near p to be a solution to the problem." Of course the condition may automatically be satisfied as long as a certain kind of continuity for f is assured.

THEOREM 8.5.1: *Let E be a locally convex space over R and let K be a compact convex subset of topological dual $(E', \sigma(E', E))$. Moreover, let $f : K \to E$ be an arbitrary function satisfying the following condition:*

(LDV*) *If $p \in K$ is such that for certain $q \in K$, $\langle f(p), p \rangle < \langle f(p), q \rangle$, then there are neighborhood U of p in K and element $q' \in K$ such that for all $p' \in U^x$, $\langle f(p'), p' \rangle < \langle f(p'), q' \rangle$.*

Then, there exists solution $x^* \in K$ of the variational inequality problem, i.e., $f(x^*)(x^* - y) \geq 0$ for all $y \in K$.

PROOF: If there is no solution to the variational inequality problem, we may define mapping $\varphi : K \to K$ as

$$\varphi(p) = \{q' \in K \mid q' \text{ satisfies condition (LDV*) for } p\}, \quad (8.2)$$

which is convex valued and has the local intersection property. Hence, by Theorem 2.1.6, φ has a fixed point, which contradicts the property $\langle f(p'), p' \rangle < \langle f(p'), q' \rangle$ for q' asserted in condition (LDV*). ∎

8.6 Equilibrium with Cooperative Concepts

It is an important direction to incorporate cooperative features in human society with the non-cooperative equilibrium concepts discussed in Chapter 3. In this section, we see two such attempts that are closely related to our arguments and that are expected to be greatly developed by the concepts and methods in this book.

8.6.1 Social systems with coordination

As a direct extension of the social equilibrium concept, equilibrium in *social systems with coordination* (*SSC*) (see Vind 1983 and Keiding 1985) is known. Consider a generalized non-cooperative strategic game form (abstract economy), $(X_i, P_i, K_i)_{i \in I}$. Denote by X product $\prod_{i \in I} X_i$ as usual. For this argument, assume each P_i and K_i correspondence not on X to X_i but on X to X (to incorporate externalities), and define social feasibility correspondence $K : X \to X$ as $K(x) = \bigcap_{i \in I} K_i(x)$. Furthermore, for each $i \in I$, consider function

$$e_i : X \times X \ni (x, y) \mapsto z = e_i(x, y) \in X$$

a *state expectation function of i*, representing player i's belief that given state $x \in X$, if society considers situation y to be an alternative to x and if player i changes his strategy from $x_i = \mathrm{pr}_i(x)$ to $y_i = \mathrm{pr}_i(y)$, then the consequence will be $z = e_i(x, y)$. We can now define a *social system with coordination* (*SSC*) as list (X_i, P_i, K_i, e_i).

An equilibrium for SSC is state $x^* \in X$ such that (1) $x^* \in K(x^*)$ (socially feasible), and (2) there is no state $y \in X$ such that $I^* = \{i \in I \mid e_i(x^*, y) \neq x^*\} \neq \emptyset$ and $\forall i \in I^*$, $e_i(x^*, y) \in P_i(x^*) \cap K_i(x^*)$. (See Vind 1983, Keiding 1985, Kim and Yuan 2001.)

I am not convinced that "$\forall i \in I^*$" in condition (2) is appropriate for an equilibrium concept. Indeed, the possible actions for each agent are restricted only by K_i, and under (subjective) expectation function e_i, each agent will deviate from an "equilibrium" basically as a private decision. Hence, in condition (2), if "there is at least one i" such that $i \in I^*$ and $e_i(x^*, y) \in P_i(x^*) \cap K_i(x^*)$, state x^* is not appropriate to be called an equilibrium. For this equilibrium condition (2), I have an additional question. If agent i expects consequent state $e_i(x^*, y)$ and if agent i asks whether $e_i(x^*, y) \in K_i(x^*)$, why does he not ask whether $e_i(x^*, y)$ itself is really feasible for him, i.e., $e_i(x^*, y) \in K_i(e_i(x^*, y))$?

Admittedly, however, for this type of problem, i.e., a description of our mind, our rationality, it is extremely hard (actually, impossible as we will discuss in Chapters 9 and 10) to obtain a final answer. The SSC concept provides an important suggestion: *even though inexhaustible ways may exist to describe our "minds" and "thoughts," by setting an appropriate framework (like expectation function e_i for SSC), we may restrict our equilibrium problems to a compact domain of our "behaviors."* In Chapter 10

(Section 10.2), I shall give one concluding theorem (Theorem 10.2.1) from this viewpoint.

8.6.2 Social coalitional equilibrium

The second attempt I introduce here is the social coalitional equilibrium approach like Ichiishi (1981b). Again let us begin with an abstract economic setting, $(X_i, P_i, K_i)_{i \in I}$. Denote by $\mathscr{I}$ the set of non-empty subsets of I and by X_J product $\prod_{i \in J} X_j$ for each $J \in \mathscr{I}$. Now extend the setting by recognizing each $K_i : X \to X_i$ as $K^{\{i\}}$ and assume that for each $J \in \mathscr{I}$, there is a corresponding $K^J : X \to X_J$: a *feasible strategy correspondence of coalition J*. Moreover, we can also extend settings for P_i in the same way, i.e., suppose that for each $J \in \mathscr{I}$, preference correspondence $P_i^J : X \to X_i$ is defined.

For this approach, we also consider the *coalition structure*. Given the family of all coalitions $\mathscr{I}$, partition $\mathscr{T}$ of I (i.e., $\mathscr{T} \subset \mathscr{I}$, $\forall J_1, J_2 \in \mathscr{T}((J_1 \neq J_2) \Rightarrow (J_1 \cap J_2 = \emptyset))$, and $\bigcup \mathscr{T} = I$) is called a *coalition structure* for society. Denote by $\mathfrak{T}$ the class of *admissible coalition structures* exogenously given as an intrinsic feature of a certain society.

Now, a society is described as the following list:

$$(X_i, P_i^J, K_J, \mathfrak{T})_{i \in I, J \in \mathscr{I}}.$$

A *social coalitional equilibrium* is a pair $((x_i^*)_{i \in I}, \mathscr{T}^*)$ of strategic profile $x^* = (x_i^*)_{i \in I}$ and admissible coalition structure $\mathscr{T}^* \in \mathfrak{T}$ satisfying the following conditions:

(SCE1) For all $T \in \mathscr{T}$, $(x_i^*)_{i \in T} \in K^T(x^*)$ (feasibility condition).
(SCE2) There are no $J \in \mathscr{N}$ and $(y_i)_{i \in J} \in K^J(x^*)$ such that for all $i \in J$, $y_i \in P_i^J(x^*)$ (stability condition).

To show the existence of equilibrium, as in cases with the core of non-sidepayment games, we may settle the problem into the existence of a balanced class that assures the existence of equilibrium.

In the above, we used a non-ordered framework for preferences (SCE2) instead of the utility-function setting like Ichiishi (1983, p. 96). The existence of equilibrium, however, may also be generalized in this case using essentially the same argument with core existence theorems for non-ordered preferences. (Indeed, the arguments in Ichiishi (1983, pp. 96–98) already include some non-ordered preferences since he defines utility functions on a graph of constraint correspondences.)

The important feature of such equilibrium approaches is that we can incorporate a certain kind of mechanism that explains how a coalition structure is formulated among the admissible coalitional structures in society. (It will especially provide a clear perspective on the classical problem of the formation of firms. See, e.g., Ichiishi 1977 and Ichiishi 1993.)

8.7 System of Inequalities and Affine Transformations

I treat here theorems on systems of real valued mappings that may be utilized for the fixed-point theory for groups of linear transformations or ergodic arguments. First, we see the modification and extension of Fan's theorem on systems of inequality. Let us begin with the next theorem, which is a modification of a theorem in Takahashi (1988, Theorem 5.1.9) in topological vector space to our general spaces.

THEOREM 8.7.1: *Let X be a compact Hausdorff space having convex structure, and $F_y(x) : X \to R$ be a real valued function for each $y \in X$. For $c \in R$, denote by Solution(c) set $\{x \in X \mid F_y(x) \geq c \text{ for all } y \in X\}$. Assume the following:*

(1-1) *For all $y \in X$, if $F_y(x) < c$, then open neighborhood U^x of x and point $y(x) \in X$ exist such that for all $z \in U^x$, $z \notin$ Solution(c) implies $F_{y(x)}(z) < c$.*
(1-2) *For all $x \in X$, set $\{y \in X \mid F_y(x) < c\}$ is convex.*
(1-3) *There is c^* such that $F_x(x) \geq c^*$ for all $x \in X$.*

Then Solution$(c^*) \neq \emptyset$.

PROOF: Assume that Solution$(c^*) = \emptyset$. Then for each $x \in X$, there is at least one y such that $F_y(x) < c^*$, so by condition (1-1), there are open neighborhood U^x of x and point $y(x) \in X$ such that for all $z \in U^x$, $F_{y(x)}(z) < c$. Since X is compact, there are finite points, $x^1, \ldots, x^n$, open covering of their neighborhoods, $U(x^1), \ldots, U(x^n)$, and functions $F_{y(x^1)}, \ldots, F_{y(x^n)}$, such that $\forall z \in U(x^t)$, $F_{y(x^t)}(z) < c^*$ for all $t = 1, \ldots, n$. Let $\beta^t, t = 1, \ldots, n$, be a partition of unity subordinate to $U(x^1), \ldots, U(x^n)$. By condition (1-2), $F_{G(\Sigma_{t=1}^n \beta^t(x) y(x^t))}(x) < c^*$ for all $x \in X$, where for the convenience of readers unfamiliar with abstract-convexity settings, I used notation $G(\sum_{t=1}^{n} \beta^t(x)y(x^t))$ instead of the precise representation $g_{\bar{A}}(\sum_{t=1}^{n} \beta^t(x)e^t)$ with $A = \{y(x^1), \ldots, y(x^n)\}$ and points $e^1, \ldots, e^n \in \Delta^A$ corresponding to $y(x^1), \ldots, y(x^n) \in X$. Note that $e \mapsto \sum_{t=1}^{n} \beta^t(g_{\bar{A}}(e))e^t$ on Δ^A to itself has fixed point e^* by Brouwer's fixed-point theorem (or, in ordinary vector spaces, $x \mapsto \sum_{t=1}^{n} \beta^t(x)y(x^t)$ restricted on co A has a fixed point x^*). Define x^* by $x^* = g_{\bar{A}}(e^*)$, and then $G(\sum_{t=1}^{n} \beta^t(x^*)y(x^t)) = g_{\bar{A}}(\sum_{t=1}^{n} \beta^t(x^*)e^t) = g_{\bar{A}}(\sum_{t=1}^{n} \beta^t(g_{\bar{A}}(e^*))e^t) = g_{\bar{A}}(e^*) = x^*$. Hence, we have $F_{x^*}(x^*) < c^*$, which contradicts condition (1-3). ∎

By using the above theorem, we can obtain the subsequent extension of Fan's theorem on system of inequalities. The next theorem is based on Theorem 5.4.2 of Takahashi (1988).

THEOREM 8.7.2: (Fan's System of Inequality) *Let X be a compact Hausdorff space having convex structure, and $f_i(x) : X \to R$ be a real valued function for each $i \in I$. Assume the following:*

(2-1) *For all $i \in I$ and $c \in R$, if $f_i(x) > c$, then open neighborhood U^x of x exists such that for all $z \in U^x$, $f_i(z) > c$.*[3]

(2-2) *For arbitrary natural number $n > 0$, finite non-negative real numbers $a^1, \ldots, a^n$, $\sum_{t=1}^{n} a^n = 1$, and finite $i^1, \ldots, i^n \in I$, the set of y such that $\sum_{t=1}^{n} a^t f_{i^t}(y) < c$ is convex.*[4]

(2-3) *For arbitrary natural number $n > 0$, finite non-negative real numbers $a^1, \ldots, a^n$, $\sum_{t=1}^{n} a^n = 1$, and finite $i^1, \ldots, i^n \in I$, $y \in X$ exists such that $\sum_{t=1}^{n} a^t f_{i^t}(y) \leqq 0$.*

Then set $SOLV = \{x \in X \mid f_i(x) \leqq 0 \text{ for all } i \in I\}$ is not empty.

PROOF: Assume that $SOLV = \emptyset$. Then for each $x \in X$, there is at least one $i \in I$ such that $f_i(x) > 0$, so by condition (2-1), open neighborhood U^x of x and $i(x) \in I$ exist such that for all $z \in U^x$, $f_{i(x)}(z) > 0$. As in the previous theorem, by the compactness of X, we have finite points, $x^1, \ldots, x^n$, open covering of their neighborhoods, $U(x^1), \ldots, U(x^n)$,

---

[3]A real valued function satisfying this property is called a *lower semi-continuous function*. If ">" is replaced by "<" in the condition, then the function is called *upper semi-continuous*.

[4]In ordinary vector spaces, this condition is automatically satisfied if all f_i's are *convex functions*, i.e., $f_i(ax + by) \leqq af_i(x) + bf_i(y)$ for all $x, y \in X$, and $a, b \in R_+$, $a + b = 1$.

8: Applications to Related Topics

functions $f_{i(x^1)}, \ldots, f_{i(x^n)}$, such that $\forall z \in U(x^t)$, $f_{i(x^t)}(z) > 0$ for all $t = 1, \ldots, n$, and partition of unity α^t, $t = 1, \ldots, n$, subordinate to $U(x^1), \ldots, U(x^n)$. Let us define function $F(x) : X \to R$ for each $y \in X$ as

$$F_y(x) = \sum_{t=1}^{n} \alpha^t(x) f_{i(x^t)}(y).$$

Then system $\{F_y(x) \,|\, y \in X\}$ satisfies conditions (1-1) and (1-2) of Theorem 8.7.1 since each α^t is continuous and condition (2-2) holds. By definition $F_y(x) > 0$ for all $x, y \in X$. Since each f_{i^t} is lower semi-continuous, and since X is compact, $c^* > 0$ exists such that $F_y(x) > c^*$ for all $x, y \in X$, so (1-3) of Theorem 8.7.1 is also satisfied. Hence, by the previous theorem, we have $x^* \in X$ such that $F_y(x^*) = \sum_{t=1}^{n} \alpha^t(x^*) f_{i(x^t)}(y) \geq c^* > 0$ for all $y \in X$, which contradicts condition (2-3). ∎

Even though conditions (1-2) and (2-2) of the above theorems are weaker than assuming f_i's to be the usual convex functions, it is also true that we do not have an appropriate concept for convex functions in our general settings since we cannot specify coefficients among points under the abstract convex-combination concept. As the third theorem in this section, I treat the Markov–Kakutani fixed-point theorem (see, e.g., Dunford and Schwartz (1966, p. 456)). For this purpose, we must extend the notion of affine transformations and operations on a class of linear mappings, so we have to develop alternative tools for particular coefficients in convex-combination arguments.

Let X be a compact Hausdorff space and denote by $(\{f_A, A \in \mathscr{F}(X)\}, C)$ the convex structure on it (see Chapter 2, Section 2.2). Let us consider class $\mathfrak{C}$ of continuous mappings $T : X \to X$ such that for all $A =\in \mathscr{F}(X)$,

$$T(C(A)) = C(T(A)). \tag{8.3}$$

(Note that if X is a convex subset of vector space, then every *affine function*, i.e., a function $T : X \to X$ satisfying $T(ax+by) = aT(x)+bT(y)$ for all a, b, $a + b = 1$, is a member of class $\mathfrak{C}$.) We call mapping $T \in \mathfrak{C}$ a $\mathfrak{C}$-*morphism* (*convex morphism*) (see Section 6.1).

Consider **Cover**(X), and for each $x, y \in X$ and $\mathfrak{N} \in$ **Cover**X, write diam $|x, y| \leq \mathfrak{N}$ if there exists $N \in \mathfrak{N}$ such that $x, y \in N$. Moreover, denote by X_∞ the set of all finite sequences in X, and assume that for each finite sequence $v = (x_0, x_1, \ldots, x_n) \in X_\infty$, of points in X, there exists point $b(v) \in X$, say the *barycenter* of v, that satisfies the following conditions:

(BC0) For two sequences v and v' such that $v = (x_0, x_1, \ldots, x_i, x_{i+1}, \ldots, x_n)$ and $v' = (x_0, x_1, \ldots, x_{i+1}, x_i, \ldots, x_n)$, $b(v) = b(v')$.
(BC1) For all $v = (x_0, x_1, \ldots, x_n)$, $A = \{x_0, x_1, \ldots, x_n\}$, $b(v) \in C(A)$.
(BC2) For closed convex subset K of X, set $\{b(v) \mid v = (x_0, x_1, \ldots, x_n) \in X_\infty, \{x_0, x_1, \ldots, x_n\} \subset K\}$ is closed.
(BC3) For all $\mathfrak{N} \in \mathbf{Cover}\, X$, a natural number n exists such that for every sequence $v = (x_0, x_1, \ldots, x_m)$ whose length m is greater than n, and two points, $x, y \in X$, $\operatorname{diam} |b(x_0, x_1, \ldots, x_m, x), b(x_0, x_1, \ldots, x_m, y)| \leq \mathfrak{N}$.
(BC4) For each $v = (x_0, x_1, \ldots, x_n)$, operation b and each morphism $T \in \mathfrak{C}$ commutes, i.e., $T(b(v)) = b(Tx_0, Tx_1, \ldots, Tx_n)$.

We call $b : X_\infty \ni v \mapsto b(v) \in X$ a *barycentric operator* on X. (In ordinary topological vector spaces, define b as $b(x_0, x_1, \ldots, x_n) = 1/(n+1)(x_0 + x_1 + \cdots + x_n)$. In this case, (BC3) is automatically satisfied since for each compact subset X and open 0-neighborhood U, we have $\frac{1}{n}X \subset U$ for all sufficiently large k.)

If T and S are of class $\mathfrak{C}$, then for each $A \in \mathscr{F}(X)$, $T \circ S(C(A)) = T(C(S(A))) = C(T \circ S(A))$, so $T \circ S$ is also a member of $\mathfrak{C}$. Since identity mapping $I : X \to X$ is clearly an element of $\mathfrak{C}$, by identifying the composition of mappings as the law of composition, $\mathfrak{C}$ may be considered a set having an associative law of composition with an identity element (a monoid). Let us consider a subclass $\mathfrak{T} \subset \mathfrak{C}$ on which the law of composition is closed and commutative, i.e., for each $T, S \in \mathfrak{T}$, $T \circ S = S \circ T$ (a semigroup).[5]

Note that if $T \in \mathfrak{T}$, then $T^2 = T \circ T, T^3 = T \circ T^2, \ldots$, are necessarily members of $\mathfrak{T}$. (Note also that identity mapping $id_X : X \to X$ is not necessarily an element of $\mathfrak{T}$.) For each $T \in \mathfrak{T}$, let us define a subset $T_n(X)$ of X as

$$T_n(X) = \{b(Q_n) \mid Q_n = (q, Tq, \ldots, T^{n-1}q), q \in X\}. \quad (8.4)$$

In ordinary vector spaces, we may identify T_n with linear transformation $n^{-1}(I + T + T^2 + \cdots + T^{n-1})$. Now we can prove the next extension of the Markov–Kakutani fixed-point theorem.

---

[5] A set with an associative law of composition is called a *semigroup*. A semigroup with an identity element is called a *monoid*.

8: Applications to Related Topics

THEOREM 8.7.3: (Extension of Markov–Kakutani Fixed-Point Theorem) *Let X be a compact Hausdorff space having a convex structure and a barycentric operator. Let $\mathfrak{T}$ be a commutative semigroup of $\mathfrak{C}$-morphisms in X. Then point $p \in X$ exists such that $Tp = p$ for all $T \in \mathfrak{T}$.*

PROOF: Denote by $\mathscr{K}$ the class of all sets of form $T_n(X)$ for $T \in \mathfrak{T}$ and natural number n. By (BC1) and (BC2), each $T_n(X)$ is a non-empty closed subset of X. Moreover, since $\mathfrak{T}$ is commutative, for each $S, T \in \mathfrak{T}$ and natural numbers m, n, $S_m(X) \cap T_n(X) = T_n \circ S_m(X) \neq \emptyset$, so $\mathscr{K}$ has the finite intersection property. Hence, by the compactness of X, $\bigcap \mathscr{K} \neq \emptyset$. Let p be an element of $\bigcap \mathscr{K}$. Assume that $p \neq Tp$ for a certain $T \in \mathfrak{T}$. Since T is continuous, there is open set U such that $p \in U$ and $Tp \notin U$. Take $\mathfrak{N} \in \mathbf{Cover}\, X$ sufficiently fine so that $\forall N \in \mathfrak{N}, (x \in N) \Rightarrow (N \subset U)$. Let us consider sequence $B = (p, Tp, T^2 p, T^3 p, \ldots)$. Then by (BC3) with (BC0), natural number m exists such that

$$\mathrm{diam}\, |b(q, Tq, \ldots, T^{m-2}q, T^{m-1}q), b(Tq, T^2 q, \ldots, T^{m-1}q, T^m q)| \leq \mathfrak{N},$$

for any $q \in X$. Note that since p is an element of $\bigcap \mathscr{K}$, p is an element of $T_m(X)$. Hence, element $q \in X$ exists such that $p = b(q, Tq, \ldots, T^{m-2}q, T^{m-1}q)$. By (BC4), however, we have also

$$Tp = T(b(q, Tq, \ldots, T^{m-2}q, T^{m-1}q)) = b(Tq, T^2 q, \ldots, T^{m-1}q, T^m q),$$

so we have $\mathrm{diam}\, |p, Tp| \leq \mathfrak{N}$. Since $Tp \notin U$, there is no $N \in \mathfrak{N}$ such that $p \in N$ and $Tp \in N$; hence we have a contradiction. ■

Chapter 9
Mathematics and Social Science

9.1 Basic Concepts in Axiomatic Set Theory

In Chapters 9 and 10, we treat problems about the mathematical modeling of human society itself. One of the most idiosyncratic features of social science is that the society we intend to describe is nothing but the world including ourselves who are describing it. An equilibrium concept in social science, therefore, is associated with certain kinds of *views on ourselves*, our *worldview* (*Weltanschauung*), implicitly or explicitly. We can find here one of the most profound sorts of fixed-point arguments (our views on ourselves), so the topics in these two chapters will be appropriate as concluding discussions.

This chapter is devoted to the study of set-theoretical limitations on a description of society to incorporate the human ability of recognition (the ability to recognize the society to which one belongs) with rational behaviors based on a consistent view of the world (our consistent worldview). This section (9.1) provides some basics of axiomatic set theory and formal logic. Sections 9.2 and 9.3 treat the set-theoretic problems on individual rationality, social validity, and consistent worldviews.

Section 9.2, "Individuals and Rationality," is a formal treatment that describes society as consisting of rational individuals. The concept of rationality, at least in the sense of the rational acceptability of sentences based on our certain worldview, cannot set-theoretically (especially finitistically) be described introspectively as long as we require it to be logically consistent. The result may be compared with Tarski's truth definition theorem, a closely related result to Gödel's Second Incompleteness Theorem. The arguments could serve as a launching pad for rigorous treatments of the problematic cognitive features of all mathematical models in social sciences based on *methodological individualism*.

Section 9.3, "Society and Values," deals with the same problems of describing human society from a macroscopic viewpoint. (We treat society as a whole and do not base our arguments on rational individuals.) A description of society, in the sense of a collection of sentences valid for descriptions of society in a certain formal language, cannot be introspectively consistent as long as we require it to be logically consistent. It follows that for logical consistency, we cannot define society in an introspectively complete manner. In other words, we cannot assure the rightness of the validity of the descriptions of society (not to speak of optimality, efficiency, etc.), except for dogmatically believing it.[1] The results may be considered a mathematical critique of *empiricism* (*logical positivism*) as a methodology of social sciences. In 1951 in his famous essay, "Two dogmas of empiricism" (reprinted in Quine 1953), W.V.O. Quine pointed out the same problem as a difficulty of defining analyticity together with dogmatic sense-data reductionism.[2]

We may relate the concepts in Sections 9.2 and 9.3 to our ordinary way of "justification" (an *internal* (*extensional*) condition for true rationality — consistency in the view of the world) and "refutability" (an *external* (*intentional*) condition for true rationality — inconsistency in the view of the world), respectively. Chapter 10 incorporates these two features into a social equilibrium argument, so that 'rationality,' as a non-refutable justification, is assured to exist as a fixed point (equilibrium) of the recognition of the world for each member of society.

9.1.1 *Formal theories, models, and theory of sets*

As stated in Chapters 1 and 6 (Sections 1.3 and 6.1), a mathematical theory is constructed by a formal language that describes (*mathematical*) *objects* as its *terms* and (*mathematical*) *relations* as its *formulas*. We are now in the position to describe the formal language, terms, and formulas.

---

[1] The problem is based on the difficulty of defining "judgments for facts" since it should also determine our value judgments concerning what our "facts" are. According to H. Putnam, the targets of such value judgments are called *epistemic values* (Putnam 2002, p. 30) whose existence expounds the collapse of the classical fact/value dichotomy. Note that mathematical models in social science always implicitly or explicitly presuppose one such value.

[2] In our case, we cannot obtain validity for descriptions of society and have to presuppose one of the epistemic values (see the previous footnote).

9: Mathematics and Social Science

All of our mathematical arguments in this book are founded on Zermelo–Fraenkel (ZF) axiomatic set theory. We need a formal language (first-order predicate logic) (1) to describe the axioms in ZF and the total system in a rigorous way, and (2) to describe the total system in a simple manner so that we can treat themselves as set theoretical (or finitistic, if possible) objects. Here, the total system means the totality of *terms* (objects in the theory), *formulas* (predicates describing relations or rigorous properties in the theory), and the *rules of inference* (rules describing how we construct relations called *theorems* based on relations called *axioms* in the theory).

We characterize a *theory in formal language* $\mathscr{L} = (L, R, T)$ by specifying *list of symbols* L, *list of syntactical rules* R, and *list of axioms* T.

List of symbols

The list of symbols is prepared as a class of names in a formal language for five types of tools: *logical signs, predicates, functions, (individual) variables,* and *(individual) constants*. For example (perhaps, as the smallest one), we can specify list of symbols L as

$$L = \{\neg, \vee, \exists, \in, =, x_0, x_1, x_2, x_3, \ldots\},$$

where $\neg$, $\vee$, and $\exists$ are logical signs intuitively meaning "not," "or," and "exist," respectively. For a language used to describe mathematical theory, it is standard to use symbols $=$ (equality) and $\in$ (elementhood) as two of the most basic predicates.[3] Infinitely (here, countably) many symbols, $x_0, x_1, \ldots$, are prepared for different individual variables.

List of syntactical rules

The list of syntactical rules, R, can be divided into the following three categories:

(R1) Rules for Construction of Terms
(R2) Rules for Construction of Formulas
(R3) Inference Rules.

---

[3]Their meanings (including the meanings of logical signs) in a rigorous sense are given only after we define a certain "believable" (credible, plausible, reliable, or trustworthy) domain of discourse and what holds — (true or false) — for each statement using these symbols on such a domain (a *model*). The *semantics* of $\mathscr{L}$, in this sense, always depend on a certain belief in ourselves. We may classify such a belief as a sort of *epistemic value* discussed in the next chapter.

These rules can be written as procedures to generate terms from the primitive terms, formulas from the atomic ones, and theorems from the axioms.

If the theorem has no individual constants or functions, the terms are nothing but the individual variables.[4] Generally, (R1-1) the individual constants and variables are terms, (R1-2) if f is a function with n variables and $t_1, \ldots, t_n$ are terms, then $f(t_1, \ldots, t_n)$ is also a term, and (R1-3) only those that we can obtain through the previous two rules are the terms.

To define the formulas in a formal language, we can use the same processes that are used for the terms. That is, (R2-1) for predicate P with n variables and terms $t_1, \ldots, t_n$, we call $P(t_1, \ldots, t_n)$ an *atomic formula*; (R2-2) if P and Q are formulas and x is a variable, then $\neg P$, $P \vee Q$, and $\exists x P$ are formulas; and (R2-3) only those that we can obtain through the above two rules are the formulas. In our special (minimal) example, there are only two types of atomic formulas, $\in (x, y)$ and $= (x, y)$, where x and y are individual variables.[5] As well as constants and functions, we can also introduce new logical signs and predicate symbols (merely as abbreviations for longer expressions) and use them to construct formulas. Especially, we use

$$P \wedge Q \text{ instead of } \neg((\neg P) \vee (\neg Q)),$$

$$P \rightarrow Q \text{ instead of } (\neg P) \vee Q,$$

$$P \leftrightarrow Q \text{ instead of } (P \rightarrow Q) \wedge (Q \rightarrow P),$$

$$\forall x P \text{ instead of } \neg(\exists x (\neg P)),$$

where P and Q are formulas and x is a variable.

Since each formula is constructed through the finite process described above (R2-1,2,3), we can always check whether variable x, which appears in a certain place of formula φ, might be seen in a certain formula with form $\exists x P$ constructing a part of φ. In such a case, the *occurrence* of x in formula φ is called *bounded*. If the occurrence of x in the formula is not bounded, the

---

[4] Even in such a theory, however, ordinarily we actually define constants and functions like natural numbers $0, 1, \ldots$, empty set $\emptyset$, addition, $+$, and use them as if they were individual constants and functions prepared from the beginning in the language. We shall identify such a new theory with the original one as long as those additional constants and functions are used merely to abbreviate long expressions.

[5] We usually write $x \in y$ instead of $\in (x, y)$ and $x = y$ instead of $= (x, y)$.

occurrence of x in the formula is said to be *free*. x is a *bounded variable* of φ if x has a bounded occurrence in φ, and x is a *free variable* of φ if x has a free occurrence in φ. A formula that has no free variable is a *closed formula*. In the same way, a term which includes no variable is called a *closed term*. A closed formula is also called a *sentence*.

For the rules of inference, we use the following two rules:

(R3-1: Modus Ponens) From two formulas, φ and $\varphi \to \psi$, we infer ψ.

(R3-2: Instantiation) From formula $\varphi \to \psi$, we infer $(\exists x \varphi) \to \psi$, where x is not a free variable of ψ.

There seems no need to refer to the first rule, modus ponens. For the second rule, note that the condition, "x is not a free variable of ψ," is essential. From $\varphi \to \psi$, it is natural to infer the contraposition, $(\neg \psi) \to (\neg \varphi)$. Since x is not a free variable of ψ, it is also natural to infer $(\neg \psi) \to (\forall x(\neg \varphi))$. Then by taking the contraposition again, we obtain formula $(\exists x \varphi) \to \psi$. Instantiation (R3-2) can also be explained as follows: We can infer from $\varphi \to \psi((\neg \varphi) \vee \psi)$ that assertion ψ (having no relation to variable x) must hold when assertion φ $(\neg(\neg \varphi))$ holds for an arbitrarily fixed value of x in φ. The instantiation rule says that in such a case, we can also infer from $\varphi \to \psi$ that ψ (saying nothing about variable x) holds even when we can assert φ for a special value of x.

List of axioms

Axioms may be classified into the following five categories: (T1) axioms for propositional logic,[6] (T2) axioms for quantifiers, (T3) axioms for equality, (T4) axioms for set theory, and (T5) others.[7] At least for (T1)–(T3), we merely fix our arguments on those axioms presented below because they form the standard *first-order predicate calculus (with equality)* with symbols in L and syntactical rules in R. For propositional logic, we use the following three types of formulas (T1-1,T1-2,T1-3) as axioms. (In (T1-1),(T1-2), and (T1-3), one may obtain explicit axioms by replacing φ, ψ, and θ with

---

[6] Precisely, as stated below, we do not directly use the system of propositional logic. Therefore, it is an abuse of language to call (T1) axioms of propositional logic. They are indeed axioms for a part of the predicate logic that may be identified with a system of propositional logic.

[7] For (T4)-type axioms, I will specifically use notation (S), so I write (S1), (S2),..., instead of (T4-1), (T4-2),..., (T5) axioms may not be specified at least for the arguments in this book.

arbitrary formulas. Each (T1-1),(T1-2), and (T1-3), therefore, is not an axiom in a precise sense but a representation of a form of axioms: an *axiom schema*.)

(T1-1) $\varphi \to (\psi \to \varphi)$
(T1-2) $(\varphi \to (\psi \to \theta)) \to ((\varphi \to \psi) \to (\varphi \to \theta))$
(T1-3) $((\neg\varphi) \to (\neg\psi)) \to (((\neg\varphi) \to \psi) \to \varphi)$

The first axiom (schema) may be rewritten as $(\neg\varphi) \vee ((\neg\psi) \vee \varphi)$, which clearly seems correct. The second and the third schemas will be easier than the first one to understand if "$\to$" is interpreted as "imply" as usual.

Under (T1)-type axiom schemas with unique inference rule, modus ponens, if we regard φ, ψ, θ in (T1) as variables for propositions, we obtain the syntactical part of the system of *propositional logic*. Moreover, by considering that we merely treat the relation among propositions with respect to values "true" or "false" that are given from outside simultaneously with the meanings (*truth tables*) of "$\neg$," "$\vee$," "$\to$," etc., we may also identify the semantical part of the propositional logic with the discussion on values ("true" or "false") of formulas that do not depend on a particular domain of discourse.[8] Indeed, under such rules (truth tables defining the meaning of each logical sign), one can check that propositions having forms in (T1-1), (T1-2), and (T1-3) are all "true" regardless what propositions φ, ψ, and θ are. In other words, all propositions of the (T1)-type are *tautologies*. One may also confirm that all consequences deduced from modus ponens on tautological formulas are also tautologies. This means that under (T1)-type axioms with modus ponens, there is no possibility to obtain two propositions P and $\neg P$ as consequences. (Indeed, if P is "true," then $\neg P$ must be "false," and vice versa.) Thus, we have the *consistency* of propositional logic. Moreover, it is also possible to show that every tautological formula (constructed merely by those processes related to $\neg$ and $\vee$ in (R2) for atomic propositional formulas) can be obtained

---

[8]In this book, we use only predicate logic, and do not directly use the system of propositional logic. Part of predicate logic, however, may be identified with the propositional logic, so we may utilize some results from the theory of propositional logic to develop theorems in predicate logic.

through (T1)-type propositions and modus ponens (*completeness* of the propositional logic).[9]

The axiom (schema) for the quantifier can be written as follows:

(T2-1) $\varphi_x[t] \to (\exists x \varphi)$,

where $\varphi_x[t]$ denotes the formula obtained by replacing every free occurence of x in φ with term t under the assumption that no variable in t is a bounded variable of φ. Usually, the system of axiom schemas (T1-1),(T1-2),(T1-3), and (T2-1) is called axioms for predicate logic. In this book, we also use the following axiom schemas for equality as basic axioms for predicate logic:

(T3-1) $\forall x(x = x)$.

(T3-2) For function f with n-variables, $\forall y_1 \cdots \forall y_n \forall z_1 \cdots \forall z_n(((y_1 = z_1) \wedge \cdots \wedge (y_n = z_n)) \to (f(y_1, \ldots, y_n) = f(z_1, \ldots, z_n)))$.

(T3-3) For predicate P with n-variables, $\forall y_1 \cdots \forall y_n \forall z_1 \cdots \forall z_n(((y_1 = z_1) \wedge \cdots \wedge (y_n = z_n)) \to (P(y_1, \ldots, y_n) \leftrightarrow P(z_1, \ldots, z_n)))$.

In our minimal example, (T3-2) is not necessary (since we have no function) and there is only one predicate, $\in$, that should satisfy (T3-3).

The totality of the axioms of (T1), (T2), and (T3) types, with formation rules (R1) and (R2), and the two formal deduction rules, modus ponens and instantiation, in (R3), is called the system of *predicate logic with equality*. We need more axioms to develop an ordinary mathematical theory. Axioms for theory of sets (T4) and other axioms (T5) will be discussed below. Before that, we discuss the concept of a *model*, which is an interpretation of a formalized theory.

Model

Let L be a list of symbols including $\{\neg, \vee, \exists, \in, =, x_0, x_1, x_2, x_3, \ldots\}$ and R be a corresponding formation rules of terms (R1) and formulas (R2) with

---

[9]To see this, for proposition P with atomic propositional formulas $\varphi_1, \ldots, \varphi_n$ show that (1) if $v_1, \ldots, v_n$ is a sequence of "true" or "false" values on $\varphi_1, \ldots, \varphi_n$ under which P is "true," then by (T1), $\hat{\varphi}_1, \ldots, \hat{\varphi}_n$, and modus ponens, we can deduce P, where $\hat{\varphi}_t$ is φ_t if v_t is "true" and $\hat{\varphi}_t$ is $\neg \varphi_t$ if v_t is "false" for each $t = 1, \ldots, n$; and (2) for a list of axioms T including all of the (T1)-type, if we can deduce P from T and φ with modus ponens, and if we can deduce P from T and $\neg \varphi$ with modus ponens, then we can deduce P from T and modus ponens. For (1), use mathematical induction on the formulas in propositional logic constructed merely by those processes related to $\neg$ and $\vee$ in (R2). If P is a tautology, an arbitrary "true" and "false" evaluation on $\varphi_1, \ldots, \varphi_n$ makes P "true," so by using (1) and (2) repeatedly, we can deduce P from T and modus ponens.

the two inference rules in (R3). Given list of formulas T including all axioms in (T1), (T2), and (T3), if we can deduce formula φ merely by applying the two formal deduction processes in (R3), modus ponens and instantiation, on formulas in T, we can write $T \vdash \varphi$ and say that T proves φ. When we call the totality of T in the formal language a *formal theory*, we can also say that φ is a *theorem in the formal theory*, (L, R, T). Since the formal deduction process is a finite process, it is obvious that if $T \vdash \varphi$, there is a list of finite formulas T' such that $T' \vdash \varphi$.[10]

A list of axioms T, in a formal language is said to be *consistent* if no φ exists such that $T \vdash \varphi$ and $T \vdash \neg\varphi$. For the consistency of a formal theory, (L, R, T), there is a well-known result called the *Gödel Completeness Theorem*. To explain it, the concept of a *model* is necessary. Formal theory (L, R, T) is said to have a *model* (or an *interpretation* of the theory) if (1) we can consider a domain of discourse in which our variables and constants may take values, on which our functions are defined, and whose predicates describe a certain relation among elements; (2) based on such an interpretation of variables, functions, and predicates, we can define a *value* as "**True**" or "**False**" of all formulas (based on the finite procedure to construct formulas); and (3) under such valuations of formulas, all axioms in T are **True**. The Gödel Completeness Theorem argues that formal theory (L, R, T) is consistent if and only if it has a model.[11] A formula in (L, R, T) which is evaluated as **True** in all models is called a *universally valid formula*. The Gödel Completeness Theorem guarantees

---

[10] This fact with the following Gödel Completeness Theorem guarantees that for a list of axioms T, (L, R, T) has a *model* if and only if for every finite subset T' of T, (L, R, T') has a *model* (*Compactness Theorem*).

[11] The sufficiency (for consistency) part is easy to check since a model is so constructed that if φ is proved then the evaluation of φ is **True**, and if the evaluation of φ is **True**, then the evaluation of $\neg\varphi$ is **False**. The necessity part is shown by directly constructing a (countable) model for a list of consistent axioms. Let us consider countably many new symbols, $a_0, a_1, \ldots$ to L and treat them as constants in new language $\mathscr{L}' = (L', R, T)$. Since the number of formulas in $\mathscr{L}$ is countable, we have a denumerable list of the closed formulas with shape $\exists x \varphi$ as $\exists x_{i(1)} \varphi^1, \exists x_{i(2)} \varphi^2, \ldots$, and we can define a sequence of increasing list of formulas, $T_0, T_1, \ldots$, as $T_0 = T$, $T_n = T_{n-1} \cup \{\exists x_{i(n)} \varphi^n \to \varphi^n x_{i(n)}[b_n]\}$, where b_n is the first symbol in $a_0, a_1, \ldots$, that does not appear in both T_{n-1} and φ^n. Let T_∞ be the union of such lists. By using countable list $\psi^1, \psi^2, \ldots$ of all closed formulas in $\mathscr{L}'$, we can further extend T_∞ as $\Delta_0 = T_\infty$, $\Delta_n = \Delta_{n-1} \cup \{\psi^n\}$ if $\Delta_{n-1} \vdash \psi^n$ and $\Delta_n = \Delta_{n-1} \cup \{\neg\psi^n\}$ if otherwise. We can verify that all Δ_n and the union Δ of such lists are consistent. Regarding the domain of discourse as the totality of closed terms in $\mathscr{L}'$ and deciding **True** and **False** for each formula through Δ by considering all statements in Δ to be **True**, we obtain a model of $\mathscr{L}$. (See, e.g., Shoenfield (1967, Section 4.2).)

that if φ is a universally valid formula in a theory (L, R, T), then $T \vdash \varphi$ (Completeness). Indeed, if T does not prove φ, then list "$T, \neg\varphi^c$" (adding $\neg\varphi^c$ to T, where φ^c denotes the *logical closure* of φ, i.e., if $x_0, \ldots, x_n$ are the free variables appearing in φ, then φ^c is $\forall x_0 \cdots \forall x_n \varphi$) is consistent.[12] Therefore, $(L, R, (T, \neg\varphi^c))$ has a model, $\mathfrak{M}$. This means, however, $\neg\varphi^c$ and $\neg\varphi$ are **True** in $\mathfrak{M}$. At the same time, φ is also **True** by the assumption that φ is a universally valid formula. Hence, in $\mathfrak{M}$, both φ and $\neg\varphi$ are evaluated as **True**, a contradiction.

Thus, as in the case with propositional logic, the system of predicate logic (with equality), ((T1), (T2), (T3), and the two inference rules, modus ponens and instantiation, in (R3)), is consistent and complete as long as we believe in the argument like the model-construction process used in Footnote 11. Note that to construct a model to show the necessity part of the Gödel Completeness Theorem, we used somewhat complicated (non-finitistic) mathematical methods and set theoretical concepts. Since we use such an argument (*metatheory*) to discuss things that are *really* true, i.e., to confirm the soundness and to assure the rightness of the formal theory, etc., it is desirable and important to know how we can restrict such methods (for metatheory) within the finite, finitistic (including recursive), ZF, or ZFC reasoning. Fortunately, in the rest of this book, the necessity (for consistency) part of the Gödel Completeness Theorem is not necessary, and we may restrict our metatheory within finitistic reasoning.[13]

Axioms for set theory

As axioms for set theory, we use here the axiomatic system of the Zermelo–Fraenkel type with the axiom of choice (ZFC). At the same time, the treatment of axioms for further developments in the theory of mathematics is also discussed. Since we treat (at least in this book) the axioms and inference rules in predicate logic with equality (axioms in (T1), (T2), and

---

[12] If not, we have $T, \neg\varphi^c \vdash \varphi \wedge \neg\varphi$. This means $T \vdash \neg\varphi^c \to (\varphi \wedge \neg\varphi)$ (Deduction Theorem). On the other hand, $(\neg\varphi^c \to (\varphi \wedge \neg\varphi)) \to \varphi^c$ is obtained through substitution for tautology $(A \to (B \wedge \neg B)) \to \neg A$. (Every tautology is known to be only obtained through axioms of the (T1)-type and modus ponens.) Hence, by modus ponens, $T \vdash \varphi^c$, so we have $T \vdash \varphi$, a contradiction.
[13] Note also that by the model-construction process used in Footnote 11, if (L, R, T) is consistent, we have a model whose domain of discourse is countable. This fact (based on non-finitistic arguments or methods) is known as the *Löwenheim–Skolem Theorem*.

(T3) with modus ponens and instantiation) as the most basic ones, it is sometimes useful to omit referring to them if there is no fear of confusion. For example, if theorem φ is deduced from the list of predicate-logic axioms T with ψ, in the following, we may write $\psi \vdash \varphi$ instead of $T, \psi \vdash \varphi$. With respect to axioms for set theory and further developments, however, we do not fail to describe the axioms on which our arguments depend.

In Zermelo–Fraenkel set theory (ZF), all the mathematical objects we treat in our formal theory are sets. Even though we sometimes use concepts like "class" that may not be treated as sets, such arguments can be recaptured by translating them into arguments on rigorously described "properties" (through formulas in formal language) that cannot be treated as conditions that characterize sets, as explained in Chapter 6, Section 6.1. The following fact, therefore, means the existence of at least one set in our theory of sets:

$$\vdash \exists x(x = x). \tag{9.1}$$

This is a theorem in predicate logic with equality.[14]

The first axiom listed here is the *axiom of extensionality*, which says that for two sets to be equal, it is sufficient that they have identical elements:

(S1) *Extensionality*
$\forall x \forall y ((\forall z (z \in x \leftrightarrow z \in y)) \to (x = y))$.

Note that by (T3-3), the necessity part automatically holds; i.e., if two sets are equal, they must have identical elements. We write $x \subset y$ (to read x is a *subset* of y) instead of $\forall z((z \in x) \to (z \in y))$. Then $(x \subset y) \wedge (y \subset x)$ implies $x = y$ by (S1).

The second axiom is the *axiom of pairing*. This axiom asserts that for sets x and y, set z exists whose elements are exactly x and y:

(S2) *Pairing*
$\forall x \forall y \exists z \forall w ((w \in z) \leftrightarrow ((w = x) \vee (w = y)))$.

When we admit extensionality (S1), such a set must be unique, so we may denote it by $\{x, y\}$. Since case $x = y$ is not excluded, it is possible that

---

[14]Indeed, by (T3-1), we have $\vdash \forall x(x = x)$. By (T2-1) and tautology $(\psi \to \theta) \to (\neg \theta \to \neg \psi)$, we have $\vdash \neg \exists x \neg \varphi \to \neg \neg \varphi_x[t]$, i.e., $\vdash \forall x \varphi \to \neg \neg \varphi_x[t]$. Hence, we have $\neg \neg (x = x)_x[t]$, which means $\vdash (x = x)_x[t]$ since $\neg \neg \varphi \to \varphi$ is a tautology. By taking x as t, we have $\vdash x = x$, which with (T2-1) gives the result.

$\{x, y\} = \{x, x\}$. Then we may write $\{x\}$ instead of $\{x, x\}$. For two sets x and y, their *ordered pair* (x, y) is defined as $(x, y) = \{\{x\}, \{x, y\}\}$. We can immediately verify that (x, y) and (z, w) are equal if and only if $x = z$ and $y = w$.

In the ordinary description of sets, the next *axiom (schema) of separation* is quite important and useful. Given set y and formula φ describing a certain property of its unique free variable, the collection of all elements of y that satisfy property φ is a set:

(S3) *Separation Schema*
Let φ be a formula with one free variable x. Then
$$\forall x \forall y \exists z ((x \in z) \leftrightarrow ((x \in y) \land \varphi)).$$

Since there are countably many formulas, this schema is indeed a countable list of axioms. Based on the fact (Equation (9.1)) that at least one set $(\exists y(y = y))$ exists, the separation axiom schema guarantees the existence of a set that has no elements (e.g., let φ be $\neg(x = x)$). Under the extensionality axiom, we may suppose such a set is unique and denote it by $\emptyset$.[15] This axiom also maintains the existence of the intersection for non-empty family of sets v as $\bigcap v = \{u|\, \forall w((w \in v) \rightarrow (u \in w))\}$ and the set theoretic difference between two sets as $y \backslash x = \{u|\, (u \in y) \land (u \notin x)\}$.

The next two axioms provide two important concepts in set theory: *union* and *power sets*, respectively:

(S4) *Union*
$$\forall x \exists y \forall z ((z \in y) \leftrightarrow \exists w((z \in w) \land (w \in x))).$$

(S5) *Power Set*
$$\forall x \exists y (\forall z ((z \in y) \leftrightarrow (z \subset x))).$$

Axiom (S4) asserts the existence of set y, whose elements are all the elements of the elements of x, and (S5) says that there is set y whose elements are the subsets of x. As before, such a set can be verified as unique by the extensionality axiom (S1). Let us denote unique y in (S4) by $\bigcup x$

---

[15]The axiom of extensionality (S1) with separation schema (S3) gives a proof of the existence of the empty set. Is there any possibility for other sets to be proved to exist under these two axioms? The answer is negative. Let us consider the model whose domain consists merely of the empty set. In this model, (S1), (S3), and "$\forall x(x = \emptyset)$" are all true. That is, the list (S1), (S3), and "$\forall x(x = \emptyset)$" is consistent because it has a model. It follows that we cannot prove "$\exists x(x \neq \emptyset)$" only by (S1) and (S3).

and call it the union of family x. Unique set y in (S5) is called the power set of x and denoted by $\mathscr{P}(x)$. (We have already defined these notations in Chapter 1, Section 1.3.1.)

The 6th axiom is the *axiom of infinity*, which asserts that there is at least one infinite set:

(S6) *Infinity*
$$\exists y((\emptyset \in y) \wedge ((x \in y) \rightarrow (\bigcup\{x,\{x\}\} \in y))).$$

In the above, notations are used that are only defined after we admit axioms (S1), (S2), (S3), and (S4). They are not essential, since we can replace $(\emptyset \in y)$ with $(\exists z_0(\forall w_0(w_0 \notin z_0)) \wedge (z_0 \in y))$ and $(\bigcup\{x,\{x\}\} \in y)$ with $(\exists z_1(\forall w_1((w_1 \in z_1) \leftrightarrow ((w_1 \in x) \vee (w_1 = x)))) \wedge (z_1 \in y))$, where z_0, z_1, w_0, w_1 are variables that do not appear in (S6). If we denote $\emptyset$ by 0, $\bigcup\{0,\{0\}\} = \{0\}$ by 1, $\bigcup\{1,\{1\}\} = \{0,1\}$ by 2, $\bigcup\{2,\{2\}\} = \{0,1,2\}$ by 3, $\bigcup\{3,\{3\}\} = \{0,1,2,3\}$ by $4,\ldots,$ [16] then axiom (S6) maintains the existence of a set that has all natural numbers $0,1,2,3,4,\ldots$ as its elements. Given set y in (S6), by considering the intersection among sets in family $\{z|\, z \subset y, (\emptyset \in z) \wedge ((x \in z) \rightarrow (\bigcup\{x,\{x\}\} \in z))\} \subset \mathscr{P}(y)$ (under S3 and S5), we obtain set of natural numbers, $N = \{0,1,2,3,\ldots\}$, which is unique under extensionality (S1).

Although the axiom of infinity (S6) is indispensable for developing almost all mathematical arguments today, note that ZF without the axiom of infinity, ZF $-$ INF, has the following model:

(I) Let $R(0)$ be the empty set.
(II) Let $R(n+1)$ be the power set of $R(n)$ for each $n = 0,1,2,\ldots$.
(III) Define $R(\omega)$ as $R(\omega) = \bigcup_{n=0}^{\infty} R(n)$.

Here we use set-theoretic concepts from an informal (realistic) standpoint. They are constructed only based on finitistic objects and methods except for the last one, $R(\omega)$.[17] When we allow $R(\omega)$ as the domain of discourse, under

---

[16] The validity to denote a set consisting of finite elements by listing its elements as $\{0,1,2\}$ and $\{0,1,2,3\}$ is given by (S1) in exactly the same reasoning for doubleton $\{x,y\}$ and singleton $\{x\}$.

[17] For $R(\omega)$ to be a domain of discourse, however, it is not necessary for us to treat $R(\omega)$ as a set (object) in the metatheory. That is, we may treat $R(\omega)$ as a certain clear assertion on x in the metatheory; e.g., we say $x \in R(\omega)$ merely in the sense that $x \in R(n)$ for some number n in the metatheory, so it may be identified with a statement that just depends on finitistic objects.

9: Mathematics and Social Science

the natural interpretation of the elementhood, we have indeed a model of ZF − INF.[18]

Let us consider formula φ that has two free variables, x and y, and variables z and w that do not appear in φ. Then we denote by $\varphi(z,w)$ the formula obtained by replacing x with z and y with w. We say (as a condition on formulas) that formula φ with free variables x and y determines a *functional relation* if $(\varphi(z_1, y_1) \wedge \varphi(z_1, y_2)) \to (y_1 = y_2)$, where three (mutually different) variables z_1, y_1, y_2 do not appear in φ. Clearly the number of such formulas is countable, and we have the next axiom (schema) of *replacement* that means that if φ is a formula with free variables x and y defining a functional relation, then for each set u, there is set v such that for all $x \in u$, every y satisfying φ with x belongs to v:

(S7) *Replacement Schema*
Let φ be a formula with two free variables, x and y. Then
$$((\varphi(z_1, y_1) \wedge \varphi(z_1, y_2)) \to (y_1 = y_2))$$
$$\to \forall u \exists v \forall x \forall y (((x \in u) \wedge \varphi) \to (y \in v)).$$

Here, variables z_1, y_1, y_2 are so taken that they do not appear in φ, as noted above. By the extensionality axiom (S1) and the separation axiom (S3), we can assume set v as exactly the set of all values of y that satisfy φ with a certain value of x in u, say, the *range* of the functional relation on *domain* u. An important role of the replacement schema is to guarantee the existence of *Cartesian product* between two sets. Given two sets u and v, for each element x of u, we may obtain (by (S7) with (S1) and (S2)) the set of all ordered pairs, $v_x = \{(x, y) | y \in v\}$, as the range of the functional relation, "z is an ordered pair of x and y," with free variables y and z and domain v for y. Take w as the set (again by (S7) and (S1)) $\{v_x | x \in u\}$, the Cartesian product, $u \times v$, is obtained as the union (under (S4)) $\bigcup w$.

As a subset (by (S3)) of Cartesian product (by (S1)(S2)(S4)(S7)) $u \times v$ a *function* f on u to v may be defined as $f = \{(z_0, y_0) | ((z_0, y_0) \in u \times v) \wedge$

---

[18]The validity of (S1) is trivial. To show the validity of other axioms, the next property of $R(\omega)$ plays an essential role. Class M is *transitive* if for all $x \in M$ we have $x \subset M$. $R(\omega)$ is transitive since an element of $R(n+1)$ is a subset of $R(n)$, so it is a subset of $R(\omega)$ for each $n = 0, 1, 2, \ldots$. Moreover, since $R(0) \subset R(1)$ holds, we can see that each element $x \in R(1)$ (a subset of $R(0)$) can also be identified with a subset of $R(1)$, so it is an element of $R(2)$; hence we have $R(1) \subset R(2)$. It follows that if we use mathematical induction, $R(n) \subset R(n+1)$ for all $n = 0, 1, 2, \ldots$. With the fact that each element of $R(\omega)$ is a finite set, one can easily verify that (S2), (S3), (S4), and (S5) hold. It is also easy to verify for axioms (S7) and (S8) introduced below (see p. 214, Footnote 20).

$\varphi(z_0, y_0)\}$, where φ with two free variable determines a functional relation and variables z_0 and y_0 do not appear in φ. Function f on u to v is said to be *one to one* if for all z_0 and z_1 in u, $(z_0 \neq z_1) \wedge ((z_0, y_0) \in f) \wedge ((z_1, y_1) \in f)$ means $(y_0 \neq y_1)$. f is said to be *onto* if for all y_0 in v, z_0 in u exists such that $(z_0, y_0) \in f$. With respect to the power set whose existence is maintained in (S5), it is known that there is no one to one and onto f on u to $\mathscr{P}(u)$.[19] To represent the fact that one to one and onto function f on set x to y exists, we say that the *cardinality* of set x is not greater than y. The cardinality of set $\mathscr{P}(u)$, therefore, is always greater than u.

The last axiom of ZF is the axiom of *foundation* that asserts that for each set, the existence of an element that is minimal in the sense of $\in$. More precisely:

(S8) *Foundation*
$$\forall x \exists y ((x \neq \emptyset) \to ((y \in x) \wedge (x \cap y = \emptyset))).$$

Here, $x \cap y$ denotes the *intersection* of sets x and y whose existence and uniqueness are guaranteed by (S3) and (S1). If (S8) is not satisfied, there is set x such that all elements of x have an element that may also be identified with an element of x.[20]

We add the next axiom of *choice* as an axiom (S9) of set theory. We have already seen it informally in Chapter 1, Section 1.3.1 (p. 15):

(S9) *Choice*
$$\forall x \forall y \exists w \forall z_1 \forall z_2 \forall z_3 (((x \neq \emptyset) \wedge ((y \in x) \to (y \neq \emptyset))) \to ((w \subset x \times \bigcup x) \wedge ((((z_1, z_2) \in w) \wedge ((z_1, z_3) \in w)) \to ((z_2 = z_3) \wedge (z_2 \in z_1))))).$$

---

[19] If $f : u \to \mathscr{P}(u)$ is one to one and onto, then for all $y \in \mathscr{P}(u)$, there exists $z \in u$ such that $(z, y) \in f$. Consider set $y_1 = \{z | (z \in u) \wedge (((z, y) \in f) \to (z \notin y))\}$. This set is a subset of u, so it is a member of $\mathscr{P}(u)$, and member z_1 in u exists such that $(z_1, y_1) \in f$. Then we have $(z_1 \in y_1) \leftrightarrow (z_1 \notin y_1)$, a contradiction.

[20] Now, we can confirm that $R(\omega)$ is a model of (S7) and (S8). With respect to (S7), since $R(0) \subset R(1) \subset \cdots \subset R(n)$ for every n, we may assume for every $u \in R(\omega)$ and function f on u, the range v of f (a finite set) is a subset of $R(n)$ for sufficiently large n. Hence, v is an element of $R(n+1)$. To see (S8), let x be an element of $R(\omega)$ such that for each element z of x, there is element y that may also be identified with an element of x. Since x is a finite set, we have a sequence of elements of x satisfying $z \ni y \ni y_1 \ni \cdots \ni y_k \ni z \ni y \ni \cdots$. For z in x, n exists such that $z \notin R(n-1)$ and $z \in R(n)$, the first n such that $z \in R(n)$. Then $y \in z$ is an element of $R(n-1)$ and the first m such that $y \in R(m)$ is less than n. By repeating this argument, the first m_k such that $y_k \in R(m_k)$ is verified to be less than n. Then $z \in y_k$ is an element of $R(m_k - 1)$, which contradicts the fact that n is the first number such that $z \in R(n)$.

9: Mathematics and Social Science

Of course, the axiom asserts the existence of *choice function* w for any non-empty class x of non-empty sets. The axiomatic system (S1)–(S8) is called ZF, and the total system (S1)–(S9) is named ZFC (the axiomatic system of ZF with axiom of choice). Since every $x \in R(\omega)$ is finite, the choice axiom is obviously true in $R(\omega)$. Hence, $R(\omega)$ is also considered a model of ZFC − INF.

The axiom of choice has many important (equivalent) theorems. One of the most famous assertions is the next theorem about well-ordering. As in Chapter 1, Section 1.3.1, let us call a subset (under (S3)) of Cartesian product (under (S1) (S2) (S4) (S7)) $x \times y$ as a (strict mathematical sense of) *relation* on x to y. Relation $\preceq$ on x to x is a *preordering* on x if $\forall z(z \preceq z)$ (reflexivity) and $\forall z_0 \forall z_1 \forall z_2(((z_0 \preceq z_1) \land (z_1 \preceq z_2)) \to (z_0 \preceq z_2))$ (transitivity) are satisfied. A preordering on x is an *ordering* on x if we have $\forall z \forall y(((z \preceq y) \land (y \preceq z)) \to (z = y))$. Ordering $\preceq$ on x is said to be a *total ordering* on x if $\forall z \forall y(((z \in x) \land (y \in x)) \to ((z \preceq y) \lor (y \preceq z)))$ holds. Relation $\preceq$ on x to x is a *well-ordering* on x if $\preceq$ is a total ordering on x and for every non-empty subset w of x, an element y of w exists such that $\forall z((z \in w) \to (y \preceq z))$, where y is called the $\preceq$-least element of w. If we define $0, 1, 2, 3, \ldots$ as before (under (S1) (S2) (S3) (S4)), we can verify on each $0 = \emptyset$, $1 = \bigcup\{0, \{0\}\} = \{0\}$, $2 = \bigcup\{1, \{1\}\} = \{0, 1\}, \ldots$ that relation $\in$ may be identified with a well-ordering.[21] Moreover, under (S5) and (S6), on the set of all natural numbers $N = \{0, 1, 2, \ldots\}$, $\in$ is also a well-ordering.

THEOREM 9.1.1: (Well-Ordering Theorem: Zermelo) *In ZFC, for every set x, there is a well-ordering $\preceq$ on x.*

PROOF: If x is a finite set, since $n \in N$ is well-ordered by $\in$, we can obtain well-ordering $\preceq$ on x by defining it as $\in$ on n through the one to one function between x and n. If x is infinite, let $\mathcal{C}$ be the class of all non-empty subsets of x (under (S5) and (S3)) and f be a choice function on $\mathcal{C}$ (under (S9)). Denote by $\mathcal{W}$ (a subset of $\mathscr{P}(x \times x)$ under (S3)) the set of relations on x such that $\preceq$ is a member of $\mathcal{W}$ if and only if $\preceq$ is a total ordering on a subset y of x, and for all $z \subset y$, $z \neq y$, if $((v \in z) \land (u \preceq v)) \to (u \in z)$ (z is an initial segment of y under $\preceq$), $f(x \backslash z)$ is an element of y and is the least upper bound of z in y under $\preceq$. Since the empty relation on the

---

[21] Precisely, we must say that $\in$ with = may be identified with a well-ordering. Generally, if $\prec$ is a relation and relation $\preceq$ defined by $\prec$ with equality = (i.e., $x \preceq y$ if and only if $x \prec y$ or $x = y$) gives a well-ordering, we say in the following that $\prec$ is a well-ordering, $\prec$-least element, etc., without referring anything to relation $\preceq$.

empty set and the identity relation on $\{f(x)\}$ belong to $\mathcal{W}$, $\mathcal{W}$ is not an empty set. Union $\bigcup \mathcal{W}$ (under (S4)) gives the well-ordering on x. Note that each member of $\mathcal{W}$ is easily seen to be a well-ordering since for each $\preceq$ on $y \subset x$ in $\mathcal{W}$ and for each non-empty $z \subset y$, intersection $z_0 = \bigcap \{\{v| (v \preceq u) \wedge (v \neq u)\} | u \in z\} \neq y$ is an initial segment of y and $f(x \backslash z_0)$ must be the least upper bound of z_0. Moreover, if $\preceq_1 \neq \emptyset$ on $y_1 \subset x$ and $\preceq_2 \neq \emptyset$ on $y_2 \subset x$ are two elements of $\mathcal{W}$, we must have $y_1 \subset y_2$ or $y_2 \subset y_1$. Indeed, let y_3 be the common initial segment of y_1 and y_2 on which $\preceq_1$ and $\preceq_2$ are identical. Since $\{f(x)\}$ is included in y_3, y_3 is not empty and must equal y_1 or y_2. (If not, $f(x \backslash y_3)$ must belong to both y_1 and y_2; hence it also clearly belongs to y_3, a contradiction.) Now it will be easy to check that union $\preceq^* = \bigcup \mathcal{W}$ itself is a member of $\mathcal{W}$. It remains for us to show that domain $y^* = \{u | \exists v ((u,v) \in \preceq^*)\}$ of $\preceq^*$ equals x. Assume not, and then $f(x \backslash y^*)$ can be added to y^* as the maximal element by extending $\preceq^*$ on $y^* \cup \{f(x \backslash y^*)\}$, and the extended relation may also be identified with an element of $\mathcal{W}$, which contradicts the fact that the domain of $\bigcup \mathcal{W}$ does not have point y^*. ∎

It is obvious that the well-ordering theorem implies the axiom of choice (S9). (Given class x of non-empty sets, by the well-ordering theorem we have well-ordering $\leq$ on $\bigcup x$, and we can define for each $y \in x$, $f(x) \in \bigcup x$ as the $\leq$-least element of y.) Therefore, we can freely replace the well-ordering theorem with the axiom of choice (S9) to obtain an equivalent axiomatic system with ZFC.

Assume (S1)–(S8) and that there is a well-ordering $\leq$ on set x. As in the case with replacement schema (S7), for formula φ with one free variable x, let us denote by $\varphi(y)$ the formula obtained by replacing variable x in φ with term y, where it is assumed that variables in y do not appear in φ except for x. The next theorem is known as the method of *transfinite induction*.

THEOREM 9.1.2: (Transfinite Induction Theorem) *In ZF, let $\leq$ be a well-ordering on x, and denote by $u < v$ relation $u \leq v \wedge u \neq v$. If $\forall u ((u < v) \rightarrow \varphi(u))$ implies $\varphi(v)$, then, we have $\forall u ((u \in x) \rightarrow \varphi(u))$.*

PROOF: If not, then there is at least one u in x such that $\neg \varphi(u)$, so $y = \{u | (u \in x) \wedge \neg \varphi(u)\}$ is a non-empty subset of x. Since x is a well-ordered set, y has the $\leq$-least element v^*. Since every $u < v^*$ satisfies condition $\varphi(u)$, v^* must satisfy $\varphi(v^*)$, so we have a contradiction. ∎

Now we utilize the above two theorems for arguments to assure the existence of maximal elements. Let X be a non-empty set and $\mathcal{W}$ be a subset (under (S3)) of the power set of X (under (S5)). $\mathcal{N} \subset \mathcal{W}$ is a *nest* in $\mathcal{W}$ if for all x and y in $\mathcal{N}$, $x \subset y$ or $y \subset x$ holds.

THEOREM 9.1.3: (Hausdorff Maximal Principle) *Let x be a non-empty set and $\mathcal{W}$ be a non-empty subset of $\mathcal{P}(x)$. Under ZFC, for each nest $\mathcal{N}$ in $\mathcal{W}$, there is a $\subset$-maximal nest $\mathcal{M}$ in $\mathcal{W}$ such that $\mathcal{N} \subset \mathcal{M}$.*

PROOF: By the Well-ordering Theorem, there is well-ordering $\leqq$ on $\mathcal{W}$. Denote by $y < z$ relation $(y \leqq z) \wedge (y \neq z)$ in $\mathcal{W}$. Let $\mathcal{L}$ be a nest in $\mathcal{W}$ satisfying the following conditions:

(1) $\mathcal{N} \subset \mathcal{L}$,
(2) $\mathcal{L} \setminus \mathcal{N}$ is well-ordered by $\leqq$, and
(3) for each $y \in \mathcal{L} \setminus \mathcal{N}$ and initial segment $I(y) = \{z | (z \in \mathcal{L} \setminus \mathcal{N}) \wedge (z \subset y) \wedge (z \neq y)\}$ of $\mathcal{L} \setminus \mathcal{N}$, the $\leqq$-least element of $(\mathcal{L} \setminus \mathcal{N}) \setminus I(y)$ is the $\leqq$-least element in $\mathcal{W}$ of $\{N^* | \mathcal{N} \cup I(y) \cup \{N^*\}$ forms a nest$\}$.

Consider class $\mathfrak{M}$ of all such $\mathcal{L}$, which is non-empty since $\mathcal{N} \in \mathfrak{M}$. Then, we may define $\mathcal{M}$ as $\mathcal{M} = \bigcup \mathfrak{M}$. ∎

A family of subsets of set x, $\mathcal{V} \subset \mathcal{P}(x)$, is of *finite character* if it satisfies the next condition: For $y \subset x$, $y \in \mathcal{V}$ if and only if each finite subset of y belongs to $\mathcal{V}$. If $\mathcal{V}$ is of finite character, for each $z_1 \in \mathcal{V}$, $y \in \mathcal{V}$ exists such that for all $z_2 \in \mathcal{V}$, $(y \subset z_2) \to (z_2 \subset y)$, i.e., y is a maximal element in $\mathcal{V}$ including z_1. This result is known as *Tuckey's Lemma*.

THEOREM 9.1.4: (Tuckey's Lemma) *Under ZFC, if $\mathcal{V} \subset \mathcal{P}(x)$ is of finite character, then for each $z_1 \in \mathcal{V}$, there exists $\subset$-maximal element y in $\mathcal{V}$ such that $z_1 \subset y$.*

PROOF: By the Hausdorff Maximal Principle, there is maximal nest $\mathcal{M}$ of $\mathcal{V}$ including $\{z_1\}$. Define $y \subset x$ as $\bigcup \mathcal{M}$. Since $\mathcal{V}$ is of finite character, each finite subset of $\bigcup \mathcal{M}$ is a finite subset of a certain member of nest $\mathcal{M} \subset \mathcal{V}$, so $y = \bigcup \mathcal{M}$ is a member of $\mathcal{V}$ under the finite character of $\mathcal{V}$. Clearly, y is a $\subset$-maximal element of $\mathcal{V}$ including z_1. ∎

Let $\precsim$ be a preordering on set x. If the restriction of $\precsim$ on subset y of x is a total ordering, y is a *totally ordered* subset of x under $\precsim$. Element $u \in x$ is an *upper bound* of subset y of x under $\precsim$ if $v \precsim u$ for all $v \in y$.

THEOREM 9.1.5: (Zorn's Lemma) *Let $\precsim$ be a preordering on x. Under ZFC, if every totally ordered subset y of x has an upper bound under $\precsim$, then for each $v^* \in x$, $\precsim$-maximal element u^* in x exists such that $v^* \precsim u^*$.*

PROOF: For each $u \in x$, denote by $y(u)$ set $\{v|\, (v \in x) \wedge (v \precsim u)\}$. Consider family $\mathcal{V} = \{y(u)|\, u \in x\} \subset \mathscr{P}(x)$. By the Hausdorff Maximal Principle, we have maximal nest $\mathcal{M}$ in $\mathcal{V}$ including $\{y(v^*)\}$. Note that with respect to $\mathcal{M} = \{y(u_m)|\, m \in \mathcal{M}\}$, for each m, n in nest $\mathcal{M}$, we have $(m \subset n) \leftrightarrow (u_m \precsim u_n)$, so $z = \{u_m|\, m \in \mathcal{M}\}$ (under (S9) and (S7)) is totally ordered by $\precsim$. Hence, z has upper bound u^*. u^* is a maximal element of x under $\precsim$. If not by taking element $\hat{u} \in x$ such that $u^* \precsim \hat{u}$ and $\neg \hat{u} \precsim u^*$, we obtain set $y(\hat{u}) \in \mathcal{V}$. Then the existence of family $\mathcal{M} \cup \{y(\hat{u})\}$ contradicts the maximality of $\mathcal{M}$. ∎

In this section, we have seen that many basic mathematical concepts, ordered pairs, products, natural numbers, relations, and functions, assumed in Chapters 1–8, can be formally constructed under ZF or ZFC. One can easily verify that such construction is also easy for other objects like the set of *integers Z rational numbers Q real numbers R* etc.[22] For our purpose in the remainder of this chapter, finite and finitistic methods are sufficient. Hence, I deliberately avoid using notions related to ordinal and cardinal transfinite arithmetic. A certain minimum level of treatment for such concepts and the notion of Continuum Hypothesis (CH) will be given in Mathematical Appendix III.

9.2 Individuals and Rationality

In this section, we see a serious constraint for set-theoretical formal descriptions of human society incorporating our quite natural and important kinds of inference (recognition) ability. There are many causes for the impossibility of obtaining a 'complete' social model in the sense that every feature of the world is completely described. Indeed, standard economic theory allows many types of 'externality.' There are many unknown structures in the real world, especially in scientific technologies, information, preferences, expectations, etc. It seems, however, that such problems have been recognized by theorists as merely the gap between an idealized

---

[22] We can identify Z as the quotient set of $N \times N$ through equivalence relation $((a, b) \sim^z (c, d)) \leftrightarrow (+(a, -(b)) = +(c, -(d))$, where $+$ and $-$ are identified with functions. In the same way Q may be considered a quotient set of $Z \times Z$ under equivalence relation $((a, b) \sim^q (c, c)) \leftrightarrow (\times(a, (b)^{-1}) = \times(c, (d)^{-1}))$, where $\times$ is treated as a function on $Z \times Z$ and $(\cdot)^{-1}$ is also treated as an operation (function) on $Z \setminus \{0\}$. Let $\mathcal{R}$ be the subset of $\mathscr{P}(Q)$ such that $A \in \mathcal{R}$ if and only if A has a $\leqq$-upper bound. Define equivalence relation $\sim^r$ on $\mathcal{R}$ as $A \sim^r B$ if and only if the set of upper bounds of A and B is equal. Then we may construct set of real numbers R as $\mathcal{R}/\sim^r$.

economic model and reality. What I am concerned with here is not the gap between them but the impossibility of the notion of an *idealized model* itself.

If the purpose of economics is to describe human society as theoretical and well-founded mechanisms of 'rational' individuals, an economic model must rigorously describe a *system of rules* that enables each agent's behavior to be called 'rational.' To formalize such (economic) 'rationality,' however, as seen in this section, we should premise a certain restricted view on individual prospects or thoughts about the entire world. If we don't, such a view of the world necessarily becomes inconsistent (hence, every action may possibly be rational). On the other hand, with such a restricted view of the world, agents are not allowed to ask whether the world is exactly as they are thinking (in their view of the model). In other words, a consistent view (description) of the world must be introspectively incomplete in the sense that every agent cannot be convinced of the rightness of the view itself.

The result in this section can be related to Gödel's second incompleteness theorem. Indeed, the main theorem in this section may be considered a transformed version of Tarski's truth definition theorem, which is another important result of Gödel's lemma for the incompleteness theorem.[23] Note, however, the important difference between the foundation of mathematics (Gödel's theorem) and the foundation of our view on society including ourselves (the social sciences). The former problem concerns what mathematics can do to formalize our "true" rationality, and the latter is an argument for formalizing rationality itself. We may change and reconstruct the rules of mathematics through our convictions and beliefs. To formalize our rationality, however, any restriction may cause our formalization to fail to characterize our "true" thoughts or recognition ability; there is no simple way or regular routine to formalize our general intelligence.[24]

In this section, *rationality* is treated as an attitude with which to accept certain kinds of formal assertions written in formal language.[25] The syntax for such a language and semantics (especially for the meanings of

---

[23]These theorems may be found in the standard literature in mathematical logic and/or theory of sets, e.g., see Kunen (1980), Jech (2003), Fraenkel *et al.* (1973), Shoenfield (1967).

[24]Hilary Putnam wrote, "The impossibility (in practice at least) of formalizing the assertibility conditions for arbitrary sentences is just the impossibility of formalizing general intelligence itself" (Putnam 1983, Introduction, p. 18).

[25]In this book, I borrow several concepts and some terminology from analytical philosophy; e.g., with respect to notions of rationality and truth, I use many from the recent works of H. Putnam (after 1980s, Putnam 1983, Putnam 1990, Putnam 1995, Putnam 2004, etc.), concepts in Kripke (1972), Kripke (1975), and works in cognitive science such as Lakoff (1987).

rationality) is given by a certain theory of sets $\mathscr{B} = (L_B, R_B, T_B)$ called an *underlying theory of sets*.[26] We assume that each person i using his/her formal language has theory $\mathscr{L}_i = (L_i, R_i, T_i)$ that is (1) sufficient to describe $\mathscr{B}$, and is (2) written by (constructed as objects in) the underlying theory of sets $\mathscr{B}$.[27] In other words, we are considering a situation where person i can treat his/her assertion θ in the language of i (theory $\mathscr{L}_i$) as set-theoretic object $\ulcorner \theta \urcorner$ through basic underlying theory $\mathscr{B}$. The problem we treat in this section is whether we can construct *formula $P_i(x)$ of person i* in one free variable x such that $P_i(\ulcorner \theta \urcorner)$ "means" that θ is a rationally acceptable assertion of i. Of course the answer depends on properties requested for the "meanings" of "rational acceptability." What we are concerned with here are *logical consistency* ($P_i(\ulcorner \theta \urcorner)$ and $P_i(\ulcorner \neg \theta \urcorner)$ never occur simultaneously), *introspective completeness* ($P_i(\ulcorner \theta \urcorner)$ means $P_i(\ulcorner P_i(\ulcorner \theta \urcorner) \urcorner)$), and *introspective consistency* ($P_i(\ulcorner \theta \urcorner)$ and $P_i(\ulcorner \neg P_i(\ulcorner \theta \urcorner) \urcorner)$ never occur simultaneously). The theorems in this section show that:

(1) If P_i is introspectively consistent, then we cannot use the fact as a rationally acceptable assertion (see Theorem 9.2.1).
(2) Under an introspectively complete view of the world, if we want to maintain the validity of using P_i's logical or introspective consistency (such as the law of excluded middle), we must understand that such an assertion causes introspective or logical inconsistency (see Theorem 9.2.2).
(3) A further reliance on the logical consistency (e.g., a presumption like $\mathscr{B} \vdash CONS(P_i)$) will cause the impossibility of defining rationality from the social scientific viewpoint (see Theorem 9.2.3).

Therefore, all economic agents in a standard economic model should dogmatically believe in their rational choices without knowing whether being rational is "truly" rational. Players in non-cooperative game theory must believe in their own rational behaviors as well as their opponents' without knowing what rationality exactly means. This seems to be a failure in all mathematical models in social science based on *methodological individualism* from the introspective viewpoint. The concept of 'rational individual' (consistency) always prevents us from providing a satisfactory answer to the question: 'What exactly is our society?' (introspection). Every

---

[26] This is not necessarily ZF nor ZFC.
[27] Of course there must be an appropriate translation between his/her formal language and the language for $\mathscr{B}$.

such agent, therefore, necessarily fails to define his/her "true" rationality (non-definability). This is not to say, however, that attempts to describe society as the totality of rational individuals are meaningless; instead it suggests that such attempts can never be completed even in an asymptotic sense. True rationality, if it exists, cannot be treated as a personal matter, and we must allow for its relation among our developing recognition ability and alternatively possible views of the world.

9.2.1 *Views of the world*

We treat each agent's reasoning to choose an action by identifying individual rationality with the consistency of actions to a certain view of the world. Let $I = \{1, 2, \ldots, m\}$ be the index set of agents. For each $i \in I$, denote by A_i the set of possible actions for agent i. Each action profile, $(a_1, a_2, \ldots, a_m) \in \prod_{i \in I} A_i$, in the economy decides consequence c_i in set C_i for each $i \in I$.

In standard economic arguments or non-cooperative game theory, stories (mathematical structures) exist about equilibrium and/or solution concepts that enable each agent i to justify his/her choices of action $a_i \in A_i$. Since there are many reasons for mutually exclusive actions to be chosen, many equilibrium and solution concepts may also exist. Rationality (reasoning) in this sense crucially depends on such views of the world (each equilibrium or solution concept). It would be possible to interpret arguments in this section as showing that such rationality is different from our 'true' rationality (thoughts or recognition abilities) and that the use (at least a part) of our true rationality may lead us to deny any such specific view of the world and rationality in a restricted sense.

In the following, we suppose that agent i has a (private) theory (written by a (private) formal language) $\mathscr{L}_i = (L_i, R_i, T_i)$ for obtaining a reason to decide action $a_i \in A_i$. L_i is the list of all symbols for the language, R_i is the list of all syntactical rules including construction rules for terms, formulas, and all inference rules (making a consequent formula from original formulas, e.g., modus ponens, instantiation, etc.), and T_i is the list of all axiomatic formulas for the theory. We assume that each element of L_i may be uniquely identified with (coded into) an object in a certain basic theory of sets, $\mathscr{B} = (L_B, R_B, T_B)$, written by the first-order predicate logic discussed precisely in the previous section. We call $\mathscr{B}$ an *underlying theory of sets* for $\mathscr{L}_i$.[28]

---

[28] The reader may identify $\mathscr{B}$ with ZF or ZFC in the previous section. Since such a coding argument may usually be treated in the domain of finitistic objects, a minimal theory may be $ZF^- - P - INF$, ZF without the axiom of foundation, the power set, and infinity.

The first important assumption of this section is that such a set theory is so basic that every agent could develop (understand) it by its own language:

(A.1) Language $\mathscr{L}_i = (L_i, R_i, T_i)$ is sufficiently rich to describe $\mathscr{B} = (L_B, R_B, T_B)$.[29] (Here, we implicitly assume an appropriate translation between the languages for $\mathscr{L}_i$ and $\mathscr{B}$. Throughout this section, such a translation is assumed to be fixed, and we suppose that each formula φ in $\mathscr{B}$ could be identified with the same formula in $\mathscr{L}_i$.)

The second assumption in this section is that the structure of theory $\mathscr{L}_i$, i.e., each symbol in L_i, the rules in list R_i, and list T_i may be treated as objects in the underlying theory of sets, $\mathscr{B}$. More precisely:

(A.2) $\mathscr{B}$ describes $\mathscr{L}_i$ in the following sense: (i) Each member of list L_i is a term (a set) in theory $\mathscr{B}$. (ii) List R_i consists of formulas in theory $\mathscr{B}$. There are formulas in one free variable, $Term_i(x)$, $Form_i(x)$, $Form_i^1(x)$, in two free variables, $Neg(x,y)$, in three free variables, $Sbst(x,y,z)$, in the language of $\mathscr{B}$, maintaining, respectively, that in $\mathscr{L}_i$, x is a term, x is a formula, x is a formula in one free variable, x is a negation formula of formula y, x is a substitution formula of term z into the single free variable of formula y, based on descriptions of construction rules for them written in theory $\mathscr{B}$.[30] Every inference rule, as a relation among formulas of i, is also written in the language of $\mathscr{B}$ as a well-defined set-theoretic procedure. (iii) $Axiom_i(x)$, which defines formulas of i belonging to list T_i, is written in the language of $\mathscr{B}$ as a well-defined set-theoretic procedure.

Assumption (A.2) is intended to be a sufficient condition in which a combination of inference procedures, such as a proof procedure in theory $\mathscr{L}_i$, may be identified with a set-theoretic procedure written in the form of a formula in theory $\mathscr{B}$. Note that each term, formula, and inference procedure (including the proof procedure) of i may not be finitistic (recursive) since

---

[29] For example, $\mathscr{B}$ may be identified with a part of $\mathscr{L}_i$ so that every theorem in $\mathscr{B}$ is a theorem in $\mathscr{L}_i$. (This is not necessarily the case, however, since the concept of the proof in $\mathscr{L}_i$ will play no role — will be replaced by concept P_i — in the following arguments.)

[30] By "based on descriptions of construction rules," I mean that the set of formulas in L_i is closed under such formation rules that are well-defined in set theory $\mathscr{B}$. That is, if θ is a formula in L_i, then $\neg\theta$ is also a formula in L_i; if $P(x)$ is a formula in one free variable x in L_i and if t is a term in L_i, then $P(t)$ is also a formula in L_i; and so forth.

the set-theoretic methods in $\mathscr{B}$ may possibly be much stronger than the finitistic method. (Descriptions for them, however, are given in the language of $\mathscr{B}$ as set-theoretical objects and processes that are well-defined in set theory $\mathscr{B}$.)

Under (A.1) and (A.2), agent i can treat assertion (formula) θ in the language of i (theory $\mathscr{L}_i$) as set-theoretic object $\ulcorner\theta\urcorner$ through the underlying theory of sets, $\mathscr{B}$,[31] as an object in theory $\mathscr{L}_i$. (It can be said, therefore, that theory $\mathscr{L}_i$ is at least assumed to be so rich that it can treat its statements as its objects.) In the following, we construct the worldview of i as theory $\mathscr{L}_i = (L_i, R_i, T_i)$, satisfying these two assumptions, (A.1) and (A.2). The worldview may include many features of the real world by adding additional axioms and syntactical rules, if necessary, and we suppose that agent i chooses a 'rational' action $a_i \in A_i$ under worldview $\mathscr{L}_i$. The third assumption concerns the possibility of such a structure in the worldview deciding 'rationality'.

(A.3) Formula $P_i(x)$ in one free variable x, in the theory of i denotes that $x = \ulcorner\theta\urcorner$ for a certain formula θ of i and θ is *rationally acceptable* for i. (Note that under (A.1) and (A.2), the theory of i can treat its formulas as its objects.) The meaning of $P_i(x)$, as a way to decide such acceptable sentences, is given as a set theoretic property in theory $\mathscr{B}$,[32] so $P_i(x)$ may be described by (replaced with) an appropriate formula in $\mathscr{B}$. In the following, by considering this fact, we always use notation P_i for such a replaced formula in $\mathscr{B}$ instead of the original formula in $\mathscr{L}_i$.

Under (A.2), one of the most typical set-theoretic properties in $\mathscr{B}$ satisfying conditions in (A.3) for $P_i(x)$ (the rational acceptability) would be the formal proof procedure in ZF under $\mathscr{B} = ZF$. (In this case, "the existence of a formal proof for a certain sentence from axioms in ZF through the inference processes" is described (could be maintained) through a formula in $\mathscr{B}$, though I have no intention to confine ourselves to this most familiar example.)

---

[31] For finitistic objects, notation $\ulcorner\ \urcorner$ is called *Quine's corner convention*.
[32] We may freely utilize the axioms in T_i to define P_i, so we may not require the method to define the rational acceptability to be finitistic even when $\mathscr{B}$ is finitistic theory $ZF^- - P - INF$. For example, we can decide that Continuum Hypothesis (CH) is rationally acceptable when CH is contained in list T_i in $\mathscr{B}$ even when CH is not an axiom in $\mathscr{B}$. When all methods in $\mathscr{B}$ are restricted as finitistic, the requirement for P_i being considered here may be restated as the "set of axioms defining property P_i is *recursive*."

In ordinary settings in economics, such a P_i may be considered the foundation of all arguments allowing for each person to choose, at least, one assertion specifying a certain character of $a_i \in A_i$ as a reason (justification) for a possible final decision rationally acceptable for agent i. For example, such arguments may be a class of assertions (based on specific R_i and T_i) compatible with "final decision $a_i \in A_i$ of i is a price taking and utility maximizing behavior," for an ordinary microeconomic view of the world; "final decision $a_i \in A_i$ of i is a best response given other agents' behaviors," for Nash equilibrium settings; and so on. It follows that agent $i \in I$ chooses action $a_i \in A_i$ if there is a sentence of i, θ that is rationally acceptable, $(P_i(\ulcorner\theta\urcorner))$, asserting that it is appropriate for agent i to choose action a_i as his/her best decision.

9.2.2 *Rationality*

The main concern of this chapter is to incorporate economic models with each agent's recognition structure for *rational* behaviors. In the previous section, such a structure is represented by formula $P_i(x)$ for agent i under a worldview constructed as theory $\mathscr{L}_i = (L_i, R_i, T_i)$ of i. We shall make in this section a further specification on the property (meaning) of $P_i(x)$ as the rationality of i.

Perhaps the most important condition for P_i to be called the rationality of i is consistency. It seems, however, that there are two kinds of consistency. One is logical (or the extensional part of) consistency and the other is introspective (or the intentional part of) consistency. $P_i(x)$ is *logically consistent* if for any sentence θ of i, $P_i(\ulcorner\theta\urcorner)$ and $P_i(\ulcorner\neg\theta\urcorner)$ do not hold simultaneously. The logical consistency of $P_i(x)$ as a fact in underlying theory of sets $\mathscr{B}$ is denoted by $CONS(P_i)$. Formally;

(D.1) $CONS(P_i)$ is a formula in $\mathscr{B}$ that is equivalent to:
$Form_i(\ulcorner\theta\urcorner) \to (P_i(\ulcorner\theta\urcorner) \to \neg P_i(\ulcorner\neg\theta\urcorner))$.[33]

The *introspective consistency* of P_i is the requirement that for any sentence θ of i, $P_i(\ulcorner\theta\urcorner)$ and $P_i(\ulcorner\neg P_i(\ulcorner\theta\urcorner)\urcorner)$ do not hold simultaneously. This condition may naturally be obtained if we request that P_i be logically consistent and *introspectively complete* in the sense that for each sentence θ, $P_i(\ulcorner\theta\urcorner) \to P_i(\ulcorner P_i(\ulcorner\theta\urcorner)\urcorner)$.

---

[33] As noted in (A.2), we assume that for each formula θ in $\mathscr{L}_i$, $\neg\theta$ is also a formula in $\mathscr{L}_i$, and that the translation process between $\ulcorner\theta\urcorner$ and $\ulcorner\neg\theta\urcorner$ may be written in a formula in $\mathscr{B}$. Note also that as stated in (A.3), $P_i(x)$ is considered a formula in $\mathscr{B}$.

(D.2) $COMP(P_i)$ is a formula in $\mathscr{B}$ that is equivalent to:
$Form_i(\ulcorner\theta\urcorner) \to (P_i(\ulcorner\theta\urcorner) \to P_i(\ulcorner P_i(\ulcorner\theta\urcorner)\urcorner))$.

Since $CONS(P_i)$ means $P_i(\ulcorner P_i(\ulcorner\theta\urcorner)\urcorner) \to \neg P_i(\ulcorner\neg P_i(\ulcorner\theta\urcorner)\urcorner)$, under $CONS(P_i)$ condition $COMP(P_i)$ automatically implies the introspective consistency. The introspective consistency, however, itself has a sufficiently clear and important meaning. One may confirm it by rewriting it in the next contrapositive form of $P_i(\ulcorner\theta\urcorner) \to \neg P_i(\ulcorner\neg P_i(\ulcorner\theta\urcorner)\urcorner)$.

(D.3) $IntrCons(P_i)$ is a formula in $\mathscr{B}$ that is equivalent to:
$Form_i(\ulcorner\theta\urcorner) \to (P_i(\ulcorner\neg P_i(\ulcorner\theta\urcorner)\urcorner) \to \neg P_i(\ulcorner\theta\urcorner))$.

In the following, logical consistency $CONS(P_i)$ and introspective completeness $COMP(P_i)$ are treated as the most desirable properties for P_i to satisfy. Logical consistency asserts the soundness in our development of theorems, especially, enabling us to exploit the *reductio ad absurdum* (or *the law of excluded middle*). On the other hand, the importance of introspective consistency stems from the natural social scientific viewpoint that we are describing the world including ourselves (so everything that we know must be known by ourselves described in the model).

Assumption (A.4) below asserts that we use $\mathscr{B}$ as the fundamental belief in constructing P_i, and (A.5) says that as conditions for P_i described under $\mathscr{B}$, we use $COMP(P_i)$ as one of the most basic requirements:

(A.4) If $\mathscr{B} \vdash \theta$, then $\mathscr{B} \vdash P_i(\ulcorner\theta\urcorner)$.[34]
(A.5) $\mathscr{B} \vdash P_i(\ulcorner COMP(P_i)\urcorner)$.

We do not directly treat the two conditions, $COMP(P_i)$ and $CONS(P_i)$, since (1) we may concentrate our attention to how these conditions are used to construct P_i, and (2) condition (A.5) is weaker than assuming $COMP(P_i)$ directly as long as we use (A.4), i.e., $\mathscr{B}$ as the basic theory constructing rationally acceptable sentences under P_i. We do not treat $COMP(P_i)$ and $CONS(P_i)$ even in parallel since our main concern is to describe logical consistency from the introspective viewpoint of social sciences. In this sense, in this section we treat $COMP(P_i)$ as a more fundamental requirement than $CONS(P_i)$.

---

[34] As stated in the previous section, symbol $\vdash$ means that the right hand side formula is deduced under the list of the left side formulas merely through the two inference rules (R3-1 and R3-2: Modus Ponens and Instantiation) together with the purely logical axioms with equation (T1–T3).

We also use the following two conditions (A.6 and A.7) for P_i as additional basic conditions, if necessary:

(A.6) $\mathscr{B} \vdash Form_i(\ulcorner\theta\urcorner) \to (P_i(\ulcorner P_i(\ulcorner\theta\urcorner)\urcorner) \to P_i(\ulcorner\theta\urcorner))$.

This condition may be identified with a sort of introspective consistency. (One may confirm it by considering the contrapositive form, $\neg P_i(\ulcorner\theta\urcorner) \to \neg P_i(\ulcorner P_i(\ulcorner\theta\urcorner)\urcorner)$.) We rarely use this property, but as the introspective consistency in (D.3), the condition has a quite natural meaning for P_i as the rational acceptability of i; i.e., the rational acceptability of the rational acceptability of θ should always mean the rational acceptability of θ.

(A.7) $\mathscr{B} \vdash (Form_i(\ulcorner\theta\urcorner) \wedge Form_i(\ulcorner\eta\urcorner)) \to (P_i(\ulcorner\theta \to \eta\urcorner) \to (P_i(\ulcorner\theta\urcorner) \to P_i(\ulcorner\eta\urcorner)))$.

If $\theta \to \eta$ and θ are rationally acceptable, then η is rationally acceptable. That is, the assumption means that rationally acceptable statements are closed under the modus ponens. This assumption is essential in our argument since under (A.4) it assures the following:

(A.8) If $\mathscr{B} \vdash \theta \leftrightarrow \eta$, then $\mathscr{B} \vdash P_i(\ulcorner\theta\urcorner) \leftrightarrow P_i(\ulcorner\eta\urcorner)$.

Note that if we define P_i as the proof procedure under $\mathscr{B}$, which is at least as strong as $ZF^- - P - INF$, then under (A.1), (A.2), (A.3), and (A.8), condition $COMP(P_i)$ is automatically satisfied.[35] (In this case, it is also obvious that condition (A.4) is automatically satisfied.) On the other hand, we never know whether $CONS(P_i)$ is satisfied.[36]

9.2.3 Rational acceptability

In the following, the main result of this section is given as three theorems, which are different aspects of the same fact; i.e., a restriction on what we

---

[35] Indeed, in this case, $P(\ulcorner\theta\urcorner)$ is equivalent to the assertion that $\mathscr{B} \vdash \theta$, so by (A.1) and (A.2), to an assertion in $\mathscr{B}$ that is equivalent to saying "there is a finite sequence of statements (as a set of objects, Z, in $\mathscr{B}$) constructing the proof in $\mathscr{B}$ to statement θ (object $\ulcorner\theta\urcorner$)." Let us denote this assertion by $\exists Z(\Psi(Z, \ulcorner\theta\urcorner))$. It is also possible to show that $\mathscr{B} \vdash \exists Z(\Psi(Z, \ulcorner\theta\urcorner))$ since Z is a finite object for any given θ and $\mathscr{B}$ is possible "to construct" and "assure the existence of" such a set of finite objects. Hence, we have $\mathscr{B} \vdash P(\ulcorner\theta\urcorner) \to \exists Z P(\ulcorner\exists Z(\Psi(Z, \ulcorner\theta\urcorner))\urcorner)$.

[36] This fact is known as the Gödel Second Incompleteness Theorem. In this book, we shall see its generalized form in Theorem 9.2.2-(2).

9: Mathematics and Social Science

can rationally accept if we require such rationality to be introspective as well as consistent. The first theorem does not depend on (A.5) and (A.6).

THEOREM 9.2.1: *Under (A.1), (A.2), (A.3), (A.4), and (A.7),*[37] *we see that*

(1) $\mathscr{B} \vdash P_i(\ulcorner IntrCons(P_i) \urcorner) \to \neg IntrCons(P_i)$, and that
(2) $\mathscr{B} \vdash \neg P_i(\ulcorner COMP(P_i) \urcorner) \vee \neg COMP(P_i) \vee \neg P_i(\ulcorner CONS(P_i) \urcorner) \vee \neg CONS(P_i)$.

PROOF: Let θ be an arbitrary formula in one free variable in $\mathscr{L}_i$. Define formula $q(x)$ in one free variable x in $\mathscr{B}$ through the set-theoretic process defining formula $q(\ulcorner \theta \urcorner)$ as an equivalent formula of $P_i(\ulcorner \neg \theta(\ulcorner \theta \urcorner) \urcorner)$. (Under condition (A.2), we can confirm that the procedure, $\ulcorner \theta \urcorner \mapsto \ulcorner \neg \theta(\ulcorner \theta \urcorner) \urcorner$, is well-defined through formulas in $\mathscr{B}$. For example, we may define this procedure as "for each x such that $Form^1(x)$, define y as the unique set satisfying $Neg(y, v)$, where v is the unique set satisfying $Sbst(v, x, x)$". Of course, P_i may be identified with a process in $\mathscr{B}$ under (A.3).) Since $\mathscr{B} \vdash q(\ulcorner q \urcorner) \leftrightarrow P_i(\ulcorner \neg q(\ulcorner q \urcorner) \urcorner)$, by defining Q as $q(\ulcorner q \urcorner)$, we have

$$\mathscr{B} \vdash Q \leftrightarrow P_i(\ulcorner \neg Q \urcorner).^{38} \tag{9.2}$$

If we define Z as $\neg Q$, we also have

$$\mathscr{B} \vdash Z \leftrightarrow \neg P_i(\ulcorner Z \urcorner). \tag{9.3}$$

By using Z we can prove the first assertion as follows:

$$\mathscr{B}, IntrCons(P_i) \vdash P_i(\ulcorner \neg P_i(\ulcorner Z \urcorner) \urcorner) \to \neg P_i(\ulcorner Z \urcorner) \tag{9.4}$$

$$\mathscr{B} \vdash P_i(\ulcorner \neg P_i(\ulcorner Z \urcorner) \urcorner) \to P(\ulcorner Z \urcorner) \quad \text{(by (9.3) and (A.7))} \tag{9.5}$$

$$\mathscr{B}, IntrCons(P_i) \vdash \neg P_i(\ulcorner \neg P_i(\ulcorner Z \urcorner) \urcorner) \quad \text{(by (9.4) and (9.5))} \tag{9.6}$$

$$\mathscr{B} \vdash IntrCons(P_i) \to \neg P_i(\ulcorner \neg P_i(\ulcorner Z \urcorner) \urcorner) \tag{9.7}$$

---

[37]More precisely, as assumptions for a metatheorem, these conditions (all facts listed) are treated as "really true" and will be used everywhere they are needed like the axioms of predicate logic.
[38]Note that by (A.1)–(A.3), q, P_i, and Q may be considered formulas in $\mathscr{B}$ as well as $\mathscr{L}_i$ though θ may not be. Since P_i is a formula in $\mathscr{B}$, it may also be possible to obtain assertion (9.2) as an application of Gödel's lemma (see, e.g., Kunen (1980, p. 40, Theorem 14.2)). I have proved it directly for the sake of completeness.

$$\mathscr{B} \vdash IntrCons(P_i) \to \neg P_i(\ulcorner Z \urcorner) \quad \text{(by (A.7))} \tag{9.8}$$
$$\mathscr{B} \vdash P_i(\ulcorner IntrCons(P_i) \urcorner \to \neg P_i(\ulcorner Z \urcorner)) \quad \text{(by (A.4))} \tag{9.9}$$
$$\mathscr{B} \vdash P_i(\ulcorner IntrCons(P_i) \urcorner) \to P_i(\ulcorner \neg P_i(\ulcorner Z \urcorner) \urcorner) \quad \text{(by (A.7))} \tag{9.10}$$
$$\mathscr{B} \vdash P_i(\ulcorner IntrCons(P_i) \urcorner) \to P_i(\ulcorner Z \urcorner) \quad \text{(by (A.7))} \tag{9.11}$$
$$\therefore \ \mathscr{B} \vdash P_i(\ulcorner IntrCons(P_i) \urcorner) \to \neg IntrCons(P_i). \tag{9.12}$$

The final line comes from the contradiction given through (9.8) and (9.11). For the second assertion, let us use formula Q. Since $\mathscr{B}, COMP(P_i) \vdash P_i(\ulcorner \neg Q \urcorner) \to P_i(\ulcorner P_i(\ulcorner \neg Q \urcorner) \urcorner)$, by (9.2) and (A.7),

$$\mathscr{B}, COMP(P_i) \vdash P_i(\ulcorner \neg Q \urcorner) \to P_i(\ulcorner Q \urcorner). \tag{9.13}$$

Hence, we have $\mathscr{B}, COMP(P_i), P_i(\ulcorner \neg Q \urcorner), CONS(P_i) \vdash \neg P_i(\ulcorner \neg \neg Q \urcorner) \wedge P_i(\ulcorner Q \urcorner)$, which also means (since by $\neg \neg Q \leftrightarrow Q$ and (A.7), the right hand side is contradictory) that

$$\mathscr{B}, COMP(P_i), CONS(P_i) \vdash \neg P_i(\ulcorner \neg Q \urcorner), \text{ and} \tag{9.14}$$
$$\mathscr{B}, COMP(P_i) \vdash P_i(\ulcorner \neg Q \urcorner) \to \neg CONS(P_i). \tag{9.15}$$

By the contrapositive form of (9.15), $\mathscr{B}, COMP(P_i) \vdash CONS(P_i) \to \neg P_i(\ulcorner \neg Q \urcorner)$. Hence, we also have by (A.4) and (A.7) that

$$\mathscr{B} \vdash P_i(COMP(P_i)) \to (P_i(\ulcorner CONS(P_i) \urcorner) \to P_i(\ulcorner \neg P_i(\ulcorner \neg Q \urcorner) \urcorner)), \tag{9.16}$$

which by considering $\neg P_i(\ulcorner \neg Q \urcorner) \leftrightarrow \neg Q$ with (9.15), means

$$\mathscr{B}, P_i(\ulcorner COMP(P_i) \urcorner), COMP(P_i) \vdash P_i(\ulcorner CONS(P_i) \urcorner) \to \neg CONS(P_i). \tag{9.17}$$

It follows that under $\mathscr{B}$, $P_i(\ulcorner COMP(P_i) \urcorner)$, $COMP(P_i)$, $P_i(\ulcorner CONS(P_i) \urcorner)$, and $CONS(P_i)$ are contradictory. Hence, we have the result. ∎

Thus, we see that: (i) *it is hard to demand our rationality to be introspectively consistent*, and (ii) *it is also difficult to accept both the introspective completeness and logical consistency* (for under (A.4) and (A.7), $P_i(\ulcorner COMP(P_i) \wedge CONS(P_i) \urcorner)$ implies $P_i(\ulcorner IntrCons(P_i) \urcorner)$). This situation actually *forbids us to have a deterministic view of the world* (in the sense that from $P_i(\ulcorner \theta \urcorner)$, we cannot exclude the possibility of $P_i(\ulcorner \neg P_i(\ulcorner \theta \urcorner) \urcorner)$), as long as we incorporate the structure of introspection into our knowledge-construction framework from the natural standpoint of social sciences.

9: Mathematics and Social Science

THEOREM 9.2.2: *Under* (A.1), (A.2), (A.3), (A.4), (A.5), *and* (A.7), *we have*

(1) $\mathscr{B} \vdash P_i(\ulcorner CONS(P_i) \urcorner) \to \neg IntrCons(P_i)$.
(2) *If* $\mathscr{B} \vdash COMP(P_i)$,[39] *then* $\mathscr{B} \vdash P_i(\ulcorner CONS(P_i) \urcorner) \to \neg CONS(P_i)$.

PROOF: By (9.15) in the previous proof, we have

$$\mathscr{B}, COMP(P_i) \vdash \neg CONS(P_i) \to \neg P_i(\ulcorner \neg Q \urcorner). \tag{9.18}$$

Then by (A.4), (A.7), and (A.5),

$$\mathscr{B} \vdash P_i(\ulcorner CONS(P_i) \urcorner) \to P(\ulcorner \neg P_i(\ulcorner \neg Q \urcorner) \urcorner). \tag{9.19}$$

Since $\neg Q \leftrightarrow P_i(\ulcorner \neg Q \urcorner)$, it follows from (9.19) and (9.18) that

$$\mathscr{B}, COMP(P_i) \vdash P_i(\ulcorner CONS(P_i) \urcorner) \to \neg CONS(P_i), \tag{9.20}$$

$$\mathscr{B}, COMP(P_i) \vdash CONS(P_i) \to \neg P_i(\ulcorner CONS(P_i) \urcorner), \quad \text{(Contraposition)} \tag{9.21}$$

$$\mathscr{B} \vdash P_i(\ulcorner CONS(P_i) \urcorner) \to P_i(\ulcorner \neg P_i(\ulcorner CONS(P_i) \urcorner) \urcorner), \quad \text{(A.4,A.7,A.5)} \tag{9.22}$$

$$\therefore \mathscr{B}, P_i(\ulcorner CONS(P_i) \urcorner) \vdash \neg IntrCons(P_i). \tag{9.23}$$

The last line comes from the contradiction obtained by $\mathscr{B}$, $P_i(\ulcorner CONS(P_i) \urcorner)$, and $IntrCons(P_i)$, hence we have proved the first assertion. The second assertion is obvious from (9.20). ∎

The above result shows that *if we base our arguments on the standpoint of social sciences* (A.5) *or a stronger assumption like* $\mathscr{B} \vdash COMP(P_i)$, *then we cannot expect P_i to have logical consistency without damaging its introspective consistency.*

THEOREM 9.2.3: *Under* (A.1), (A.2), (A.3), (A.4), *and* (A.7), *we have*:

(1) $\mathscr{B}, P_i(\ulcorner IntrCons(P_i) \urcorner) \vdash \neg IntrCons(P_i)$.
(2) $\mathscr{B}, COMP(P_i), CONS(P_i) \vdash \neg P_i(\ulcorner IntrCons(P_i) \urcorner)$.
(3) *If* $\mathscr{B} \vdash P_i(\ulcorner COMP(P_i) \urcorner)$ *and* $\mathscr{B} \vdash P_i(\ulcorner CONS(P_i) \urcorner)$, *then* $\mathscr{B} \vdash \neg CONS(P_i)$.

---

[39] In this case, (A.5) may be reduced under (A.4).

PROOF: Assertion (1) follows directly from (1) of Theorem 9.2.1. Moreover, as stated before, $CONS(P_i), COMP(P_i) \vdash IntrCons(P_i)$. Hence, the second assertion also follows from (1) of Theorem 9.2.1. By assertion (2) of Theorem 9.2.1, if $\mathscr{B} \vdash P_i(\ulcorner COMP(P_i)\urcorner)$ and $\mathscr{B} \vdash P_i(\ulcorner CONS(P_i)\urcorner)$, then $\mathscr{B}, COMP(P_i) \vdash \neg CONS(P_i)$ and $\mathscr{B}, CONS(P_i) \vdash \neg COMP(P_i)$. Hence, we have by (A.7),

$$\mathscr{B} \vdash P_i(\ulcorner COMP(P_i)\urcorner) \to P_i(\ulcorner \neg CONS(P_i)\urcorner), \text{ and} \qquad (9.24)$$

$$\mathscr{B} \vdash P_i(\ulcorner CONS(P_i)\urcorner) \to P_i(\ulcorner \neg COMP(P_i)\urcorner). \qquad (9.25)$$

Therefore, we have $\mathscr{B} \vdash P_i(\ulcorner \neg CONS(P_i)\urcorner)$ and $\mathscr{B} \vdash P_i(\ulcorner \neg COMP(P_i)\urcorner)$ under the conditions $\mathscr{B} \vdash P_i(\ulcorner CONS(P_i)\urcorner)$ and $\mathscr{B} \vdash P_i(\ulcorner CONS(P_i)\urcorner)$. It follows that we have $\mathscr{B} \vdash \neg CONS(P_i)$. ∎

Theorem 9.2.3 shows the inconsistency among natural requirements for P_i under $\mathscr{B}$, especially between $COMP$ and $CONS$. Since i believes $\mathscr{B}$ to be an underlying theory of sets, it would be hard to interpret such an inconsistent P_i as representing rationality for i. Therefore, if we really think that such conditions (like logical and introspective consistency) are indispensable, we must treat Theorem 9.2.3 as an impossibility theorem in defining P_i through methods that can be described (in the sense of (A.3)) in $\mathscr{B}$.

We can sum up the three theorems as follows. With respect to the introspective consistency, we can believe it (introspective consistency) but we cannot use it as a rationally acceptable assertion (Theorem 9.2.1). If we want to use part of its consistency like the *reductio ad absurdum* (the law of excluded middle), we must understand that its use also means (automatically) the inacceptability of consistency $\neg P_i(CONS(P_i))$ (Theorem 9.2.2). A further reliance on logical consistency (e.g., a presumption like $\mathscr{B} \vdash CONS(P_i))$[40] may cause the impossibility of defining rationality from introspective or consistent viewpoints (Theorem 9.2.3-(2) and -(3)).

The impossibility result clearly shows the different standpoints between the foundation of mathematics (or our knowledge) and the social sciences.

---

[40] Of course, this is one of the simplest examples of $\mathscr{B} \vdash P(\ulcorner CONS(P_i)\urcorner)$. Since $CONS(P_i)$ immediately means the consistency of $\mathscr{B}$ under (A.4), the result for such cases as $\mathscr{B} = ZF$ may directly be shown through Gödel's Second Incompleteness Theorem. Condition $\mathscr{B} \vdash P(\ulcorner CONS(P_i)\urcorner)$ may include more impressive cases, however, such that to be $P_i(\ulcorner \varphi \urcorner)$, we only request finitistic $\mathscr{B}$ to describe axioms in $\mathscr{L}_i = ZFC$ and consider the proof process for φ under $\mathscr{L}_i = ZFC$. In such a case, it may be possible for $\mathscr{B}$ to prove $P_i(\ulcorner \varphi \urcorner)$ even when $\mathscr{B}$ cannot prove φ.

9: Mathematics and Social Science

For mathematics (or to be the foundation of our knowledge), we can resign ourselves to referring to a certain kind of introspective consistency as seen in the standpoint of formalists over the Gödel's theorems. From the social scientific introspective viewpoint, however, since we cannot stop arguing whether our view of the world is "really" true, we have to make allowances for the existence (consider some possibility) of inconsistency among our assertions to define what our "true" rationality is.

REMARK 9.2.4: (Non-definability of Rationality) If we change (A.3) to (A.3′) so that it states all the properties of P_i except for the existence in (A.3), the above theorem asserts that *under* (A.1) *and* (A.2), *no* P_i *satisfies* (A.3′), (A.4), (A.5), *and* (A.7) together with full logical and introspective consistency.

REMARK 9.2.5: (Tarski's Truth Definition Theorem) The procedure to define Q for P_i in the proof of Theorem 9.2.1 merely depends on conditions (A.1), (A.2), and (A.3). Hence, for arbitrary formula $P(x)$ in one free variable x in $\mathscr{L}_i$, maintaining a certain property about formulas in $\mathscr{L}_i$ (under (A.1) and (A.2)), and being possible to be described as a set theoretic process in $\mathscr{B}$ (instead of (A.3)), we may construct Q satisfying $\mathscr{B} \vdash Q \leftrightarrow \neg P(\ulcorner Q \urcorner)$ in exactly the same way as before (Gödel's lemma).[41] The special case that $\mathscr{B} = \mathscr{L}_i = ZF$ and P_i is considered a definition of "truth" (which is assumed to be described as a process in $\mathscr{B}$) is easily seen to be contradictory. Indeed, if P is a definition of truth, there is a sentence Q that is "true" if and only if "not true"; hence, we cannot have such a definition of truth. This result is known as Tarski's truth definition theorem (see Kunen (1980, p. 41)).

REMARK 9.2.6: (Non-definability of Rational Common Knowledge) Let us return to the argument in Remark 9.2.4. If two agents, i and j, have the same rationality described in common underlying set theory $\mathscr{B}$, ($\ulcorner P_i \urcorner = \ulcorner P_j \urcorner$), then (D.2) is a necessary condition for their rationality to be *common knowledge*, i.e., $P_j(\ulcorner \theta \urcorner) \to P_i(\ulcorner P_j(\ulcorner \theta \urcorner) \urcorner)$ and $P_i(\theta) \to P_j(\ulcorner P_i(\ulcorner \theta \urcorner) \urcorner)$. Hence, the non-definability of rationality may also be interpreted as a *non-definability theorem of rationality* as *logically consistent common knowledge*.[42]

---

[41] To obtain this, replace P_i and $\ulcorner \neg \theta(\ulcorner \theta \urcorner) \urcorner$ with $\neg P$ and $\ulcorner \theta(\ulcorner \theta \urcorner) \urcorner$, respectively, throughout the process described in the theorem.

[42] Since we do not restrict $\mathscr{B}$ to be finitistic, the problem cannot be improved merely by strengthening the infinitary methods for the defining process.

9.3 Society and Values

In this section, we continue to analyze formal set-theoretical constraints for describing the world. The results in the previous section may be characterized as the failure to formalize human society as the whole of 'rational' individuals (*methodological individualism*). This section shows that the problem does not vanish even when we look for a structure describing the world as a whole by allowing for many other normative criteria without specifying a rigorous micro foundation.

A description of society without micro foundations needs other mechanisms of *verification* to assure the validity of the description itself. A fundamental attitude of *logical positivism* considers the world (society) as the whole of logical sentences that may or may not hold, and the purpose of social science (if it may be called a science), is to find assertions that are true (or at least may be called adequate) for a description of society. If we require such a verification for validity, however, there always exists the problem of introspective and logical consistency, as is the case with structures for rational individuals. That is, such social validity cannot be introspectively consistent as long as we require it to be logically consistent.

As before, denote here by $P(x)$ the assertion in certain formal language $\mathscr{L}$, meaning that "the society is such that assertion x holds." Here, the meaning of $P(x)$ is more directly related to the "truth" of the world than personal $P_i(x)$ in the previous section. In other words, we are more interested in the foundation of our knowledge (like the foundation of mathematics) than the foundation of social sciences (like the existence of introspectively consistent rationality). Language $\mathscr{L} = (L, R, T)$ is supposed to be treated as a list of objects in a certain theory of sets, $\mathscr{B} = (L_B, R_B, T_B)$, which is also written by formulas in language $\mathscr{L}$. We consider $\mathscr{B}$ a set theory under first-order predicate logic. (For simplicity, one may identify $\mathscr{B}$ with ZF, Zermelo–Fraenkel set theory, under first-order predicate logic.)[43] Hence, we may deal with each formula θ in $\mathscr{L}$ as a set-theoretical object, $\ulcorner\theta\urcorner$, in $\mathscr{B}$. Moreover, assume that formula $P(x)$ in one free variable x is a set-theoretically well-defined property (i.e., we may also replace $P(x)$ with a formula in $\mathscr{B}$), or an object in $\mathscr{B}$. These assumptions ((B.1),(B.2),(B.3) below) are completely parallel to those in the previous section ((A.1),(A.2),(A.3)). The meaning of $P(x)$, however, is not given

---

[43] This is the same setting as in the previous section, except that $\mathscr{L}$ and $\mathscr{B}$ are not private but public language and theory, respectively.

9: Mathematics and Social Science

from the previous social scientific viewpoint, i.e., we take logical consistency (instead of introspective consistency) as the most basic requirement for $P(x)$. Then, with several natural conditions, we have the following results:

(1) A mathematical truth ($\mathscr{B} \vdash \theta$) exists that isn't socially valid ($\mathscr{B} \vdash \neg P(\ulcorner \theta \urcorner)$) (see Theorem 9.3.1).
(2) We cannot verify the introspective consistency of description $P(x)$ itself (see Theorem 9.3.2).
(3) We cannot define (formally describe) society as long as we require it to be introspectively consistent (see Theorem 9.3.3).

These arguments may also be restated as follows: if we identify the description of society by deciding what is valid in society, then social validity (a value judgment in society) is always restrictive in the sense that we are not allowed to ask exactly what society is (as long as we require it to be logically and/or introspectively consistent). Of course, the result may also be interpreted as a general statement on various social values, i.e., we cannot completely describe social norms, justice, and/or validities as well-defined structures (mechanisms) as long as we require them to be logically and/or introspectively consistent (non-definability of social values).

These results are closely related to the arguments in the previous section in which the logically consistent *rationality of individuals* makes the description of society introspectively inconsistent. In this section, the logically consistent *values in society* makes verification of society introspectively inconsistent. Although truth and/or rationality in our society are determined by ourselves, no single (consistent) mind is sufficient to define or construct them from an introspective (social scientific) viewpoint.

9.3.1 *Society*

As in Section 9.2, we assume that all mathematical arguments and theorems are supposed to be given in a certain basic formal set theory, $\mathscr{B} = (L_B, R_B, T_B)$, where L_B is the list of symbols, R_B is the list of syntactical rules, and T_B is the list of axioms. Moreover, we also assume that in describing society, language $\mathscr{L} = (L, R, T)$ is used, where L (list of symbols), R (list of syntactical rules), and T (list of axioms) are sufficient for developing theory $\mathscr{B}$ under first-order predicate logic in the sense that every formula in $\mathscr{B}$ may be identified with a formula in $\mathscr{L}$. Intuitively, $\mathscr{B}$ is our basic belief. We require it to include finitistic objects and methods to encode the first-order predicate logic into it. $\mathscr{L}$ is a theory with more axioms and methods that we do not require to be finitistic to describe the world and define concept $P(x)$ to mean that "society is such that assertion x holds."[44] Precisely, we assume the following:

(B.1) $\mathscr{B}$ is a set theory under first-order predicate logic with equality.[45] All symbols, terms, formulas, inference rules, and logical (non-mathematical) axioms in $\mathscr{B}$ can be written by the symbols and formulas in $\mathscr{L}$. ($\mathscr{L}$ need not, however, be a stronger theorem than $\mathscr{B}$.)

Moreover, we assume that $\mathscr{L}$ is formalized under $\mathscr{B}$:

(B.2) $\mathscr{B}$ describes $\mathscr{L} = (L, R, T)$ in the following sense: (i) Each member of list L may be identified with a set in theory $\mathscr{B}$. (ii) List R consists of formulas constructed as objects in theory $\mathscr{B}$. Especially, there are formulas in $\mathscr{B}$ in one free variable, $Term(x)$, $Form(x)$, $Form^1(x)$, $Neg(x, y)$, and $Sbst(x, y, z)$ that describe, respectively, "x is a term of $\mathscr{L}$," "x is a formula of $\mathscr{L}$," "x is a formula in one free variable," "x is a negation of y," and "y is a formula in one free variable, and x is the formula obtained by substituting term z into y." Every inference rule, as a relation among formulas in $\mathscr{L}$, can also be described as set-theoretic objects in $\mathscr{B}$. (iii) $Axiom(x)$, which defines the formulas of $\mathscr{L}$ belonging to list T, can also be described as objects in $\mathscr{B}$.

Assumption (B.2) enables us to treat each assertion θ in $\mathscr{L}$ as a set-theoretical object $\ulcorner\theta\urcorner$ in theory $\mathscr{B}$. Since all the terms and formulas in $\mathscr{B}$ are also in $\mathscr{L}$ by (B.1), through theory $\mathscr{B}$, language $\mathscr{L}$ can be formalized into $\mathscr{L}$ itself. Language $\mathscr{L}$ is rich enough to treat each of its formulas as objects. Hence, as in the previous section, we can assume that the concept of *society* is given in logical formula $P(\ulcorner\theta\urcorner)$ in one free variable $\ulcorner\theta\urcorner$ in $\mathscr{L}$, maintaining that "assertion θ in $\mathscr{L}$ is valid as a description of society." That is, we identify the problem "what is society?" with "what assertions hold in society?" Hence, if a complete description of society exists, we may

---

[44] As $P_i(x)$, $P(x)$ is a formula in $\mathscr{L}$. To describe the property of x, it may use many axioms that do not belong to $\mathscr{B}$, e.g., $P(x)$ in $\mathscr{L}$ may use the axiom of choice that may not be included in $\mathscr{B}$. We may assume, however, that $\mathscr{B}$ can describe the process in which $P(x)$ uses the choice axiom to define the property "society is such that assertion x holds."

[45] In exactly the same way as before, we may suppose that the minimum requirement for $\mathscr{B}$ is $ZF^- - P - INF$.

obtain all the relevant assertions on what society is, what we are in society, what we should do in society, etc. We suppose (in exactly the same way as P_i in the previous section) that such a structure representing "validity" in the world (society), i.e., the meanings of P, is given through a set-theoretic procedure on the formulas in $\mathscr{L}$ that can be described in the underlying theory of sets $\mathscr{B}$. Formally:

> (B.3) Formula $P(x)$ in one free variable x in $\mathscr{L}$ asserts that "x is a formula in $\mathscr{L}$ that is *valid* for a description of society." We further assume that P as a property in formula x in $\mathscr{L}$ can also be described as a set-theoretic procedure in $\mathscr{B}$ as an appropriate method for formula x to define whether it is "valid" based on notions and axioms in $\mathscr{L}$ as well as $\mathscr{B}$; so it is also described by (can be replaced with) a formula in $\mathscr{B}$. In the following, we always treat $P(x)$ as a formula in $\mathscr{B}$ in this sense.

Of course, by (B.1), every formula in $\mathscr{B}$ is also in $\mathscr{L}$, so formula $P(x)$ in $\mathscr{B}$, replaced by the original one in $\mathscr{L}$, is also identified with a formula in $\mathscr{L}$. Such $P(x)$ (as identified with a formula in $\mathscr{L}$), however, is a completely different formula from the original one.[46] The "validity" stated above will be discussed axiomatically in the next subsection. However, assumption (B.3) at least maintains the standpoint that we identify the world with all valid logical formulas regardless how validity is defined.[47] Hence, in this sense, we identify society with all valid sentences in it.

9.3.2 *Social validity and mathematical truth*

We are considering in assumption (B.3) that to describe *society* is simply to decide what its valid descriptions are; hence, it is merely to decide what *validity* in society is. Such thoughts are not new but quite common. Indeed, such attitudes toward the truth of the world are nothing but the standpoint of logical positivists. Our result in this section, therefore, has an important relation with the methodology of social sciences as well as the formal (or mechanical) definition of "values" in society.

---

[46] As the case with P_i in the previous section, formula P is more desirable to be referred to as the original formula in $\mathscr{L}$ than the replaced one in $\mathscr{B}$. Even if it is written in $\mathscr{B}$, the meaning of P is given in theory $\mathscr{B}$.

[47] Or, at least, we are considering that a complete description of society should decide (in the sense of $\mathscr{B}$) a set of logical formulas that are valid views of society.

As a mechanism that defines validity in society, we will naturally expect P to have the following properties, even though we do not directly assume all of them:

(C.1) (Logical Consistency) *CONS* is a sentence equivalent to:
$Form(\ulcorner \theta \urcorner) \to (P(\ulcorner \theta \urcorner) \to \neg P(\ulcorner \neg \theta \urcorner))$.

(C.2) (Introspective Completeness) *COMP* is a sentence equivalent to:
$Form(\ulcorner \theta \urcorner) \to (P(\ulcorner \theta \urcorner) \to P(\ulcorner P(\ulcorner \theta \urcorner) \urcorner))$.

(C.3) (Introspective Consistency) *IntrCons* is a sentence equivalent to:
$Form(\ulcorner \theta \urcorner) \to (P(\ulcorner \neg P(\ulcorner \theta \urcorner) \urcorner) \to \neg P(\ulcorner \theta \urcorner))$.

For P, we treat *CONS* as a more fundamental property than *COMP*. The following assumptions will be used to supply more general conditions for P than (A.4) and (A.7) in the previous section.

(B.4) If $\mathscr{B} \vdash \theta \leftrightarrow \eta$, then $\mathscr{B} \vdash P(\ulcorner \theta \urcorner) \leftrightarrow P(\ulcorner \eta \urcorner)$.
(B.5) If $\mathscr{B} \vdash P(\ulcorner \theta \urcorner)$, then $\mathscr{B} \vdash P(\ulcorner P(\ulcorner \theta \urcorner) \urcorner)$.
(B.6) If $\mathscr{B} \vdash \neg P(\ulcorner \theta \urcorner)$, then $\mathscr{B} \vdash P(\ulcorner \neg P(\ulcorner \theta \urcorner) \urcorner)$.
(B.7) If $\mathscr{B} \vdash \theta \to P(\ulcorner \eta \urcorner)$, then $\mathscr{B} \vdash P(\ulcorner \theta \urcorner) \to P(\ulcorner P(\ulcorner \eta \urcorner) \urcorner)$.
(B.8) If $\mathscr{B} \vdash \theta \to \neg P(\ulcorner \eta \urcorner)$, then $\mathscr{B} \vdash P(\ulcorner \theta \urcorner) \to P(\ulcorner \neg P(\ulcorner \eta \urcorner) \urcorner)$.

Note that condition (B.4) is much weaker than assuming the conditions of (A.4)- with (A.7)-type. We also use some conditions (B.5)–(B.8) that are weaker than (A.4)- with (A.7)-type conditions.[48]

Now we have the following result.

THEOREM 9.3.1: *Under* (B.1), (B.2), (B.3), (B.4), *COMP, and CONS, we have a mathematical theorem that is not socially valid.*

PROOF: Let θ be an arbitrary formula in one free variable in $\mathscr{L}$. As in the proof of Theorem 9.2.1, we can define formula q in one free variable in $\mathscr{B}$ through a process that identifies $q(\ulcorner \theta \urcorner)$ with an equivalent formula of $P(\ulcorner \neg \theta(\ulcorner \theta \urcorner) \urcorner)$ (see also the discussion in Remark 9.2.5, Gödel Lemma). Let Q be formula $q(\ulcorner q \urcorner)$ and Z be $\neg Q$. Then, as before, we have:

$$\mathscr{B} \vdash Q \leftrightarrow P(\ulcorner \neg Q \urcorner), \qquad (9.26)$$

$$\mathscr{B} \vdash Z \leftrightarrow \neg P(\ulcorner Z \urcorner). \qquad (9.27)$$

---

[48] If we use a condition of the (A.7) type, e.g., "if $\mathscr{B} \vdash \theta \to \eta$, then $\mathscr{B} \vdash P(\ulcorner \theta \urcorner) \to P(\ulcorner \eta \urcorner))$," we can verify that if there is at least one sentence ψ such that $\mathscr{B} \vdash \psi$ and $\mathscr{B} \vdash P(\ulcorner \psi \urcorner)$ (socially valid), then all mathematical sentences become socially valid.

9: Mathematics and Social Science

By COMP, we have $\mathscr{B} \vdash P(\ulcorner \neg Q \urcorner) \to P(\ulcorner P(\ulcorner \neg Q \urcorner) \urcorner)$. Hence, by (9.26) with (B.4), we have

$$\mathscr{B} \vdash P(\ulcorner \neg Q \urcorner) \to P(\ulcorner Q \urcorner). \tag{9.28}$$

By CONS, we have $\mathscr{B} \vdash P(\ulcorner Q \urcorner) \to \neg P(\ulcorner \neg Q \urcorner)$. Therefore, by considering the fact that $\mathscr{B}, P(\ulcorner \neg Q \urcorner) \vdash P(Q) \wedge \neg P(\ulcorner Q \urcorner)$, a contradiction, we have

$$\mathscr{B} \vdash \neg P(\ulcorner \neg Q \urcorner). \tag{9.29}$$

Again by considering (9.26) and (B.4), we obtain

$$\mathscr{B} \vdash \neg P(\ulcorner \neg P(\ulcorner \neg Q \urcorner) \urcorner). \tag{9.30}$$

Let ψ be a sentence that is equivalent to $\neg P(\ulcorner \neg Q \urcorner)$, then by (9.29) and (9.30), ψ satisfies the condition of the theorem. ∎

It may not be clear that a mathematical truth, which cannot be socially valid in the above theorem, may have some crucial meanings in the view of social science. The sentence, however, has important relations with many kinds of assertions on the structure of P itself as seen in the following theorems.

THEOREM 9.3.2: *Assume* (B.1), (B.2), (B.3), (B.4), *and* (B.8). *Then we have the following results:*

(1) $\mathscr{B} \vdash IntrCons \to \neg P(IntrCons)$.
(2) $\mathscr{B} \vdash P(\ulcorner CONS \wedge COMP \urcorner) \to \neg IntrCons$.
(3) $\mathscr{B}, CONS, COMP \vdash \neg P(\ulcorner COMP \wedge CONS \urcorner)$.

PROOF: Let Q and Z be the same formulas defined in the previous proof of Theorem 9.3.1. Note that (9.26) and (9.27) also hold under the setting of Theorem 9.3.2. For the first assertion consider these formulas:

$$\mathscr{B}, IntrCons \vdash P(\ulcorner \neg P(\ulcorner \neg Q \urcorner) \urcorner) \to \neg P(\ulcorner \neg Q \urcorner) \text{ (by } IntrCons\text{)}, \tag{9.31}$$
$$\mathscr{B} \vdash P(\ulcorner \neg P(\ulcorner \neg Q \urcorner) \urcorner) \to P(\ulcorner \neg Q \urcorner) \text{ (by (9.26))}, \tag{9.32}$$

so we have

$$\mathscr{B}, IntrCons \vdash \neg P(\ulcorner \neg P(\ulcorner \neg Q \urcorner) \urcorner). \tag{9.33}$$

Then by (B.8),

$$\mathscr{B} \vdash P(IntrCons) \rightarrow P(\ulcorner \neg P(\ulcorner \neg P(\ulcorner \neg Q \urcorner) \urcorner) \urcorner), \qquad (9.34)$$

$$\mathscr{B}, P(IntrCons) \vdash P(\ulcorner \neg P(\ulcorner \neg P(\ulcorner \neg Q \urcorner) \urcorner) \urcorner), \qquad (9.35)$$

$$\mathscr{B}, P(IntrCons) \vdash P(\ulcorner \neg P(\ulcorner \neg Q \urcorner) \urcorner) \quad \text{(by (9.26))}. \qquad (9.36)$$

Hence, by (9.33) and (9.36), conditions $P(IntrCons)$ and $IntrCons$ are contradictory in $\mathscr{B}$, and we have assertion (1). Using $CONS$ and $COMP$, we obtain the following assertion in exactly the same way as with (9.30):

$$\mathscr{B}, CONS, COMP \vdash \neg P(\ulcorner \neg P(\ulcorner \neg Q \urcorner) \urcorner). \qquad (9.37)$$

Hence, we have

$$\mathscr{B} \vdash (CONS \wedge COMP) \rightarrow \neg P(\ulcorner \neg P(\ulcorner \neg Q \urcorner) \urcorner). \qquad (9.38)$$

It follows by (B.8) that

$$\mathscr{B} \vdash P(\ulcorner CONS \wedge COMP \urcorner) \rightarrow P(\ulcorner \neg P(\ulcorner \neg P(\ulcorner \neg Q \urcorner) \urcorner) \urcorner). \qquad (9.39)$$

Since $\neg P(\ulcorner \neg Q \urcorner) \leftrightarrow \neg Q$, we also have

$$\mathscr{B} \vdash P(\ulcorner CONS \wedge COMP \urcorner) \rightarrow P(\ulcorner \neg P(\ulcorner \neg Q \urcorner) \urcorner). \qquad (9.40)$$

Then, (9.39) and (9.40) show that $\mathscr{B}, P(\ulcorner CONS \wedge COMP \urcorner)$, and $IntrCons$ prove $P(\ulcorner \neg P(\ulcorner \neg P(\ulcorner \neg Q \urcorner) \urcorner) \urcorner) \wedge \neg P(\ulcorner \neg P(\ulcorner \neg P(\ulcorner \neg Q \urcorner) \urcorner) \urcorner)$, a contradiction. Hence, (2) follows. Since $CONS, COMP \vdash IntrCons$, (3) follows from (2). ∎

Last, we see the inconsistency among many properties in (B.1)–(B.8) and (C.1)–(C.3) with underlying theory of sets $\mathscr{B}$. It may also be possible to understand the theorem as a non-definability theorem of the concept of "social validity."

THEOREM 9.3.3: *Under* (B.1), (B.2), (B.3), *and* (B.4), *we have the following results*:

(1) *If* $\mathscr{B} \vdash IntrCons$, *condition* (B.6) *does not hold*.
(2) *Under* (B.8), *if* $\mathscr{B} \vdash CONS$, *we have* $\mathscr{B}, IntrCons \vdash \neg P(COMP)$.
(3) *Under* (B.8), *if* $\mathscr{B} \vdash COMP$, *we have* $\mathscr{B} \vdash CONS \rightarrow \neg P(CONS)$.

PROOF: Let Q and Z be the same formulas defined in the previous proof of Theorem 9.3.1. Assume that $\mathscr{B} \vdash IntrCons$; then

$$\mathscr{B} \vdash P(\ulcorner \neg P(\ulcorner Z \urcorner) \urcorner) \to \neg P(\ulcorner Z \urcorner), \tag{9.41}$$

$$\mathscr{B} \vdash P(\ulcorner \neg P(\ulcorner Z \urcorner) \urcorner) \leftrightarrow P(\ulcorner Z \urcorner) \quad \text{(by (B.4))}, \tag{9.42}$$

$$\therefore \quad \mathscr{B} \vdash \neg P(\ulcorner \neg P(\ulcorner Z \urcorner) \urcorner). \tag{9.43}$$

If (B.6) holds, we have by (9.43), $\mathscr{B} \vdash P(\ulcorner \neg P(\ulcorner \neg P(\ulcorner Z \urcorner) \urcorner) \urcorner)$, which also means by (B.4), $\mathscr{B} \vdash P(\ulcorner Z \urcorner)$. On the other hand, (9.43) means by (B.4), $\mathscr{B} \vdash \neg P(\ulcorner Z \urcorner)$, so we have a contradiction and result (1) follows. Assume that $\mathscr{B} \vdash CONS$. Under (B.8), we may consider that (9.37) holds. Note that $Q \leftrightarrow P(\ulcorner \neg Q \urcorner)$. Under (B.4) and CONS, we have by (9.37),

$$\mathscr{B} \vdash COMP \to \neg P(\ulcorner \neg Q \urcorner), \tag{9.44}$$

$$\mathscr{B} \vdash P(\ulcorner COMP \urcorner) \to P(\ulcorner \neg P(\ulcorner \neg Q \urcorner) \urcorner) \quad \text{(by (B.8))}, \tag{9.45}$$

$$\therefore \quad \mathscr{B} \vdash P(\ulcorner COMP \urcorner) \to P(\ulcorner \neg Q \urcorner) \quad \text{(by (B.4))}. \tag{9.46}$$

On the other hand, (9.45) with IntrCons means that

$$\mathscr{B}, IntrCons \vdash P(\ulcorner COMP \urcorner) \to \neg P(\ulcorner \neg Q \urcorner), \tag{9.47}$$

which with (9.46) we have (2). For (3), assume that $\mathscr{B} \vdash COMP$. Again by (9.37), $Q \leftrightarrow P(\ulcorner \neg Q \urcorner)$, and (B.4),

$$\mathscr{B} \vdash CONS \to \neg P(\ulcorner \neg Q \urcorner), \tag{9.48}$$

$$\mathscr{B} \vdash P(\ulcorner CONS \urcorner) \to P(\ulcorner \neg P(\ulcorner \neg Q \urcorner) \urcorner) \quad \text{(by (B.8))}, \tag{9.49}$$

$$\mathscr{B} \vdash P(\ulcorner CONS \urcorner) \to P(\ulcorner \neg Q \urcorner) \quad \text{(by (B.4))}. \tag{9.50}$$

Hence, we have $\mathscr{B} \vdash CONS \to \neg P(\ulcorner CONS \urcorner)$. ■

One may believe that conditions like $\mathscr{B} \vdash IntrCons$ and $\mathscr{B} \vdash CONS$ are too strong. Note, however, that we base all our arguments on $\mathscr{B}$ and intend to construct P under methods that can be described in $\mathscr{B}$. Hence, we naturally demand that P have such strong provability conditions.[49] In this sense, the meaning of (1) of Theorem 9.3.3 is critical since it argues that as long as we intend to construct P to satisfy introspective consistency, we have to discard condition (B.6). That is, we have to admit that the proof

---

[49] Of course, when $P(\ulcorner \theta \urcorner)$ necessarily follows from $\mathscr{B} \vdash \theta$, condition $\mathscr{B} \vdash CONS$ would be contradictory (too strong, perhaps, for meaningful arguments) under Gödel's Second Incompleteness Theorem. Note that even in such cases, condition $\mathscr{B} \vdash P(\ulcorner CONS \urcorner)$ still has meaning since we allow $\mathscr{B} \vdash P(\ulcorner CONS \urcorner)$ even when $\mathscr{B} \vdash \neg CONS$.

under $\mathscr{B}$ of the assertion that θ is not classified as valid cannot be classified as valid even when the validity is clearly defined under $\mathscr{B}$. (Here, note in (B.6) the difference between $\mathscr{B} \vdash \neg P(\ulcorner\theta\urcorner)$ and $\mathscr{B} \nvdash P(\ulcorner\theta\urcorner)$.) Perhaps, there is a way to renounce condition (B.6), but it seems more appropriate to regard Theorem 9.3.3 as an impossibility theorem.

REMARK 9.3.4: (Non-definability of Social Validity) If we change (B.3) to (B.3′) so that it merely asserts the property of P without maintaining its existence, the above results insist that there is no set-theoretical possibility (in $\mathscr{B}$ under (B.1), (B.2), and (B.3′)) for defining the concept of social validity P in a natural sense that satisfies important properties (C.1)–(C.3). For example, we can see by (1) of Theorem 9.3.3:

Non-definability Theorem: Under (B.1), (B.2), (B.3′), (B.4), and (B.6), no P satisfies *IntrCons*.

Indeed, if there is such a P, we have "$\mathscr{B} \vdash P(\ulcorner\neg P(\ulcorner\neg Q\urcorner)\urcorner) \to \neg P(\ulcorner\neg Q\urcorner)$" by IntrCons, and "$\mathscr{B} \vdash P(\ulcorner\neg P(\ulcorner\neg Q\urcorner)\urcorner) \to P(\ulcorner\neg Q\urcorner)$" by the definition of Q. It follows that we have $\mathscr{B} \vdash \neg P(\ulcorner\neg P(\ulcorner\neg Q\urcorner)\urcorner)$. Again, by IntrCons and the definition of Q, however, we may obtain "$\mathscr{B} \vdash \neg P(\ulcorner\neg Q\urcorner)$" and "$\mathscr{B} \vdash P(\ulcorner\neg Q\urcorner)$," a contradiction.

REMARK 9.3.5: (Other Conditions for P) The discussions we have treated in this chapter are merely an introduction to rigorous set-theoretic arguments about the basic methodology of social sciences. More natural conditions and general treatments may exist for various aspects of P. For example, we have not used "provability" in $\mathscr{B}$ itself as a property to characterize P. It would be natural to request that P satisfy

(WeakIntrCons) If $\mathscr{B} \vdash P(\ulcorner\theta\urcorner)$, then $\mathscr{B} \nvdash P(\ulcorner\neg P(\ulcorner\theta\urcorner)\urcorner)$.

The condition may be considered a *weak* version of introspective consistency. If we use this condition for Theorem 9.3.3-(3), we obtain $\mathscr{B} \nvdash P(\ulcorner CONS\urcorner)$.[50] It is also true that we have not treated many assumptions in a comprehensive way. For example, in Theorem 9.3.3, we can obtain the same result of (2) under $\mathscr{B} \vdash COMP$ and of (3) under $\mathscr{B} \vdash CONS$. Further developments would be meaningful since these arguments, even when they are impossibility results, are directly related to the knowledge about this world itself, so they construct the basic knowledge of social sciences in a truly introspective manner.

---

[50] For $\mathscr{B} \vdash P(\ulcorner CONS\urcorner) \to P(\ulcorner\neg P(\ulcorner CONS\urcorner)\urcorner)$ by Theorem 9.3.3-(3).

9: Mathematics and Social Science

REMARK 9.3.6: (Relation among Conditions) The following classification will be useful for our assumptions in this chapter. At first, we may classify them into two categories: *logical* (conditions between sentences described under P) and *introspective* (conditions between sentences described under P and $P(\ulcorner P \urcorner)$). Second, we may further classify them into two categories: *consistency* (a requirement for the non-existence of a sentence based on the existence of its negation) and *completeness* (a requirement for the existence of a sentence based on the non-existence of its negation). The following are the six most basic categories:[51]

1. (Logical Consistency) $P(\ulcorner \theta \urcorner) \to \neg P(\ulcorner \neg \theta \urcorner)$ (CONS)
2. (Logical Completeness) $\neg P(\ulcorner \neg \theta \urcorner) \to P(\ulcorner \theta \urcorner)$
3. (Introspective Consistency 1) $P(\ulcorner \theta \urcorner) \to \neg P(\ulcorner \neg P(\ulcorner \theta \urcorner) \urcorner)$ (IntrCons)
4. (Introspective Completeness 1) $\neg P(\ulcorner \neg P(\ulcorner \theta \urcorner) \urcorner) \to P(\ulcorner \theta \urcorner)$
5. (Introspective Consistency 2) $\neg P(\ulcorner \theta \urcorner) \to \neg P(\ulcorner P(\ulcorner \theta \urcorner) \urcorner)$ (A.6)
6. (Introspective Completeness 2) $\neg P(\ulcorner P(\ulcorner \theta \urcorner) \urcorner) \to \neg P(\ulcorner \theta \urcorner)$ (COMP)

As noted above, we have already seen conditions 1, 3, 5, 6 in this and the previous sections. Condition 2 is the ordinary sense of completeness when P is identified with the proof procedure. If we use condition 4, say IntrComp1, under (B.1), (B.2), (B.3), and (B.4), we have the following immediate contradiction between COMP and CONS. As (9.37), we have

$$\mathscr{B}, CONS, COMP \vdash \neg P(\ulcorner \neg P(\ulcorner \neg Q \urcorner) \urcorner), \quad (9.51)$$

$$\mathscr{B} \vdash \neg P(\ulcorner \neg P(\ulcorner \neg Q \urcorner) \urcorner) \leftrightarrow \neg P(\ulcorner \neg Q \urcorner) \text{ (by (B.4))}, \quad (9.52)$$

$$\mathscr{B} \vdash \neg P(\ulcorner \neg P(\ulcorner \neg Q \urcorner) \urcorner) \to P(\ulcorner \neg Q \urcorner) \text{ (by IntrComp1)}, \quad (9.53)$$

$$\mathscr{B}, CONS, COMP \vdash \neg P(\ulcorner \neg Q \urcorner) \wedge P(\ulcorner \neg Q \urcorner). \quad (9.54)$$

Thus, we have the next theorem.

Theorem (IntrComp1): Under (B.1), (B.2), (B.3), (B.4), and IntrComp1, we have $\mathscr{B}, COMP \vdash \neg CONS$, and $\mathscr{B}, CONS \vdash \neg COMP$.

Bibliographic Notes

The contents in Chapter 9 (and the last part of Chapter 10) are based on earlier papers written in 2002–2006 (Urai 2002a, Urai 2002b, Urai 2002c,

---

[51] With respect to introspective conditions, the position of the negation sign provides two more categories.

Urai 2006). For these researches I was greatly inspired by Mamoru Kaneko (Tsukuba University) for his enlightening work on the foundation of game theory including the concept of common knowledge (Kaneko 1996), game logic (Kaneko and Nagashima 1996), and inductive games (Kaneko and Kline 2008). The method and standpoint of my works (including this book) are completely different from his approach in the sense that the arguments are based on an *axiomatic set theory* that is *strong enough to code itself into its objects* (sets). At least for describing the methodology of social sciences, I believe that this *recursive feature* is a minimal requirement that the basic theory must have. The axiomatic set theory given in Section 9.1 is quite standard, but it is also intended to be a minimal basic theory for the social sciences. As noted, I have borrowed many notions and methods from Kunen (1980), Jech (2003), and Fraenkel *et al.* (1973).

In writing this and subsequent chapters, I have also gleaned much from philosophical writings, e.g., the recent works of H. Putnam (Putnam 1981, Putnam 1983, Putnam 1990, Putnam 1995, Putnam 2004) for rationality and truth, concepts in Kripke (1972) and Kripke (1975) for a fixed-point view of the world, and works in cognitive science such as Lakoff (1987) for problems about objectivism.

Chapter 10
Concluding Discussions

10.1 Fixed Points and Economic Equilibria

10.1.1 *Equilibrium concepts in economics*

From Adam Smith's *The Wealth of Nations* (1776) to recent works in the theory of general equilibrium, economic theorists have attempted to view human society as an integrated system whose behavior follows scientifically determinable laws. Though the word "scientific" (or the notion of "scientific objectivity") has been severely attacked since the 1950s, we have treated the world as a systematic structure whose properties may gradually but eventually be clarified through mathematical or physical investigations.[1]

In economics, the state of the world is mainly captured through the concept of *equilibrium*. The concept of equilibrium, however, is usually based not only on the *physical* sense relating to what we call facts (e.g., market clearing condition, feasibility conditions under budgets, production sets, and strategy sets) but also on the *mental* (*moral*) or *epistemic* sense, including our value judgments (e.g., utility maximization under given prices as our "rational" behaviors, "rational expectations" compatible with underlying models, the condition for Nash equilibrium strategies considering other players' strategies as given, and many kinds of stability conditions in arguments for the refinement of Nash equilibria).

---

[1] After World War II, the methodology of economics as a *social science* was founded on the philosophy of *empiricism (logical positivism)*. Even today, we often say that we can avoid *metaphysical problems* based on the distinction between *facts* and *values*. As Hilary Putnam argues (2002), such a fact/value dichotomy is closely related to the analytic/synthetic dichotomy criticized by W.V.O. Quine in his famous essay "Two dogmas of empiricism" in 1951 (reprinted in Quine (1953)), which has historical importance to the streams of philosophy pursued in the last century. One purpose of this book (as seen in the previous chapter) is to show (by purely mathematical arguments) that to see the "real" human world, we cannot "stand outside it and take the stance of an observer with perfect knowledge," since "we are organisms functioning as part of reality" (Lakoff 1987, p. 261).

General equilibrium theory after the 1950s (from Debreu 1959, Arrow and Hahn 1971, and Hildenbrand 1974 to Mas-Colell 1985, Balasko 1988, etc.) has treated such epistemic conditions as a given hypothesis or a matter outside of theory. Of course, the moral or epistemic side of equilibrium conditions are nothing but our *value judgments* or *beliefs* concerning *the society to which we belong*. Therefore, such a viewpoint had to gradually change the meaning of economic equilibrium based on the rigorous and rigid beliefs from that of human society as a whole (as in traditional works from Walras (1874) to Hicks (1939)) to that of a small and partial market mechanism.[2]

Value judgments (the moral or epistemic side of equilibrium conditions) have a special importance, however, since they inherently include our judgments regarding the difference between "facts" and "values," "necessities," "contingencies," and "possibilities," and so on. In other words, they determine our concept of "world," i.e., our basis for recognition or "rationality." Putnam calls the targets of such value judgments *epistemic values* (2002, p. 30).[3]

Implicitly or explicitly, an economic equilibrium "concept" always presumes such epistemic values.[4] Based on one such belief, *a view of the world*, we claim that our actions and plans are "rational" and that we can explain our motivation (incentive) to follow an equilibrium choice as rationality based on this view of the world. For example, the *general market equilibrium* is a physical market-clearing condition with incentive (mental or moral) conditions of individual profit/utility maximization under the (epistemic) assumption (a special view of the world) that *all prices are given*. Furthermore, the *Nash equilibrium* of an n-person game is a physical list of available personal strategies with the incentive (mental or moral) part

---

[2] As rightly pointed out by Balasko (1988, p. 2), "A widespread but rather unfortunate practice of economic theory is to consider the market as universal, in the sense that it is unique and that every commodity is traded there, an interpretation that cannot be seriously defended. ··· we would first have to understand the workings of every single market, each of which is already quite complex because each involves numerous different commodities."

[3] The collapse of the classical fact/value dichotomy caused by such epistemic values is argued in Putnam (2002) with an intimate relation to the collapse of the analytic-synthetic dichotomy, which is an essential part of Quine's critique of empiricism.

[4] Of course, this is by no means a new way of thinking but rather a traditional one. For example, we can see the same thought in Max Weber's emphasis on the "empirically impossible proof of the validity of the evaluative ideas" (1904, paragraph 66) in his argument about the objectivity of social sciences as well as his famous advocacy of "Wertfreiheit" and "Idealtypus."

of each person's best response condition under the (epistemic) supposition (a view of the world) that *no other person changes his/her current strategy*. As a description of human society, the epistemic condition (or a moral condition in the sense of a condition describing what our society is or should be) is far more important than the physical condition.

10.1.2 Fixed points and social validity

Unfortunately, however, we have little to say about the validity of such a basic belief (a view of the world) itself. It is even possible that such beliefs (at least in the ordinary sense) seem too restrictive or false, as in the concepts of general market and Nash equilibria. Indeed, agents can change prices as long as they have positive market powers (positive measures in the general equilibrium model), and other players are allowed freely to change their strategies under the settings of an n-person game. The criteria for the rationality of such a belief, if it even exists in any rigorous sense, cannot be given by reference to "rationality" based on the belief itself.[5] On the other hand, we can always freely compare the consequence of our actions (experiences in the world) with those in our presupposed beliefs (views of the world). Since the world (society) includes ourselves, our beliefs or views of the world also shape part of it. (Needless to say, this is the most idiomatical feature of social sciences.) As a minimum requirement for equilibrium concepts in social sciences under an epistemic viewpoint, therefore, it is quite natural to suppose that such a belief (a view of the world) is not *self-refuting*; that is, the consequences of "rational" actions under the belief do not contradict any supposition that constructs a part of the belief. Here we can see a profound relation between the equilibrium concepts in social sciences and mathematical *fixed points*.

Hence, we can relate the fixed-point argument in economic equilibrium theory with one of our beliefs: a view of the world that is "stable" in the sense that it is not self-refuting. As stated above, this self-irrefutable requirement is a minimum requirement, which is merely to say that the view happens to be *self-irrefutable*, not that it is justified under a particular

---

[5] "夫知有所待而後當, 其所待者, 特未定也, Knowledge must wait for something before it can be applicable, and that which it waits for is never certain" — 荘子 Chuang Tzu — (Six, The Great and Venerable Teacher, (Watson 1968)). Rationality in the above sense cannot prove its validity by itself, since for such arguments we have to take such validity for granted without any proof. (See also footnote 6 below.)

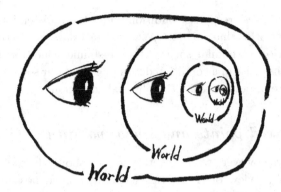

Figure 49: View of world: Most idiomatical feature of social science

Figure 50: A corner of a cube or a ceiling?

rigorous criterion (just as the three halflines in Figure 50 might be viewed as forming either the corner of a cube or a room). It is even possible (or perhaps most likely) that an equilibrium is based on an incorrect view of the world as long as we have no chance or ability to observe it. We can support the concept of Nash equilibrium since, in the equilibrium, no player has an incentive to change his/her own strategy. For the general market equilibrium, it is even possible to consider a converging path of prices to an equilibrium as long as agents continue to regard their price-taking behaviors as rational. The meaning of such "stability," therefore, fluctuates based on what kind of knowledge (or ability to think) is presupposed for each member of the model of the world. Of course, it should really change based on the development of our real knowledge in the future, especially with advances in ways to describe our knowledge (knowledge about our knowledge).

10.1.3 *Rationality and economic equilibria*

Consider, however, that the most important knowledge obtained in the previous century was that such developments in our knowledge *never end* from a highly *rigorous* viewpoint. This may partly be observed through purely mathematical results like Gödel's Incompleteness Theorem (1931), partly by works in analytical philosophy such as Wittgenstein's *Tractatus Logico-Philosophicus* (1919) and *Philosophische Untersuchungen* (1953) as well as Quine's "Two Dogmas" (*ibid.*) and "Indeterminacy of Translation" (1960), and by developments in many other academic fields related to the problems of the mind, such as psychology, linguistics, anthropology, computer science, and cognitive science. A complete description of human society, in a rigorous sense, can never be accomplished because we cannot produce a complete description of our knowledge.[6] The situation seems to make our standpoint on objectivity in the social science to be erected on a platform equivalent to that of Max Weber, where the word objectivity is merely used as a tool of research and is not considered a research target.[7]

To know what a society is, we have to know the view of the world, the *value* judgments, of each member of that society. On the other hand, if we have obtained a certain *fact* of the world based on some value judgments, our knowledge development of the fact may also affect those epistemic *value* judgments. In particular, knowing such a special fact about epistemic values as "we cannot have a complete description of our knowledge" forces us to appraise the idea that there is no satisfactory way to define "rational" behavior. In other words, to be "truly" rational, we must reconcile our irrationality with any actual choice. However, the problem is not so simple that we can resolve it with the word "irrational," since admitting irrationality, even in part of an argument, usually introduces a contradiction into our logic and renders the totality of our logic useless. (Indeed, in standard logical arguments, a contradiction is a proof applicable to every statement, so that we obtain rigorous proofs for all unreasonable statements like "$1 = 0$," "You and I are the same person," and "Every square is a triangle.") The situation is just like Epimenides's liar paradox in ancient Greek philosophy (Figure 51). Needless to say, it would be absurd to base an argument on economic or game-theoretic settings in which agents

---

[6] In relation to Gödel's theorem, we encountered this problem in the previous chapter. In the social sciences, it follows that epistemic values like "individual rationality" and "social validity" are impossible to be described as purely mathematical objects in a satisfactory manner.

[7] Weber's "objectivity" based on "Idealtypus" and "Wertfreiheit" is in this regard quite different from the scientific objectivism of logical positivists. He insists that "the 'objectivity' of the social sciences depends rather on the fact that the empirical data are always related to those evaluative ideas which alone make them worth knowing and the significance of the empirical data is derived from these evaluative ideas" (1904, *ibid.*).

Figure 51: Circle of irrationality: To be rational one must recognize one's own irrationality

have such impractical logical abilities. It remains important, however, to describe the world as it is, based on the knowledge that "we cannot have a complete description of our knowledge," since this condition is related to the truth of our world.

In economics, if we are concerned merely with the classical static framework, the above paradoxical feature of rationality may not cause any problem, since rationality may simply be identified with behaviors that are justified as optimal under a certain formalized (possibly personal) fixed view of the world. (Of course, this type of argument is all that we have developed so far in the theory, and Chapters 3–5 of this book may also be considered one consequence of this tradition from equilibrium or fixed-point arguments.) However, if the word rationality is used in the dynamic context of our ability to recognize, like "rational expectation," it may cause problems, since "true" rationality (as long as the word is used as something related to general human intelligence) should recognize that every model-specific view of the world is not complete. The paradox may also have serious consequences in game-theoretic settings, such as agents' common knowledge of rationality.[8]

10.1.4 Fixed points and view of the world

The problem is how we treat truth, rationality, and value in relation to our developing knowledge and recognition. Of course, there are impossible

---

[8] In the previous chapter, we saw the undefinability of common knowledge (Remark 9.2.6), which means that in ordinary (n-person) game theory, to describe agents' rationality as a common knowledge, the players of the game must be limited to those who cannot use ZF set theory as their basic knowledge.

results in the previous chapter such as non-definability theorems in Remarks 9.2.4 and 9.3.4 for this problem. In describing the world, therefore, we have to forsake at least one of the following requests: (a) to treat our general intelligence (like ZF), (b) to treat our developing knowledge (like social recognition), and (c) to describe the world in a deterministic way (through ZF as a process to determine what happens). Although this is not the type of problem that has a final answer, a fixed-point concept provides an important method for constructing views of the world. To define "truth" in a mathematical model of language, for example, Kripke (1975) used a fixed-point argument to incorporate the necessary infinite-order of regression in the ordinary usage of our sentences. He changes the problem of defining truth to a fixed-point problem for a developing language (*fixed-point language*).

We may treat problems like *individual rationality* and the many kinds of *social values* (morals, conventions, institutions, credits, money, etc.) in the same way, not as given constants but as variables for fixed-point arguments, including all of our actions and developing knowledge. The significance of a fixed point as a way to grasp the world in the above sense will be discussed and formulated further in the next section. The argument will be methodologically founded on new thoughts and directions for social sciences directly based on the *impossibility* of freeing the basic theory of social science from discussions on the epistemic problem.[9]

Even in the hard sciences, we are not allowed to recognize the total world as an absolutely unique and complete readymade system. We can recognize a "chair" or a "stone" as a simple collection of molecules or as a wave form from the point of view of wave equations.[10] In the social sciences, the situation is more serious, and we are not allowed (at least in a purely ideal theoretical sense) to recognize the total system as a unique artifactual

---

[9] As seen in Section 10.2, one may relate some part of our arguments with the *new theory of references*, the concept of *metaphysical necessity* or *possible worlds* (c.f. Kripke (1972), Putnam (1983)), *internal realism*, *cognitive semantics* (see Putnam (1981), Putnam (1983), Lakoff (1987)), and, moreover, to a certain pragmatic standpoint like Quine (1951), Putnam (1995), or James (1912).

[10] "Physical objects, small and large, are not the only posits. Forces are another example; and indeed we are told nowadays that the boundary between energy and matter is obsolete" (Quine (1951) *Two Dogmas of Empiricism*). "Moreover, the quantum mechanical Weltanschauung that emerged from the 2nd Solveg Conference — the 'Copenhagen Interpretation' — remains controversial even today. A substantial minority of cosmologists have deserted it for the so-called 'Many Worlds Interpretation' — an interpretation which implies, among other things, that there are 'Parallel Worlds,'..." (Putnam (1995) *Pragmatism*).

object. At the same time, we have to choose our actions as long as we live. Consequently, we have to choose our rationality, including many kinds of epistemic values, under the condition of *self-reliance*. The significance of fixed-point arguments is the ability to describe this *self-reliance*.

10.2 Rationality and Fixed-Point Views of the World

In this section, by incorporating the arguments in Sections 9.2 and 9.3 of the previous chapter, rationality or social validity may be determined as an equilibrium of the social model in which the cognitive features of members are treated explicitly. Intuitively, the model in this section describes the situation in which each member can choose an arbitrary finite number of models of the society, *one's possible worldviews*, to approximate the 'real' world. Even though the candidates for such models of society for each member may not be finite, and members are not convinced by their approximation to be complete, we can expect the existence of an equilibrium for the list of each person's view of the world and 'rational' behaviors based on it, as long as the total space for behaviors is not so large, e.g., a compact Hausdorff space. In the equilibrium, those behaviors are compatible with each person's view of the world; i.e., actions are tested in the light of their experiences and beliefs for the validity of the model.

The results in Chapter 9 testify to the difficulty of describing 'individual rationality' and/or 'social validity' as artificial (well-defined) objects. In the following, by considering the space of the action configurations of all members, $X = \prod_{i \in I} X_i$, as the set of *rigid designators* that are identified across all possible worldviews, we characterize a pair of *rational behaviors* ("justified" for each agent under a view of the world) and possible worldviews ("compatible" for each agent with such behaviors) as an equilibrium situation for recognition of society.

By such arguments in Section 9.2, a "justification" procedure for 'rational' actions (under a view of the world like an economic model) may be consistent though it may not be complete enough to justify what is every "possibility" for all views of the world. Judgments for "compatibility" (to determine what is appropriate to be called possible worlds), therefore, should be another social validity based on the same view of the world that recognizes it as valid. Section 9.3 also argues, however, that we cannot expect such "validity" to be strong enough to maintain the validity of itself as long as we require it to be consistent. It follows that any

10: Concluding Discussions

attempt to rigorously determine the "possibility of the world" through a class of necessary and sufficient conditions will fail. We have to leave the extension of such a concept somewhat receptive to the situation of the 'self-reliance' of our minds, to our 'willingness' to seek another possibility of the world, or certain 'beliefs,' if such 'really' exist.[11]

To describe and verify the existence of equilibria, we use a certain kind of mixed strategy (with expected utility) settings for the simplicity and familiarity of arguments. A *view of the world* in this section, therefore, is taken as a mixture of possible worldviews (e.g., a probability measure on the set of possible worldviews). The justification for behaviors and the judgment for compatibility, first associated merely with each possible world (worldview), are supposed to be extended over all such mixtures.[12] The essential feature of this section's approach, however, does not depend on such a special framework. The central ideas discussed here characterize human's "rationality" (at least in the sense of "rational behaviors" for game-theoretic settings in social science) and "validity" for a society, not as terms or objects fixed by a set of criteria laid down in advance but as *references* or equilibrium fixed points in determining the extensions of the terms that refer to them using classes of laws not all of which we exactly know.[13]

---

[11] Note, however, that it is inappropriate to think that we can further formulate such 'self-reliance,' 'willingness,' or 'beliefs' as a part of our rationality, since such an attempt (to formalize our rationality) should be treated as a possible worldview itself. In this sense, there is no room for our 'rationality' to play over the list of all possible worldviews or the 'self-reliance' of our minds. It is natural and interesting to treat the totality of such possible worldviews as a black-box mechanism for each person from the *radical empiricist's* standpoint (James 1912).

[12] One may ask why such a view of the world (in the above sense of mixture) itself is not classified into one of the possible worlds (worldviews). Of course, we may call it a possible world and select it from the list as a candidate for possible worldviews among which further mixtures (views of the world) are taken.

[13] The concepts may be related to *natural kind words* in the sense of Kripke (1972) and Putnam (see, e.g., Putnam (1983, 4. Reference and Truth)). Their approaches (independently advanced in the 1960s and 1970s) are called the *new theory of reference*; at least in this section, the concept of "possible world(views)" is used in relation to this context. I am not going to say, however, that we should (or we can) "define" rationality as a natural kind word, but merely to state that as long as we want to treat it under mathematical (and social scientific) arguments, mathematics should (and can) treat it properly through characters fixed by considering all possible views of the world. Note also that the term, "natural kind," is sometimes used under the tradition of classical view of objectivism or metaphysical realism (c.f. Lakoff (1987, Chapters 16–17)). The approach I have taken here does not favor such *externalism* but the *internalism* (internal realism) of Putnam (1981). For this, as stated in footnote 11, we should not (and cannot) formalize the totality of possible worldviews.

10.2.1 Individual and society

Let $I = \{1, 2, \ldots, m\}$ be the index set of members of society. For each $i \in I$, denote by X_i the set of possible *behaviors* for individual i. For simplicity, we assume that each X_i is a non-empty compact convex subset of a Hausdorff topological vector space. We also assume that X_i covers all behaviors observable by others for each $i \in I$ and that each *behavior profile* $(x_1, x_2, \ldots, x_m) \in \prod_{i \in I} X_i$ is sufficient to decide *consequence* $c_i \in C_i$ for each $i \in I$.

10.2.2 Languages and possible worlds

Each member i is assumed to have a set of *logical formulas* T_i, *inference rules* R_i, and *language* $\mathscr{L}_i$ that may be considered a list of symbols describing the terms and formulas. Triplet $(\mathscr{L}_i, R_i, T_i)$ is called a *possible world (worldview)* of i. Member i may have many (we suppose possibly denumerably many) possible worlds, $W_i^0 = (\mathscr{L}_i{}^0, R_i^0, T_i^0), W_i^1 = (\mathscr{L}_i{}^1, R_i^1, T_i^1), W_i^2 = (\mathscr{L}_i{}^2, R_i^2, T_i^2), \ldots$. We assume they may not mutually be consistent or may not even be translatable to one another.

Let us consider the inductive limit under inclusion (with respect to non-empty finite subsets of natural numbers directed by inclusion) of abstract simplices, $W_i = \varinjlim_A \overline{W_i^{a(0)} \cdots W_i^{a(n)}}$, where $A = \{a(0), a(1), \ldots, a(n)\}$ and each $\overline{W_i^{a(0)} \cdots W_i^{a(n)}}$ is identified with n-dimensional standard unit simplex Δ^n. In the following, we denote by W_i^A simplex $\overline{W_i^{a(0)} \cdots W_i^{a(n)}}$, so we write $W_i = \varinjlim_A W_i^A$. Point $w \in W_i$ may be considered to represent a special standpoint of i's thought. We call it i's (*extended*) *view of the world* (*worldview*).

Let W be set $\prod_{i \in I} W_i$. In the following, we describe the situation where each $i \in I$ cannot specify or grasp totality W_i of one's own thoughts. (It does not mean, however, that each i cannot understand the abstract level discussion we are going to have below since we do not prevent each i from understanding ZF. In this sense, the discussion in this section is not an argument from God's eyes.)

10.2.3 Possible worldviews and behaviors

For each $i \in I$, the possible worldviews (in the narrow sense) of i, $W_i^0, W_i^1, W_i^2, \ldots$, represent various kinds of reasoning for a certain behavior, $x_i \in X_i$, to be considered better than other others. For example, W_i^n

may be a possible world of a Nash Equilibrium; i.e., under W_i^n, i is convinced that his/her choice of behavior x_i is reasonable since it is part of a Nash equilibrium strategy profile for a certain game-theoretic model of society that is completely described and treated as valid in $(\mathscr{L}_i{}^n, R_i^n, T_i^n)$. (In such a case, i's estimation on the thoughts of other players are also described sufficiently for the equilibrium concept and treated as valid for individuals in $(\mathscr{L}_i{}^n, R_i^n, T_i^n)$.) It is also possible to consider W_i^n a world of cooperation equilibrium (i.e., i thinks that his/her behavior may be considered reasonable since everyone's behavior may be classified as choices to decide a consequence in the core of a game defined in W_i^n), a world of an incomplete information game, a world of an abstract economy (in which the constraint correspondence is described as a rule in T_i^n), and so on. Note that we are assuming that all behaviors that are possible for i are completely listed in X_i. Hence, each X_i is defined to include a mixed strategy if such behaviors are allowed to exist in the formalized model.

10.2.4 *Justification and refutability*

For each possible worldview (in the narrow sense) $W_i^n \in W_i$ of $i \in I$, define for a given profile of behaviors, $x = (x_j)_{j \in I} \in X = \prod_{j \in I} X_j$, the set of *improved behaviors*, $\Phi_i(W_i^n, x) \subset X_i$, as behaviors that are better than x_i under worldview W_i^n, and the set of *compatible* profiles of behaviors $\Xi_i^A(W_i^n) \subset X$ for each finite set of natural numbers $A \ni n$. Behavior x_i is not *justified* under worldview W_i^n, if it can be improved ($\Phi_i(W_i^n, x) \neq \emptyset$). When (observable state) x is not compatible with W_i^n, W_i^n is *refutable* for i under x and A.[14] In the following, we deal with the concept of the *rationality* of i based on a justification (reasoning) for behavior x_i of profile $x = (x_j)_{j \in I} \in \prod_{j \in I} X_j$ under a certain possible worldview that is not refutable under x. For each non-empty finite subset A of natural numbers,[15] define $\Psi_i^A(x)$ as $\Psi_i^A(x) = \{W_i^n | n \in A, x \in \Xi_i^A(W_i^n)\}$. (Note that for each i, it is possible to treat mappings Ξ_i^A, a restriction of Φ_i, and Ψ_i^A in a certain formal theory $W_i^{n'}$, although it may not be possible to specify Φ_i in a certain $W_i^{n'}$. Even so, it is possible for i to consider correspondences Φ_i under a certain theory at least in the same abstraction level with our arguments in this section as long as we use ZF and i is allowed to use ZF.)

---

[14] Here, we consider that whether worldview W_i^n is refutable under an observable state x or not is dependent on A, i.e., some properties of i outside theory W_i^n.
[15] Let us use notation $A \in \mathscr{F}(N)$ as in cases with spaces having convex structure.

As stated before, we consider that i's view of the world (in the extended sense) is a point, $w_i \in W_i = \varinjlim_A \overline{W_i^{a(0)} \cdots W_i^{a(n)}}$. Each i has no formalized theory on the rightness among all possible worldviews in W_i.[16] However, we do not prevent i from having formal treatments among finite possible worldviews $W_i^0, \ldots, W_i^n$ in a certain $W_i^{n'}$. For example, we may interpret point $w_i \in \overline{W_i^0 \cdots W_i^n} \subset W_i$ as a representation of the state of i's thought as a degree of confidence among possible worldviews, $W_i^0 \cdots W_i^n$, and i has a certain theory integrating those views using concepts like expected utility. In this sense, we may suppose (for simplicity) that Φ_i and Ξ_i^A may adequately be extended as correspondences on $W_i \times X$ ($W_i^A \times X$ for each A) and W_i (W_i^A for each A), respectively, and we may redefine Ψ_i^A as $\Psi_i^A(x) = \{w_i \in W_i^A \mid x \in \Xi_i^A(w_i)\}$ for each $A \in \mathscr{F}(N)$.

Of course, a certain $w_i = (w_i^0, \ldots, w_i^{\ell(i)}) \in W_i^A$ may justify or refute behaviors and other worldviews in various ways based on $W_i^0, \ldots, W_i^{\ell(i)}$ or through other unknown factors. We may assume, however, that it would be natural for $\Phi_i : W_i \times X \to X_i$, $\Xi_i^A : W_i^A \to X$, and $\Psi_i^A : X \to W_i^A$ to satisfy the following conditions.[17]

(A-0) $\Phi_i : W_i \times X \to X_i$ has the local intersection property at each (v_i, x) such that $\Phi_i(v_i, x) \neq \emptyset$ in $W_i \times X$ and satisfies the irreflexivity condition, i.e., for all $x = (x_i)_{i \in I}$, and $v_i \in W_i$, $x_i \notin \Phi_i(v_i, x)$.

(A-1) $x \in \Phi_i(v_i, z)$ and $y \in \Phi_i(w_i, z)$ imply that for all $\lambda \in [0, 1]$, there exists $\hat{\lambda} \in [0, 1]$, such that $\lambda x + (1 - \lambda)y \in \Phi_i(\hat{\lambda} v_i + (1 - \hat{\lambda})w_i, z)$.

(A-2) $\Psi_i^A : X \to W_i^A$ is convex valued.

Three conditions may be naturally interpreted. No explanation is needed for the irreflexivity in Condition (A-0). Condition (A-1) says that a mixture of better strategies will be supported by a certain mixture of two possible worlds. Note here that $\hat{\lambda}$ may be different from λ. The convexity condition in (A-2) asserts that any mixture of worldviews compatible with x is also a worldview compatible with x.

By considering that no complete description of the world exists (results from earlier sections), it would be appropriate to treat such a standpoint

---

[16]The non-definability of such a complete rightness is indeed the consequence of the arguments in the previous chapter.

[17]We treat these structures on $W_i^A \times X$ as fundamental features of i whose totality (except for the abstract properties (A-0)–(A-3) below) cannot be known by anyone, or more precisely, does not affect the abstract argument (on the existence of equilibrium) we discuss below. See footnotes 11, 12, and 13.

10.2.5 Equilibrium under social recognition

We assume possible worldviews $W_i^0, W_i^1, W_i^2, \ldots$ for each $i \in I$, which means that person i does not have any formalized ideas on the relation among 'all' such possible views. In other words, $W_i^0, W_i^1, W_i^2, \ldots$ are all formalized ideas of i with respect to society. At the same time, we suppose that for each $i \in I$, relations Φ_i and Ψ_i^A exist on subsets of $W_i = \varinjlim_A \overline{W_i^{a(0)} \cdots W_i^{a(\sharp A-1)}}$, based on justifications and refutabilities.

Let $X = \prod_{i \in I} X_i$, $W^A = \prod_{i \in I} W_i^A$, and $W = \prod_{i \in I} W_i$. Given behavior profile $x = (x_1, \ldots, x_m) \in X = \prod_{i \in I} X_i$ and views of the world $w \in W^A$, we denote by $\varphi_i^A(x) \subset W_i^A \times X_i$, the set of pairs, (v_i, y_i) such that $v_i \in \Psi_i^A(x) \subset W_i^A$ and $y_i \in \Phi_i(v_i, x)$, i.e., the set of pairs of compatible view v_i of the world with x under A and improved behavior y_i better than x_i under v_i. When we restrict our attention merely to the behavioral aspects, we may call state $x^* = (x_i^*)_{i \in I} \in X$ an *equilibrium (under social recognition)* for society $((W_i, X_i, \Phi_i, \Psi_i^A)_{i \in I, A \in \mathscr{F}(N)})$ if $\varphi_i^A(x^*) = \emptyset$ for all $i \in I$ and $A \in \mathscr{F}(N)$. Equilibrium $x^* = (x_i^*)_{i \in I}$ is said to be *rationally accessible* if for every open neighborhood U of x^* and $A \in \mathscr{F}(N)$, we have $A' \supset A$ and $x^{A'} \in U$ such that $\varphi_i^{A'}(x^{A'}) = \emptyset$ for all i.[18]

THEOREM 10.2.1: *Society $((W_i, X_i, \Phi_i, \Psi_i^A)_{i \in I, A \in \mathscr{F}(N)})$ has equilibrium if (A-0), (A-1), and (A-2) are satisfied.[19] Moreover, assume the following condition:*

(A-3) *For each $i \in I$ and $x \in X$, open neighborhood U of $x \in X$ and $A \in \mathscr{F}(N)$ exist such that for all $A' \supset A$, we have $\Psi_i^A(z) \subset \Psi_i^{A'}(z)$ for all $z \in U$.*

Then we may also confirm the existence of rationally accessible equilibria.

---

[18] In this definition, it is possible (and, indeed, more desirable) to take A' differently for each i. We use the above setting merely for notational simplicity.

[19] We use (A-0) (the (Product K*)-type condition in Chapter 3) to assure the existence of equilibrium. Of course, it is also possible to base our arguments directly on other conditions like (Product B), (Product K), (Product K1–V4*), (Product K1–V4), and (Product B or K1).

PROOF: For each $i \in I$, define $\varphi_i^X : X \to X_i$ as $\varphi_i^X(x) = \{y_i \in X_i | y_i \in pr_X(\varphi_i^A(x)), A \in \mathscr{F}(N)\}$. By definition, the existence of equilibrium may be verified by showing the existence of $x^* = (x_i^*)_{i \in I}$ such that $\varphi_i^X(x^*) = \emptyset$ for all $i \in I$. Therefore, under (A-0), to show the existence of equilibrium, it is sufficient by Theorem 3.2.1, to verify that each φ_i^X satisfies conditions (P1)–(P3) in Chapter 3. By (A-0), each φ_i^X satisfies (P1). By (A-1) and (A-2), each φ_i^X is convex valued. Moreover, the local intersection property of each Φ_i (under (A-0)) is sufficient to assure the local intersection property of each φ_i^X. Hence, there is equilibrium point $x^* \in X$. To show the existence of rationally accessible equilibria, for each $A \in \mathscr{F}(N)$, consider the equilibrium problem for product mapping:

$$\varphi^A : W^A \times X \ni ((w_i)_{i \in I}, (x_i)_{i \in I}) \mapsto \prod_{i \in I} \varphi_i^A((x_i)_{i \in I}) \subset W^A \times X.$$

Since $W^A \times X$ is compact, we may obtain (through exactly the same argument with the first assertion under conditions (A-0), (A-1), (A-2), and Theorem 3.2.1) $x^A = (x_i^A)_{i \in I} \in X$ such that $\varphi_i^A(x^A) = \emptyset$ for every i for each $A \in \mathscr{F}(N)$. Since X is compact, we may obtain a converging subnet of $\{x^A\}_{A \in \mathscr{F}(N)}$ that converges to $x^* \in X$. By the subnet definition, to see that x^* is a rationally accessible equilibrium, it is sufficient to check that x^* is an equilibrium. Indeed, x^* is an equilibrium. If not, for some i and A, we have $\varphi_i^A(x^*) \neq \emptyset$, which means by the local intersection property of Φ_i, for a certain $v_i \in W_i^A$, $v_i \in \Psi_i^A(x^*)$ and an open neighborhood U of x^*, $\Phi_i(v_i, z) \neq \emptyset$ for all $z \in U$. Here, under (A-3), we can take U to satisfy $z \in \Xi_i^{A'}(v_i)$ $(v_i \in \Psi_i^{A'}(z))$ for all $z \in U$ for every sufficiently large A'. Therefore, for each $A' \supset A$ (since $v_i \in W_i^{A'}$ and $v_i \in \Psi_i^{A'}(z)$ for all $z \in U$), we have $\Psi_i^{A'}(v_i, z) \neq \emptyset$, so $\varphi^{A'}(z) \neq \emptyset$, which contradicts that $x^A \in U$ for every sufficiently large A in the subnet. ∎

Mathematical Appendix I

P-adic Expansion of a Real Number

As stated in Chapter 1, Section 3, we identify the set of real numbers, R, with the conditionally complete ordered field having the usual order topology. This view gives us an axiomatically clear standpoint on the domain of discourse for the calculus. From this viewpoint, however, the relation between a real number in this book and a real number in the ordinary sense under the decimal system is not clear. Precisely speaking, except for 0 (identity element under addition) and 1 (identity element under multiplication), elements of R do not have any previously given symbolic forms.

In the following, we see that for any natural number $P \geq 2$, real number $x \in R$ has an expression through a sequence of natural numbers less than P, the P-adic expansion of x. Therefore, the set R may be identified with the set of such sequences of natural numbers less than P.[1]

Let $P \in N = \{0, 1, 2, \ldots\}$ be an arbitrary natural number greater than or equal to 2.[2] Since $1+1, 1+1+1, 1+1+1+1, \ldots$ are elements of R, we may identify P with the element of R obtained by adding 1 for P times. There are elements of R which are obtained by multiplying P for $2, 3, 4, \ldots$ times, which will be denoted by $P^2, P^3, P^4, \ldots$. There are also multiplicative inverses of $P, P^2, P^3, \ldots$ in R, denoted by $P^{-1}, P^{-2}, P^{-3}, \ldots,$

---

[1] Of course, such an expression may fail to be unique as real number 1 has two decimal expressions like $1.000\cdots$ and $0.999\cdots$. The argument below ((a)(b) and (1)–(3)) describes a process to specify one of those alternatives (e.g., in decimal case, we use $0.999\cdots$ to 1).

[2] Here, the set of natural numbers $N = \{0, 1, 2, \ldots\}$ is an object defined directly through the set-theoretic axioms (like Infinity), and it is independent of R. It is also assumed that on set N we have the ordinary addition and multiplication that are also independent but can be identified with those in R when N is naturally embedded in R.

respectively. The definition of ordered field is sufficient to ensure the next relation among these elements:

$$0 < \cdots < P^{-3} < P^{-2} < P^{-1} < P^0 < P^1 < P^2 < P^3 < \cdots, \quad (10.1)$$

where $P^0 = 1$ and $P^1 = P$. Moreover, we can verify from the definition of ordered field that for any real number $y > 0$, there is a sufficiently large n such that $P^{-n} < y$ and $y < P^n$. Note also that in the above sequence, each element (except for 0) is obtained by adding its immediate predecessor for P times.

For a positive real number $z > 0$, we can define an integer $m(z)$ and a natural number $a(z)$ in $\{1, \ldots, P-1\}$ as follows:

(a) If $z \geq 1$, let m be the first natural number such that $z < P^{m+1}$ and n be the first natural number such that $z < (n+1)P^m$. Let $a(z) = n$ and $m(z) = m$.
(b) If $z < 1$, let m be the first natural number such that $P^{-m} < z$ and n be the first natural number such that $z < (n+1)P^{-m}$. Let $a(z) = n$ and $m(z) = -m$.

Thus we have defined $P^{m(z)}$ and $a(z)$ satisfying

$$a(z)P^{m(z)} \leq z < (a(z)+1)P^{m(z)}.$$

Define for $t = 0, 1, 2, 3, \ldots$, natural number $a_{m(z)-t}(z)$ in $\{0, 1, 2, \ldots, P-1\}$ by the following process:

(1) Let $a_{m(z)}(z) = a(z)$.
(2) If we have $P^{m(z)-1} \leq z - a_{m(z)}(z)P^{m(z)}$, then let n be the first natural number such that $z < (n+1)P^{m(z)-1}$, and define $a_{m(z)-1}(z)$ as $a_{m(z)-1}(z) = n$. If $P^{m(z)-1} > z - a_{m(z)}(z)P^{m(z)}$, let $a_{m(z)-1}(z)$ be 0.
(3) Suppose that $a_{m(z)}(z), a_{m(z)-1}(z), \ldots, a_{m(z)-(t-1)}(z)$ are defined. If we have $P^{m(z)-t} \leq z - \sum_{i=0}^{t-1} a_{m(z)-i}(z)P^{m(z)-i}$, then let n be the first natural number such that $z - \sum_{i=0}^{t-1} a_{m(z)-i}(z)P^{m(z)-i} < (n+1)P^{m(z)-t}$, and define $a_{m(z)-t}(z)$ as $a_{m(z)-t}(z) = n$. If $P^{m(z)-t} > z - \sum_{i=0}^{t-1} a_{m(z)-i}(z)P^{m(z)-i}$, let $a_{m(z)-t}(z)$ be 0.

By the continuity axiom (the conditionally completeness) for R, we have

$$z = \lim_{t \to \infty} \sum_{i=0}^{t-1} a_{m(z)-i}(z)P^{m(z)-i},$$

i.e., the series $\sum_{i=0}^{\infty} a_{m(z)-i}(z)P^{m(z)-i}$ converging to z.

Mathematical Appendix I

Based on the above process (1)–(3), let us define for a non-negative real number $x \geq 0$, *P-adic expansion of* x. Let $n(x) = m(x)$ if $m(x)$ is well-defined and $m(x) \geq 0$. If $m(x) < 0$ or is not defined, let $n(x) = -1$ and let all undefined $a_{n(x)-t}$ be equal to 0. Then, P-adic expansion of $x \geq 0$ is the following expression for x by the series converging to x,

$$x = \sum_{t=0}^{\infty} a_{n(x)-t}(x) P^{n(x)-t}.$$

For negative real number $x < 0$, the P-adic expansion of x is defined through its opposite $-x$ as $-\sum_{t=0}^{\infty} a_{n(-x)-t} P^{n(-x)-t}$.

THEOREM AI.1: *The conditionally complete ordered field, R, is an uncountable set having a countable dense subset.*

PROOF: Assume that R is countable. Then the subset $Z = \{z | z \in R, 0 < z < 1\}$ of R is also countable. Enumerate the elements of Z as $\{z_0, z_1, z_2, \ldots\}$. Take a natural number $P \geq 4$ arbitrarily, and consider P-adic expansions of elements of Z.

$$z_0 = \sum_{t=0}^{\infty} a_{n(z_0)-t} P^{n(z_0)-t}$$

$$z_1 = \sum_{t=0}^{\infty} a_{n(z_0)-t} P^{n(z_1)-t}$$

$$z_2 = \sum_{t=0}^{\infty} a_{n(z_0)-t} P^{n(z_2)-t}$$

$$\vdots$$

Note that since $0 < z_n < 1$ for all $n = 0, 1, 2, \ldots$, $P^{n(z_n)-t}$ is a member of $\{P^{-1}, P^{-2}, P^{-3}, \ldots\}$ for every $t = 0, 1, 2, \ldots$. By using the above list, let us define a sequence $\{a_t\}_{t=0}^{\infty}$ in $\{0, 1, \ldots, P-1\}$ as follows.

$$a_t = \begin{cases} 2 & (\text{if } a_{n(z_t)-t} = 1) \\ 1 & (\text{if } a_{n(z_t)-t} \neq 1) \end{cases}$$

Then, the series $\sum_{t=0}^{\infty} a_t P^{-1-t}$ converges to non-negative real number $z_* \leq 3P^{-1} < 1$. Moreover, the P-adic expansion of z_* is precisely $\sum_{t=0}^{\infty} a_t P^{-1-t}$. However, $z_* \neq z_t$ for each $t = 0, 1, 2, \ldots$, since $a_t \neq a_{n(z_t)-t}$

for each $t = 0, 1, 2, \ldots$ (Cantor's Diagonal Argument), so we have a contradiction.

R has a countable dense subset since for natural number $P > 2$, the set of all real numbers whose P-adic expansion $z = \sum_{t=0}^{\infty} a_{n(z)-t} P^{n(z)-t}$ satisfies $a_{n(z)-t} = 0$ for all sufficiently large t is countable and clearly dense in R. ∎

Urysohn's Lemma

LEMMA AI.2: (Urysohn) *Let X be a normal space and A and B be two non-empty disjoint closed subsets of X. Then, there is a continuous function f on X to $[0, 1]$ such that $f(x)$ is 0 on A and 1 on B.*

PROOF: If we consider the dyadic expansion (P-adic expansion for $P = 2$) of each element in the interval $[0, 1]$, it is easy to verify that the following set,

$$D = \left\{ \frac{n}{2^m} \,\middle|\, m \in N, n \in N, n \leqq 2^m \right\},$$

is a countable dense subset of $[0, 1]$. Let us define a function g on a subset of X to D by the following process. (In this proof, for each subset U of X, the complement of U in X, $X \backslash U$, is denoted by U^c. It would be helpful in the next process to interpret $U_{n/m}$ as the set preserved to have values less than n/m under f and $V_{n/m}$ as the set preserved to have values greater than n/m under f.)

(1) For $m = 0$, define $g(x) = 0 = 0/2^m$ on $x \in A$ and $g(x) = 1 = 1/2^m$ on $x \in B$. Let us define $V_0 = A^c$, $U_1 = B^c$, $C_0 = V_0^c = A$, and $C_1 = U_1^c = B$.

(2) For $m = 1$ and $n = 1$, define closed set $C_{1/2^1} = Cn/2^m$ as $C_{1/2^1} = U_{1/2^1}^c \cap V_{1/2^1}^c$, where $U_{1/2^1}$ and $V_{1/2^1}$ are two disjoint open sets such that $U_{1/2^1} \supset A = V_0^c$ and $V_{1/2^1} \supset B = U_1^c$, and define $g(x) = 1/2^1$ on $C_{1/2^1}$.

(3) For $m = 2$ and $n = 1, 3$, let us define closed sets $C_{1/2^2}$ and $C_{3/2^2}$ as $C_{1/2^2} = U_{1/2^2}^c \cap V_{1/2^2}^c$ and $C_{3/2^2} = U_{3/2^2}^c \cap V_{3/2^2}^c$, where $U_{1/2^2}$ and $V_{1/2^2}$ are two disjoint open sets such that $U_{1/2^2} \supset A = V_0^c$ and $V_{1/2^2} \supset U_{1/2^1}^c$, and $U_{3/2^2}$ and $V_{3/2^2}$ are two disjoint open sets such that $U_{3/2^2} \supset V_{1/2^1}^c$ and $V_{3/2^2} \supset B = U_1^c$. Define $g(x) = 1/2^2$ on $C_{1/2^2}$ and $g(x) = 3/2^2$ on $C_{3/2^2}$.

(4) Generally, suppose that for each natural number $m < k$ and $n \leq 2^m$, the closed set $C_{n/2^m}$ is defined as $C_{n/2^m} = U^c_{n/2^m} \cap V^c_{n/2^m}$ and $g(x)$ is defined to be equal to $n/2^m$ on $C_{n/2^m}$. For $m = k$ and $2n+1 = 1, 3, \ldots, 2^k - 1$, let us define closed sets $C_{(2n+1)/2^k}$ as $C_{(2n+1)/2^k} = U^c_{(2n+1)/2^k} \cap V^c_{(2n+1)/2^k}$, where $U_{(2n+1)/2^k}$ and $V_{(2n+1)/2^k}$ are two disjoint open sets such that $U_{(2n+1)/2^k} \supset V^c_{n/2^{k-1}}$ and $V_{(2n+1)/2^k} \supset U^c_{(n+1)/2^{k-1}}$. Define $g(x) = (2n+1)/2^k$ on $C_{(2n+1)/2^k}$ for each $2n+1 = 1, 3, \ldots, 2^k-1$.

We verify in the above process that all $C_{(2n+1)/2^k}$'s are non-empty and mutually disjoint, so g is indeed well-defined.

Now let us define function $f : X \to [0,1]$ as an extension of g as follows. For each $x \in X$, if $x \in C_{n/2^m} = U^c_{n/2^m} \cap V^c_{n/2^m}$ for some $m = 0, 1, \ldots$ and $n = 0, 1, \ldots, 2^k$, let $f(x) = g(x)$. If $x \notin C_{n/2^m} = U^c_{n/2^m} \cap V^c_{n/2^m}$ for all $m \in N$ and $n = 0, 1, \ldots, m$, we can divide set D exclusively into two sets $D_U = \{n/2^m \mid x \in U_{n/2^m}\}$ and $D_V = \{n/2^m \mid x \in V_{n/2^m}\}$ as $D = D_U \cup D_V$. Note that for each $r \in D_U$ and $s \in D_V$, we have $s < r$. Since D is dense in $[0,1]$, the partition $D_U \cup D_V$ uniquely defines a real number $t_x \in [0,1]$ such that $\sup D_V = t_x = \inf D_U$. Let $f(x) = t_x$ for such x. By definition, we have $f(x) = 0$ for $x \in A$ and $f(x) = 1$ for $x \in B$, so it remains for us to show that f is continuous. Since the class of all half open intervals forms a subbase for the topology of R, it is sufficient for our purpose to show that inverse images of intervals, $f^{-1}((r,1])$ and $f^{-1}([0,r))$, are closed for each $r \in (0,1)$. Let x be an element of $f^{-1}((r,1])$. Then, we have $f(x) > r$, so $\sup\{n/2^m \mid x \in V_{n/2^m}\} > r$. Therefore, we can take m' and n' so that $n'/2^{m'} > r$ and $x \in V_{n'/2^{m'}}$. For each point z in the open neighborhood $V_{n'/2^{m'}}$ of x, $\sup\{n/2^m \mid z \in V_{n/2^m}\}$ is greater than r, so z is an element of $f^{-1}((r,1])$ and $f^{-1}((r,1])$ is open. On the other hand, if x is an element of $f^{-1}([0,r))$, we have $f(x) < r$, so $\inf\{n/2^m \mid x \in U_{n/2^m}\} < r$. Therefore, we can take m' and n' so that $n'/2^{m'} < r$ and $x \in U_{n'/2^{m'}}$. Again, for each point z in the open neighborhood $U_{n'/2^{m'}}$ of x, $\inf\{n/2^m \mid z \in U_{n/2^m}\}$ is less than r, so z is an element of $f^{-1}([0,r))$ and $f^{-1}([0,r))$ is open. ∎

Partition of Unity Theorem

We can now prove the partition of unity theorem for normal spaces.

THEOREM AI.3: (Partition of Unity) *Let X be a normal space, and let $\mathscr{U} = \{U_1, \ldots, U_n\}$ be a finite open covering of X. There is a family of non-negative real valued continuous functions, $f_1 : X \to R_+, \ldots, f_n : X \to$*

R_+, such that $f_i(x) = 0$ for all $x \in X\backslash U_i$ for each i, and $\sum_{i=1}^n f_i(x) = 1$ for all $x \in X$.

PROOF: We may assume that X is non-empty, $X\backslash U_i \neq \emptyset$ for each $i = 1,\ldots,n$, and that we have a closed refinement $\mathscr{C} = \{C_1,\ldots,C_n\}$ of $\mathscr{U}$ (whose existence is assured by Theorem 3.1.3 in Chapter 3) such that $\operatorname{int} C_i \neq \emptyset$ for all $i = 1,\ldots,n$.[3] For each $i = 1,\ldots,n$, since C_i and $X\backslash U_i$ are two non-empty disjoint closed subsets of X, there is a continuous function $F_i : X \to [0,1]$ such that $F_i(x) = 0$ for all $x \in X\backslash U_i$ and $F_i(x) = 1$ for all $x \in C_i$ by Urysohn's Lemma AI.2. Note that $\{C_1,\ldots,C_n\}$ covers X, so for each $x \in X$ there is at least one i such that $F_i(x) > 0$. For each $i = 1,\ldots,n$, define $f_i(x)$ as $f_i(x) = F_i(x)/\sum_{j=1}^n F_j(x)$. ∎

---

[3] If $X\backslash U_i = \emptyset$ for some i, we may define f_i as $f_i(x) = 1$ for all $x \in X$ and $f_j(x) = 0$ for all $j \neq i$ and x. If $\operatorname{int} C_i = 0$ for some i, then corresponding U_i is not necessary for family $\mathscr{U}$ to cover X, so we may define f_i as $f(x) = 0$ for all $x \in X$.

Mathematical Appendix II

Vector-Space Topology

A *topological vector space* over R is a vector space having a topology on which the addition and the scalar multiplication are continuous (Chapter 1, Section 1.3.3). More precisely, on vector space X over R,

(LT1) $(x,y) \mapsto x+y$ is continuous on $X \times X$ to X, and
(LT2) $(\lambda, x) \mapsto \lambda x$ is continuous on $R \times X$ to X.

If $A = \{x_1, \ldots, x_\ell\}$ is a linearly independent subset of topological vector space over R, the bijective linear mapping $f : R^\ell \ni (a_1, \ldots, a_\ell) \mapsto a_1 x_1 + \cdots + a_\ell x_\ell \in L(A)$ is continuous, where $L(A)$ denotes the subspace of L spanned by the set of all linear combinations among points in A. In this section, we see that if the topology of L is Hausdorff, f^{-1} is also continuous, so $L(A)$ may be identified with $\sharp A$-dimensional Euclidean space. The discussion includes many important basic properties and concepts on vector-space topology.

THEOREM AII.1: (Translation Invariance) *If X is a topological vector space over R, for each $x^* \in X$ and for each $\lambda^* \in R \setminus \{0\}$, mappings $X \ni x \mapsto x + x^* \in X$ and $X \ni x \mapsto \lambda x \in X$ are homeomorphisms.*

PROOF: Since $x \mapsto x - x^*$ and $x \mapsto \lambda^{-1} x$ are inverse translations of $x \mapsto x + x^*$ and $x \mapsto \lambda x$, respectively, the theorem holds by the continuity of vector addition and scalar multiplication. ∎

This assertion implies that in topological vector space X, the convergence of net $\{x^\nu\}$ to point x^* is equivalent to the convergence of net

$\{x^\nu - x^*\}$ to 0, and thus the vector-space topology of X is completely determined by the class of 0-neighborhoods. Furthermore, as the next theorem asserts, we can take a more specialized class of 0-neighborhoods as a 0-neighborhood base. We say that a 0-nhd, U, is *circled* if for all $\lambda \in R$ such that $|\lambda| \leq 1$, $\lambda U \subset U$. A 0-nhd U is said to be *radial* if for all $x \in X$, $x \in \lambda U$ for all λ such that $|\lambda|$ is sufficiently large.

THEOREM AII.2: (**Circled and Radial 0-nhd Base**) *If X is a topological vector space over R, there is a 0-neighborhood base $\mathfrak{V}$ such that (1) every $V \in \mathfrak{V}$ is circled, (2) every $V \in \mathfrak{V}$ is radial, and (3) for each $V \in \mathfrak{V}$, there exists $U \in \mathfrak{V}$ such that $U + U \subset V$. Conversely, for a vector space X over R, by defining $G \subset X$ open iff for each $z \in G$ there exists $V \in \mathfrak{V}$ such that $z + V \subset G$, a 0-neighborhood base $\mathfrak{V}$ satisfying (1)–(3) gives a vector-space topology.*

PROOF: If W is a 0-neighborhood in X, (LT2) at $(0,0) \in R \times X$ means that there is a 0-neighborhood U and $\epsilon > 0$ such that for all $\lambda \in (-\epsilon, +\epsilon)$, we have $\lambda U \subset W$. Define $U(W)$ as $U(W) = \bigcup_{\lambda \in (-\epsilon, +\epsilon)} \lambda U$. Clearly, $U(W)$ is circled. $U(W)$ is radial by the partial continuity assured by (LT2) at $(0, x) \in R \times X$ near $0 \in R$ for each $x \in X$, since λU is also a 0-neighborhood by Theorem AII.1. The continuity (LT1) at $(0,0) \in X \times X$ means that there is a 0-neighborhood V such that $V + V \subset U(W)$, so if we take $V(U(W))$ for $U(W)$ in exactly the same way as $U(W)$ for W, we have $V(U(W)) + V(U(W)) \subset V + V \subset U(W)$. Hence, if we define $\mathfrak{V}$ as $\mathfrak{V} = \{U(W)|\ W$ is a 0-neighborhood in $X\ \}$, $\mathfrak{V}$ is a 0-neighborhood base satisfying (1), (2), and (3). To see the converse, we have to check (LT1) and (LT2) for the topology given by a certain $\mathfrak{V}$ satisfying (1)–(3). Since for each $x, y \in X$ and open neighborhood $x + y + W$ of $x + y$, there is an element $U \in \mathfrak{V}$ such that $U + U \subset W$ by (3), we have (LT1). In order to check (LT2), assume that $\lambda^\nu \to \lambda^*$ and $x^\nu \to x^*$. Take $V \in \mathfrak{V}$ arbitrarily and $V_1 \in \mathfrak{V}$ such that $V_1 + V_1 \subset V$. Let $\alpha^* = |\lambda^*| + 1$ and take $U \in \mathfrak{V}$ such that $alpha^* U \subset V_1$ by using condition (3) repeatedly. There exists $\bar{\nu}$ such that $\forall \nu \geq \bar{\nu}$, $|\lambda^\nu| < \alpha^*$, $x^\nu - x^* \in U$, and $(\lambda^\nu - \lambda^*)x^* \in U$ since U is radial. Then, for all $\nu \geq \bar{\nu}$, we have $\lambda^\nu x^\nu - \lambda^* x^* = \lambda^\nu(x^\nu - x^*) + (\lambda^\nu - \lambda^*)x^* = (\lambda^\nu/\alpha^*)(\alpha^* x^\nu - \alpha^* x^*) + (\lambda^\nu - \lambda^*)x^* \in (\lambda^\nu/\alpha^*)V_1 + U \subset V_1 + V_1 \subset V$, where we use $(\lambda^\nu/\alpha^*)V_1 \subset V_1$ since V_1 is circled. Hence, (LT2) holds. ∎

THEOREM AII.3: (**1-dimensional Subspace**) *Every 1-dimensional subspace of a Hausdorff topological vector space X over R may be identified* with R. We can take the isomorphism as $\lambda x_0 \mapsto \lambda$ for an arbitrarily fixed element x_0 in the subspace.

PROOF: Let $L = \{\lambda x_0 | \lambda \in R\}$, where $x_0 \in X$ and $x_0 \neq 0$, be a 1-dimensional subspace of X and let $f : L \to R$ be the function $\lambda x_0 \mapsto \lambda$. Since $f^{-1} : R \ni \lambda \to \lambda x_0 \in L$ is clearly a continuous bijection, it is sufficient for our purpose to show that f is continuous. By Theorem AII.1, it is sufficient to check the continuity of f at $0 \in L$. Assume that net $\{x^\nu\}$ in L converges to 0. Since the vector space topology relativized on L is Hausdorff, for all $\epsilon > 0$, there is a circled and radial 0-neighborhood V of L such that $\epsilon x_0 \notin V$ and $-\epsilon x_0 \notin V$ by Theorem AII.2. There is no $\lambda x \in V$ such that $|\lambda| \geq \epsilon$, since V is circled. Moreover, since V is a 0-neighborhood, there is a $\bar{\nu}$ such that for all $\nu \geq \bar{\nu}$, $x^\nu \in V$, and thus $f(x^\nu) \in (-\epsilon, +\epsilon)$. This means that $f(\{x^\nu\})$ converges to $0 \in R$. ∎

Let M be a subspace of topological vector space X over R. Define an equivalence relation $\mathscr{R}$ on X through M as $x\mathscr{R}y$ iff $x - y \in M$ for each $x, y \in X$. On the quotient set, $X/\mathscr{R}$, we may also define the vector space structure over R as $[x] + [y] = [x + y]$ and $\lambda[x] = [\lambda x]$. (Note that for each $x \in X$, $[x] = x + M$ and the addition and the scalar multiplication in $X/\mathscr{R}$ may be identified with those among scalars and subsets in X.) Denote by ϕ the quotient map, $x \mapsto [x] = x + M$. Since for each open set G in X, $G + M = \bigcup_{m \in M} m + G$ is also open in X, by considering the relation $\phi^{-1}(G + M) = G + M$, ϕ is an *open map* (images of open sets are open). This also means that if $\mathfrak{V}$ is a 0-neighborhood base of X, then $\{\phi(V) | V \in \mathfrak{V}\}$ is a 0-neighborhood base for the quotient topology on $X/\mathscr{R}$. By exactly the same reasoning, if $\mathfrak{V}$ is a 0-neighborhood base, then for each $y \in X$, $\{\phi(y + V) = y + V + M | V \in \mathfrak{V}\}$ forms the neighborhood base at $[y] = y + M \in X/\mathscr{R}$ of the quotient topology. We denote by X/M the quotient set $X/\mathscr{R}$ with the vector space structure and the quotient topology.

THEOREM AII.4: (Quotient Vector Space) *Let M be a subspace of topological vector space X over R. Then X/M is a topological vector space. If M is a closed subspace, X/M is a Hausdorff topological vector space.*

PROOF: To see that X/M is a topological vector space, we have to check (LT1) and (LT2). Suppose that $[x] + [y] = [x + y]$ and $\lambda[x] = [\lambda x]$. Let $\mathfrak{V}$ be a 0-neighborhood base of X satisfying (1)–(3) of AII.2. As stated above, $\{\phi(x+V) | V \in \mathfrak{V}\}$ forms a neighborhood base at $[x]$ of the quotient topology for each $x \in X$. By (LT1) for X, for each open neighborhood

$x + y + W$ of $x + y$ such that $W \in \mathfrak{V}$, there are open sets U_x and U_y in $\mathfrak{V}$ such that $(x+U_x)+(y+U_y) \subset (x+y+W)$. Therefore, $\phi(x+U_x)+\phi(y+U_y) \subset \phi(x+y+W)$, so (LT1) holds in X/M. In exactly the same way, by (LT2) for X, for each open neighborhood $\lambda x + W$ of λx such that $W \in \mathfrak{V}$, there are open sets U_λ of $\lambda \in R$ and $U_x \in \mathfrak{V}$ such that $a(x + U_x) \subset \lambda x + W$ for each $a \in U_\lambda$. Therefore, we have $\phi(a(x + U_x)) = a\phi(x + U_x) \subset \phi(\lambda x + W)$ for each $a \in U_\lambda$, so (LT2) holds in X/M. Lastly, if M is a closed subspace of X, each element $[x] = x + M$ of X/M is also closed in X. For two elements $[x] = x + M$ and $[y] = y + M$ such that $[x] \neq [y]$, there is an element V of $\mathfrak{V}$ such that $x + V \cap y + M = \emptyset$. If we take $U \in \mathfrak{V}$ such that $U + U \subset V$, we have $x + M + U \cap y + M + U \neq \emptyset$. This is because $x + m_1 + u_1 = y + m_2 + u_2$ for some $m_1, m_2 \in M$ and $u_1, u_2 \in U$ implies that $x + u_1 + (-u_2) = y + m_2 + (-m_1)$, so $x + U + U \cap y + M \neq \emptyset$. Hence, we can take two disjoint open neighborhoods of $[x]$ and $[y]$ in X/M as $\phi(x + U) = x + M + U$ and $\phi(y + U) = y + M + U$, respectively, and thus X/M is a Hausdorff topological vector space. ∎

We call X/M the *quotient (vector) space* (the quotient set under the quotient vector and topological structures) of X relative to M.

Net $\{x^\nu\}$ in topological vector space X over R is a *Cauchy net* if for all 0-neighborhood U there exists $\bar\nu$ such that for all $\mu, \nu \geq \bar\nu$ we have $x^\mu - x^\nu \in U$. It is obvious that every converging net is a Cauchy net. Vector space X is said to be *complete* if every Cauchy net $\{x^\nu\}$ in X converges to a point in the space. In R^n, every Cauchy net is a converging net, hence R^n is complete. Indeed for $\epsilon > 0$, there exists $\bar\nu$ such that for all $\mu, \nu \geq \bar\nu$, $\|x^\mu - x^\nu\| \leq \epsilon$, so all $\mu \geq \bar\nu$ belong to the compact set $B = \{y \in R^n | \|y - x^{\bar\nu}\| \leq \epsilon\}$. Then there is a converging subnet of $\{x^\nu\}$ converging to a point $x^* \in B$. It is easy to check that $\{x^\nu\}$ converges to x^*.

THEOREM AII.5: **(Finite Dimensional Subspace)** *Every n-dimensional subspace of a Hausdorff topological vector space X over R may be identified with the n-dimensional Euclidean space R^n.*

PROOF: For $n = 1$ the theorem holds by AII.3. Assume that the theorem holds for $n = 1, \ldots, k$. We see the case for $n = k + 1$ in the following. Let $\{x^1, \ldots, x^k, x^{k+1}\}$ be a linearly independent subset of X. Denote by $L(k)$ and $L(k + 1)$ the subspace of X spanned by $\{x^1, \ldots, x^k\}$ and $\{x^1, \ldots, x^k, x^{k+1}\}$, respectively. By assumption, we may identify $L(k)$ with R^k. $L(k + 1)$ is a Hausdorff topological vector space over R, and $L(k)$ is a subspace of $L(k + 1)$. Assume that a net $\{x^\nu\}$ in $L(k)$ converges to a point $x^* \in L(k + 1)$. Let $\mathfrak{V}$ be a 0-neighborhood base satisfying

(1)–(3) in Theorem AII.2. Then for all $V \in \mathfrak{V}$, we have $U \in \mathfrak{V}$ such that $U + U \subset V$ and $\bar{\nu}$ such that for all $\nu \geq \bar{\nu}$, $x^\nu \in x^* + U$. Since for all $\mu, \nu \geq \bar{\nu}$, $x^\mu - x^\nu = (x^\mu - x^*) + (x^* - x^\nu) \in U + U \subset V$, $\{x^\nu\}$ is a Cauchy net in $L(k)$. Since $L(k)$ is identified with R^n, net $\{x^\nu\}$ converges to a point $x' \in L(k)$. Moreover, since X is a Hausdorff space, $x' \neq x^*$ is impossible. It follows that $L(k)$ is closed in $L(k+1)$. Then by AII.4, $L(k+1)/L(k) = \{\lambda x^{k+1} + L(k)|\, \lambda \in R\}$ is a 1-dimensional Hausdorff topological vector space, and thus by AII.3 it may be identified with R. Let L be the subspace of X defined as $L = \{\lambda x^{k+1}|\, \lambda \in R\}$. As a 1-dimensional subspace of X, L is identified with R, so $L \times L(k)$ is identified with R^{k+1}. Define $g : L \times L(k) \ni (\lambda x^{k+1}, z) \mapsto \lambda x^{k+1} + z \in L(k+1)$. We can see easily that g is a continuous bijection. Theorem AII.3 gives an isomorphism v on $L(k+1)/L(k) \ni \lambda x^{k+1} + M \mapsto \lambda x^{k+1} \in L$. Let us consider $v \circ \phi$, where ϕ is the quotient map on $L(k+1)$ to $L(k+1)/L(k)$. Then $L(k+1) \ni x \mapsto (v(\phi(x)), x - v(\phi(x))) \in L \times L(k)$ gives the inverse function of g, which is clearly continuous, so g is an isomorphism between $L \times L(k)$ and $L(k+1)$. ∎

Hahn–Banach Theorem

LEMMA AII.6: *Let L be a Hausdorff topological vector space over R with dimension greater than or equal to 2. If there is a convex open set $A \subset L$ such that $0 \notin A$, there is a 1-dimensional subspace of L which does not intersect with A.*

PROOF: If $A = \emptyset$, the assertion is obviously true. Let us consider the case with $A \neq \emptyset$. Denote by $K(A)$ the cone with vertex 0 generated by A, i.e., $K(A) = \{x \in L|\, x = \lambda y, \lambda \in R_{++}, y \in A\}$. Since for each $x \in K(A)$, $-x$ cannot be an element of $K(A)$ (if $\lambda a = -\lambda' a'$ for some a, a' in A, we have $\lambda a + \lambda' a' = 0$), there is a 2-dimensional subspace $L(2)$ of L such that the convex open cone, $K(A) \cap L(2)$, is pointed (i.e., if $x \in K(A) \cap L(2)$, $-x \notin K(A) \cap L(2)$). By identifying $L(2)$ with R^2 (Theorem AII.5), we can find that the intersection of $K(A) \cap L(2)$ and the unit circle (i.e., the range of the projection $A \ni x \mapsto 1/\|x\|$) is a connected open arc having no diametrical points. Therefore, there is an element, x_0, in the unit circle such that both x_0 and $-x_0$ are not elements of $K(A)$. Hence, the 1-dimensional subspace given by x_0 does not intersect with A. ∎

The above lemma together with Zorn's lemma provides the following geometric form of the Hahn–Banach theorem.

THEOREM AII.7: (Hahn–Banach: Geometric Form) *Let L be a topological vector space over R and M be an affine subspace of L such that there is a non-empty convex open subset A of L which does not intersect with M. Then, there is a closed hyperplane H in L which contains M and does not intersect with A.*

PROOF: By using a translation, if necessary, we may assume $0 \in M$, so M is a subspace of L without loss of generality. Since A is open and M is a subset of closed set $L \setminus A$, we have $\operatorname{cl} M \cap A = \emptyset$. Denote by $\mathfrak{M}$ the family of all closed subspace of L which does not intersect with A. $\mathfrak{M}$ is non-empty since $\operatorname{cl} M \in \mathfrak{M}$. If $\mathfrak{M}$ is ordered by inclusion, every totally ordered subset $\{M_i \mid i \in I\}$ of $\mathfrak{M}$ has an upper bound $\operatorname{cl}(\bigcup_{i \in I} M_i)$ in $\mathfrak{M}$, and thus by Zorn's lemma, for $\operatorname{cl} M$, there is a maximal closed subspace $M_* \supset \operatorname{cl} M$ in $\mathfrak{M}$. Let us consider the quotient space L/M_*. Since M_* is closed, L/M_* is a Hausdorff topological vector space by AII.4. The dimension of L/M_* is greater than 1 since A is non-empty. If it is greater than 1, there is a 1-dimensional subspace of L/M_* not intersecting $A + M_*$, i.e., an element $x_0 \in X$, $x_0 \notin M_*$, such that $\lambda x_0 + M_* \cap A + M_* = \emptyset$ for all $\lambda \in R$. This implies that the closure of the linear subspace including M_* and x_0 does not intersect with A, which contradicts the maximality of M_* as a closed subspace of X not intersecting A. Therefore, we have a closed linear subspace M_* of X such that X/M_* is 1-dimensional, i.e., there is an element $x_1 \in X$ such that $X/M_* = \{\lambda x_1 + M_* \mid \lambda \in R\}$, so if we define $f : X \to R$ as $f = v \circ \phi$, where $\phi : X \to X/M$ is the quotient map and v is the map $v : X/M \ni \lambda x_1 + M_* \mapsto \lambda$, we have $M_* = \{x \in X \mid f(x) = 0\}$. ∎

In the above proof, it is easy to see that $f = v \circ \phi$ is continuous since v is (by AII.2) nothing but the isomorphism between the 1-dimensional Hausdorff vector space X/M and R. Generally, for a topological vector space X over R, the existence of a closed subspace M such that X/M is of dimension 1 is equivalent to the existence of a non-zero continuous real valued linear function f on X. It is also important to note that the existence of a non-zero continuous real valued linear function on X may be characterized (by the geometric Hahn–Banach theorem) through the existence of non-empty convex open set $A \neq X$.

COROLLARY AII.8: *If X is a topological vector space, the existence of a non-zero continuous real-valued linear function on X is equivalent to the existence of non-empty convex open subset $A \neq X$.*

Usually, the Hahn–Banach theorem is given in a form that maintains the existence of a continuous linear extension to the total space of a continuous linear form on a subspace. In the next form of the Hahn–Banach theorem, no topological structure is presupposed on the vector space. It essentially asserts, however, the existence of a continuous extension of linear forms defined on its subspaces.

THEOREM AII.9: (Hahn–Banach: Analytic Form) *Let L be a vector space over R and M be a subspace of L. Assume that there is a linear form g on M such that for some $\alpha \in R$ and non-empty radial circled convex subset A of L, $g(z) < \alpha$ for all $z \in A \cap M$. Then, there is a linear form f on L extending g and satisfying $f(z) \leqq \alpha$ for all $z \in A$.*

PROOF: If $g(x) = 0$ for all $x \in L$, the assertion is obvious. Assume that $g(x) \neq 0$ for some $x \in L$. Let us define a 0-neighborhood base $\mathfrak{V}$ of L as $\mathfrak{V} = \{(1/n)A \mid n = 1, 2, \ldots\}$. Since $\mathfrak{V}$ satisfies (1)–(3) in Theorem AII.2, it gives a vector space topology such that $G \subset L$ is open iff for each $z \in G$ there is an element $V \in \mathfrak{V}$ such that $z + V \subset G$. Then $\operatorname{int} A \ni 0$ is a non-empty convex open subset of L not intersecting $\{x \in M \mid g(x) = \alpha\}$, where $\{x \in M \mid g(x) = \alpha\} \neq \emptyset$ since $g(x) \neq 0$ for some $x \in L$. By Theorem AII.7, we have a closed hyperplane H of L that contains $\{x \in M \mid g(x) = \alpha\}$ and does not intersect with $\operatorname{int} A$. Given a linear form $F : L \to R$ such that $F(x) = \beta$ for each $x \in H$, we can define $f : L \to R$ as $f(x) = (\alpha/\beta)F(x)$. ∎

COROLLARY AII.10: (Hahn–Banach: Locally Convex) *In topological vector space L having 0-neighborhood base $\mathfrak{V}$ consisting of convex sets (locally convex topology), every continuous linear form g on subspace M of L to R has a continuous linear extention f on L to R.*

PROOF: If $g(x) = 0$ for all $x \in L$, the assertion is obvious. If $g(x) \neq 0$ for some $x \in L$, since g is continuous on M, for each $\epsilon > 0$, there is a 0-neighborhood V in $\mathfrak{V}$ such that on $V \cap M$, $g(x) < \epsilon$. Hence, by Theorem AII.9, g has an extension f such that $f(x) \leqq \epsilon$ for all $x \in V$. By considering the fact that V is circled ($\lambda V \subset V$ whenever $|\lambda| \leqq 1$), we have $|f(x)| \leqq \epsilon$ for all $x \in V$. Linear form f is continuous since $|f(x)| \leqq \epsilon/n$ for all $x \in (1/n)V$ for all $n = 1, 2, \ldots$. ∎

Given a radial circled convex subset U of vector space L, mapping $p_U : L \to R_+$ defined as

$$p_U(x) = \inf\{\lambda \in R_+ \mid x \in \lambda U\}$$

is called the *gauge function* of U. A gauge function of radial circled convex subset of vector space L is also called a *semi-norm* on L. One can prove that if p is a semi-norm on L, it satisfies the following two conditions:

(i) $p(x+y) \leq p(x) + p(y)$ for all $x, y \in L$.
(ii) $p(\lambda x) = |\lambda| p(x)$ for all $\lambda \in R$ and $x \in L$.

Since $0 = |0|p(x) = p(0x) = p(0) = p(-x+x) \leq p(-x) + p(x) = 2p(x)$ for each $x \in L$, $p(x) \geq 0$ automatically follows from (i) and (ii). It is also easy to see that p is a gauge of $\{x \in L \mid p(x) < 1\}$. A *norm* is a semi-norm such that $p(x) = 0$ means $x = 0$. It is worth remarking that the Hahn–Banach theorem has the next general form based on the concept of semi-norm.

THEOREM AII.11: (**Hahn–Banach: Semi-norm Form**) *Let L be a vector space over R and M be a subspace of L. Assume that there is a linear form g on M such that for some semi-norm p on L, $g(x) \leq p(x)$ for all $x \in M$. Then, there is a linear form f on L extending g and satisfying $f(x) \leq p(x)$ for all $x \in L$.*

PROOF: Let A be the radial circled convex set $\{x \in L \mid p(x) < 1\}$. Then, we have $g(z) < 1$ for all $z \in A \cap M$. Hence, by AII.9, there is a linear form f on L extending g satisfying $f(z) \leq 1$ for all $z \in A$. Since p is a gauge function of A, we have $f(x) \leq p(x)$ for all $x \in L$. ∎

Separation Theorem

Now we can prove the first separation theorem as an immediate consequence of the Hahn–Banach theorem (AII.7).

THEOREM AII.12: (**First Separation Theorem**) *In topological vector space L, if A is a convex set whose interior, int A, is non-empty and B is a non-empty convex set such that $\text{int } A \cap B = \emptyset$, then closed hyperplane H exists that separates A and B. If both A and B are open, we may choose H so that A and B are strictly separated.*

PROOF: Let C be the set $-A + B$. Since $-\text{int } A + B$ is an open subset of C, the interior of C, int C, is non-empty and $0 \notin \text{int } C$. The interior of C is convex since $C = -A + B$ is convex. Therefore by Theorem AII.7, there is a closed hyperplane H_0 in L which contains 0 and does not intersect with int C. Let f be a continuous linear form on L such that $H_0 = \{x \in L \mid f(x) = 0\}$ and $f(c) > 0$ for all $c \in \text{int } C$. Then, for each $a \in \text{int } A$ and $b \in B$, we

Mathematical Appendix II   271

have $f(-a+b) > 0$, so $f(b) > f(a)$. If we take $c = \sup\{f(a)|\, a \in \text{int}\, A\}$, we have $f(a) \leqq c$ for all $a \in A$ and $f(b) \geqq c$ for all $b \in B$, and thus $H = \{x \in L|\, f(x) = c\}$ is the closed hyperplane separating A and B. If both A and B are open, we can see that there is no $a^* \in A$ (or $b^* \in B$) such that $f(a^*) = c$ (or $f(b^*) = c$). Indeed, by AII.3 and AII.4, the quotient 1-dimensional Hausdorff topological vector space L/H_0 can be identified with R, where by taking $x_1 \in X$ such that $f(x_1) = 1$, we can define an isomorphism $h: L/H_0 \to R$ as $\lambda[x_0] \mapsto \lambda$. Since for each $x \in X$, $[x] = (f(x)/f(x_0))[x_0]$ in L/H_0, the isomorphism h is nothing but the mapping $[x] \mapsto h(x)$. Since the quotient map $\phi: L \to L/H_0$ is an open map, both $h \circ \phi(A)$ and $h \circ \phi(B)$ are open subsets in R. If there exists $a^* \in A$ (or $b^* \in B$) such that $f(a^*) = c$ (or $f(b^*) = c$), $h \circ \phi(a^*) = f(a^*) = c$ (or $h \circ \phi(b^*) = f(b^*) = c$) is an interior of the open set $h \circ \phi(A) = f(A)$ (or $h \circ \phi(B) = f(B)$) in R, which contradicts the fact that $f(a) \leqq c$ for all $a \in A$ (or $f(b) \geqq c$ for all $b \in B$). ∎

Mathematical Appendix III

In Chapter 9, the set-theoretic methods necessary for logical arguments in Sections 9.2 and 9.3 are restricted to finitistic ones. It would be appropriate, however, to refer to some transfinite arithmetical concepts like ordinals, cardinals, and Continuum hypothesis at least to the minimum extent, because they are fundamental tools for standard axiomatic set theory.

Ordinal

Set x is called an *ordinal* if (1) relation $\in$ restricted on x is transitive, i.e., $\forall y \forall z(((y \in x) \land (z \in y)) \to z \in x)$, (2) relation $\in$ restricted on x is irreflexive, i.e., $\forall y((y \in x) \to (y \notin y))$, and (3) x is well-ordered by $\in$ (precisely, under the relation $(y \in z) \lor (y = z)$). For example, all natural numbers are ordinals. (Indeed, if x is an ordinal, then $\bigcup \{x, \{x\}\}$, denoted henceforth by $x + 1$, is an ordinal.) We can also verify that if X is a set of ordinals, then $\bigcup X$ is also an ordinal. The above condition (1) is equivalent to saying that every element of x is a subset of x, so every element of an ordinal is also an ordinal. No ordinal can be an element of itself. (For if $x \in x$, relation $\in$ on x fails to be irreflexive.)[1] The set of all natural numbers, N, (under (S6)) is an ordinal. We usually denote it by ω instead of N if we treat N as an ordinal. If $x \neq \emptyset$ is an ordinal, $\in$-least element (more precisely, "$\in$ or $=$"-least element) of x must be the empty set $\emptyset$. (By

---

[1] If we treat the class of all ordinals, **ON**, as a set, we have a contradiction. Indeed, if **ON** is a set, **ON** itself also satisfies (1) and (2), and thus it is verified to be an ordinal. Accordingly, we have $\mathbf{ON} \in \mathbf{ON}$, which is impossible, since no ordinal x can be a member of itself (Burali–Forti Paradox).

using this, we may assure that if x and y are ordinals, one (and only one) relation of the following holds: $x \in y$, $x = y$, $y \in x$.) Ordinal ω is the $\in$-least infinite ordinal. Moreover, if we denote by $\omega+1, \omega+2, \ldots$ the ordinals $\bigcup\{\omega, \{\omega\}\}, \bigcup\{\omega+1, \{\omega+1\}\}, \ldots$, respectively (under (S2)), and for the limit by taking the union (under (S4) with (S7)), we have $\omega + \omega = 2 \cdot \omega$. Then, $3 \cdot \omega, 4 \cdot \omega, \ldots, \omega \cdot \omega$. We refrain from denoting $\omega \cdot \omega$ by ω^2 since we want to reserve notation A^B for the set of all functions on B to A.[2] For each set x, under ZFC, there is a well-ordering $\leqq$ on x (Well-ordering Theorem). With respect to well-ordered set $(x, \leqq)$ (set x having well-ordering $\leqq$ on it), we have the following theorem.

THEOREM AIII.1: (Ordinals and Well-ordered Sets) *Under ZF, if $(x, \leqq)$ is a well-ordered set, then there is a one to one, onto, monotonic function f on x to a certain ordinal $(y, \in)$.*

PROOF: If x is empty, there is nothing to be proved. Assume that $x \neq \emptyset$. Since x has the $\leqq$-least element and clearly has a one to one, onto, monotonic relation with the ordinal $\{0\} = 1$, the subset z of x, which consists of all $u \in x$ such that the initial segment $y(u) = \{v| v \leqq u\}$ has a one to one, onto, monotonic function on it to a certain ordinal, is not empty. If $z \neq x$, since x is a well-ordered set, there is a $\leqq$-least element u^* in $x \setminus z$. Note that since each $v \in z$ has the one to one function f^v on $y(v)$, we can see that $z = \bigcup\{y(v)| v \in z\}$ itself has a one to one, onto, monotonic function f^z on z to the ordinal $\bigcup\{f^v(y(v))| v \in z\}$. We can extend f^z to f^* on u^* by defining $f^*(u^*) = \bigcup\{f^v(y(v))| v \in z\} + 1$, which contradicts the definition of z. ∎

Cardinal

Since two ordinals y and z are equal if there is a one to one, onto, and monotonic function between them (consider the least ordinal in $y \cup z$ that fails to satisfy the required property), the ordinal assured to exist in the above theorem is unique. By considering the above result together with the well-ordering theorem, we may correspond each set x with a certain ordinal. Note, however, that the possible well-ordering on x is not unique. (For example, on N, we may consider well-orderings like those of $\omega + 1$,

---

[2]Unfortunately, we cannot reconcile the latter usage with the former. Note that $\omega \cdot \omega$, $\omega \cdot (\omega \cdot \omega), \ldots$, including the limit $\cdots (\omega \cdot (\omega \cdot \omega))$, may be considered countable unions of countable sets. The latter is not countable even in case 2^N.

Mathematical Appendix III 275

$\omega + 2$.) In general, if set x has a one to one and onto function f^x on x to a certain ordinal y, then the $\in$-least ordinal $Card(x)$ (e.g., as an element of $y + 1$) satisfying the same property (i.e., x has a one to one and onto function on x to it) exists, and this is called the *cardinality of x*. In this case, $Card(Card(x)) = Card(x)$ is satisfied. Ordinal z, satisfying $z = Card(x)$, is called a *cardinal*. All natural numbers are cardinals, and ω is the first infinite cardinal. In this sense, we write $\aleph_0$ or ω_0 instead of ω to emphasize the feature. For each infinite ordinal x, consider the set $O(x) = \{\preceq \mid \preceq$ is a well-ordering on $x\}$. Each $\preceq \in O(x)$ defines one ordinal $x(\preceq)$ by the previous theorem, and by replacement axiom (S7), we have set $A = \{x(\preceq) \mid \preceq \in O(x)\}$ of ordinals. Then, we cannot have a one to one and onto function f between two ordinals $\bigcup A$ and x. (See the proof of the next theorem.) Hence, we have the following assertion, which is obvious if we use power set axiom (S5) and well-ordering theorem (or equivalently, choice axiom (S9)). In the next theorem, we do not require the axiom of choice.

THEOREM AIII.2: (**Existence of Arbitrarily Large Successor Cardinals**) *Under ZF, for each ordinal x, there is a cardinal x^+, the least cardinal which is greater than x.*

PROOF: Indeed, if a one to one and onto function $f : x \to \bigcup A$ exists, we have a well-ordering on x corresponding to the well-ordering on $\bigcup A$ through f in $O(x)$, so $\bigcup A \in A$. Since x is an infinite ordinal, by replacing the least element of x as the last element of x, we can see that $x+1 \in A$. Then, since $\bigcup A$ is the set of elements of elements of A, and since $x \in x + 1$, x is an element of $\bigcup A$. Hence, $\bigcup A$ is also an infinite ordinal. Again, by replacing the first element of $\bigcup A$ as the last element of it, we can also see $\bigcup A+1 \in A$. Then, since $\bigcup A \in (\bigcup A+1)$, and since $\bigcup A$ is a set of element of element of A, we have $\bigcup A \in \bigcup A$, a contradiction. (No ordinal is a member of itself.) It follows that $Card(\bigcup A)$ is greater than x, so under the well-ordering on $\bigcup A$, there must exist x^+, the least ordinal whose cardinality is greater than x, which is nothing but the least cardinal greater than x. ∎

Continuum Hypothesis

We call x^+ the *successor cardinal* of x. We denote by $\aleph_1$ the successor cardinal $(\aleph_0)^+$. Under the well-ordering theorem (hence, under ZFC), power set $\mathscr{P}(\aleph_0)$ can be well-ordered, so we can define $\aleph = Card(\mathscr{P}(\aleph_0))$. The next assertion is known as the *Continuum Hypothesis*.

(CH) *Continuum Hypothesis.*

The cardinality $\aleph$ of $\mathscr{P}(\aleph_0)$ (defined under ZFC) is equal to $\aleph_1$.

It can be proved through a finitistic method called *forcing* that if ZFC + CH (or ZFC + $\neg$ CH) is inconsistent, then ZFC is also inconsistent. In other words, as long as ZFC is consistent, there is no proof in ZFC for CH or $\neg$ CH.[3]

---

[3] If ZFC + $\neg$ CH (or ZFC + CH) is inconsistent, then by the Compactness Theorem there is a list of finite axioms in ZFC + $\neg$ CH (or ZFC + CH, respectively) proving the contradiction. The method of forcing, however, gives a general finitistic way of constructing a model in which the list of axioms holds, so we have a contradiction under ZFC. See, e.g., Kunen (1980, Chapter VII).

References

Aliprantis, C. D. (1996): *Problems in Equilibrium Theory*. Springer-Verlag, Berlin.
Aliprantis, C. D. and Brown, D. J. (1983): "Equilibria in markets with a Riesz space of commodities," *Journal of Mathematical Economics 11*, 189–207.
Aliprantis, C. D., Brown, D. J., and Burkinshaw, O. (1989): *Existence and Optimality of Competitive Equilibria*. Springer-Verlag, New York/Berlin.
Arrow, K. and Hahn, F. (1971): *General Competitive Analysis*. Holden-Day, San Francisco.
Bagh, A. (1998): "Equilibrium in abstract economies without the lower semicontinuity of the constraint maps," *Journal of Mathematical Economics 30*(2), 175–185.
Balasko, Y. (1988): *Foundations of the Theory of General Equilibrium*. Academic Press, New York.
Begle, E. G. (1942): "Locally connected spaces and generalized manifolds," *Amer. J. Math. 64*, 553–574.
Begle, E. G. (1950a): "The Vietoris mapping theorem for bicompact spaces," *Annals of Mathematics 51*(3), 534–543.
Begle, E. G. (1950b): "A fixed point theory," *Annals of Mathematics 51*(3), 544–550.
Ben-El-Mechaiekh, H., Chebbi, S., Florenzano, M., and Llinares, J. V. (1998): "Abstract convexity and fixed points," *Journal of Mathematical Analysis and Applications 222*, 138–150.
Bewley, T. (1972): "Existence of equilibria in economies with infinitely many commodities," *Journal of Economic Theory 4*, 514–540.
Border, K. C. (1984): "A core existence theorem for games without ordered preferences," *Econometrica 52*(6), 1537–1542.
Border, K. C. (1985): *Fixed Point Theorems with Applications to Economics and Game Theory*. Cambridge University Press, Cambridge.
Bourbaki, N. (1939): *Eléments de Mathématique*. Hermann, Paris. English translation: Springer-Verlag.
Browder, F. (1968): "The fixed point theory of multi-valued mappings in topological vector spaces," *Mathematical Annals 177*, 283–301.

Čech, E. (1932): "Theorie générale de l'homologie dans une espace quelconque," *Fund. Math.* 19, 149–183.

Debreu, G. (1952): "A social equilibrium existence theorem," *Proceedings of the National Academy of Sciences of the U.S.A.* 38, 886–893. Reprinted as Chapter 2 in G. Debreu, *Mathematical Economics*, Cambridge University Press, Cambridge, 1983.

Debreu, G. (1956): "Market equilibrium," *Proceedings of the National Academy of Sciences of the U.S.A.* 42, 876–878. Reprinted as Chapter 7 in G. Debreu, *Mathematical Economics*, Cambridge University Press, Cambridge, 1983.

Debreu, G. (1959): *Theory of Value.* Yale University Press, New Haven, CT.

Ding, X. P. (2000): "Existence of solutions for quasi-equilibrium problems in non-compact topological spaces," *Computers & Mathematics with Applications* 39, 13–21.

Dixit, A. (2005): "Paul Samuelson as teacher," in *Paul A. Samuelson, The Economics Wunderkind* (Gottesman, A. and Szenberg, M. ed.), Pinto Books, New York.

Dunford, N. and Schwartz, J. T. (1966): *Linear Operators Part I: General Theory.* A Wiley-Interscience Publication, New York.

Eaves, B. C. (1974): "Properly labeled simplexes," in *Studies in Optimization*, Volume II (Dantzig, G. B. and Eaves, B. C. ed.), Vol. 10 of *Studies in Mathematics*, Mathematical Association of America.

Eilenberg, S. and Montgomery, D. (1946): "Fixed point theorems for multi-valued transformations," *American Journal of Mathematics* 68, 214–222.

Eilenberg, S. and Steenrod, N. (1952): *Foundations of Algebraic Topology.* Princeton University Press, Princeton, New Jersey.

Fan, K. (1952): "Fixed-point and minimax theorems in locally convex topological linear spaces," *Proceedings of the National Academy of Sciences of the U.S.A.* 38, 121–126.

Fan, K. (1969): "Extensions of two fixed point theorems of F.E. Browder," *Mathematische Zeitschrift* 112, 234–240.

Florenzano, M. (1983): "On the existence of equilibria in economies with an infinite dimensional commodity space," *Journal of Mathematical Economics* 12, 207–219.

Fraenkel, A. A., Bar-Hillel, Y., and Levy, A. (1973): *Foundations of Set Theory*, Second edn. Elsevier, Amsterdam.

Gale, D. (1955): "The law of supply and demand," *Math. Scad.* 3, 155–169.

Gale, D. and Mas-Colell, A. (1975): "An equilibrium existence theorem for a general model without ordered preferences," *Journal of Mathematical Economics* 2, 9–15. (For some corrections see *Journal of Mathematical Economics*, Vol. 6: 297–298, 1979.)

Giraud, G. (2001): "An algebraic index theorem for non-smooth economies," *Journal of Mathematical Economics* 36, 255–269.

Glicksberg, K. K. (1952): "A further generalization of the Kakutani fixed point theorem, with application to Nash equilibrium points," *Proceedings in the American Mathematical Society* 3, 170–174.

Granas, A. and Dugundji, J. (2003): *Fixed Point Theory*. Springer-Verlag, New York.

Grandmont, J. M. (1977): "Temporary general equilibrium theory," *Econometrica* 45(3), 535–572.

Gödel, K. (1931): "Über formal unentscheidbare sätze del Principia Mathematica und verwandter system I," *Monatsch. Math. Ph. 38*, 173–198. See *Anzeiger Akad. Wien* 67, 214–215, 1930.

Górniewicz, L. (1976): "Homological methods in fixed-point theory of multivalued maps," *Dissertationes Mathematicae 129*.

Hayashi, T. (1997): "A generalization of the continuity condition for excess demand functions," mimeo, Osaka University.

Hicks, J. (1939): *Value and Capital*. Clarendon Press, Oxford.

Hildenbrand, W. (1974): *Core and Equilibria of a Large Economy*. Princeton University Press, Princeton.

Hocking, J. G. and Young, G. S. (1961): *Topology*. Dover Publications, Inc., New York.

Horvath, C. D. (1991): "Coincidence theorems for the better admissible multimaps and their applications," *Journal of Mathematical Analysis and Applications* 156, 341–357.

Hurewicz, W. and Wallman, H. (1948): *Dimension Theory*. Princeton University Press, Princeton.

Ichiishi, T. (1977): "Coalition structure in a labor-managed market economy," *Econometrica* 45(2), 341–360.

Ichiishi, T. (1981a): "On the Knaster–Kuratowski–Mazurkiewicz–Shapley Theorem," *Journal of Mathematical Analysis and Applications* 81(2), 297–299.

Ichiishi, T. (1981b): "A social coalitional equilibrium lemma," *Econometrica* 49(2), 369–377.

Ichiishi, T. (1983): *Game Theory for Economic Analysis*. Academic Press, New York.

Ichiishi, T. (1988): "Alternative version of Shapley's theorem on closed coverings of a simplex," *American Mathematical Society* 104(3), 759–763.

Ichiishi, T. (1993): *The Cooperative Nature of the Firm*. Cambridge University Press, Cambridge.

Ichiishi, T. and Idzik, A. (1990): "Theorems on closed coverings of a simplex and their applications to cooperative game theory," *Journal of Mathematical Analysis and Applications* 46(1), 259–269.

Ichiishi, T. and Idzik, A. (1991): "Closed covers of compact convex polyhedra," *International Journal of Game Theory* 20, 161–169.

Ichiishi, T. and Idzik, A. (1999a): "Equitable allocation of divisible goods," *Journal of Mathematical Economics* 32, 389–400.

Ichiishi, T. and Idzik, A. (1999b): "Market allocation of indivisible goods," *Journal of Mathematical Economics* 32, 457–466.

James, W. (1912): *Essays in Radical Empiricism*.

Jaworowski, J. W. (1958): "Some consequences of the Vietoris mapping theorem," *Fundamental Mathematicae* 45, 261–272.

Jech, T. (1997): *Set Theory*, Second edn. Springer-Verlag, Berlin.
Jech, T. (2003): *Set Theory*, Third edn. Springer-Verlag, Berlin.
Kahn, M. A. and Yannelis, N. C. ed. (1991): *Equilibrium Theory in Infinite Dimensional Spaces*. Springer-Verlag, Berlin.
Kajii, A. (1988): "Note on equilibrium without ordered preferences in topological vector spaces," *Economic Letters 27*, 1–4.
Kakutani, S. (1941): "A generalization of Brouwer's fixed point theorem," *Duke Math. J. 8*(3).
Kaneko, M. (1996): "Structural common knowledge and factual common knowledge," RUEE Working Paper No. 87-27, Hitotsubashi University.
Kaneko, M. and Kline, J. J. (2008): "Inductive game theory: A basic scenario," *Journal of Mathematical Economics 44*, 1332–1363.
Kaneko, M. and Nagashima, T. (1996): "Game logic and its applications I," *Studia Logica 57*, 325–354.
Keiding, H. (1985): "On the existence of equilibrium in social systems with coordination," *Journal of Mathematical Economics 14*, 105–111.
Kelley, J. L. (1955): *General Topology*. Springer-Verlag, New York/Berlin.
Kim, W. K. and Yuan, G. X.-Z. (2001): "Existence of equilibria for generalized games and generalized social systems with coordination," *Nonlinear Analysis 45*, 169–188.
Komiya, H. (1981): "Convexity on a topological space," *Fundamenta Mathematicae 111*, 107–113.
Komiya, H. (1999): "Fixed point theorems and related topics in abstract convex spaces," *Kokyuroku No. 1108*, 1–11. Research Institute for Mathematical Sciences, Kyoto University.
Kreps, D. M. (1990): *A Course in Microeconomic Theory*. Princeton University Press, New Jersey.
Kripke, S. A. (1972): *Naming and Necessity*. Harvard University Press, Cambridge, Massachusetts.
Kripke, S. A. (1975): "Outline of a theory of truth," *Journal of Philosophy 72*.
Kristály, A. and Varga, C. (2003): "Set-valued versions of Ky Fan's inequality with application to variational inclusion theory," *Journal of Mathematical Analysis and Applications 282*, 8–20.
Kunen, K. (1980): *Set Theory: An Introduction to Independence Proofs*. North Holland, Amsterdam.
Lakoff, G. (1987): *Women, Fire, and Dangerous Things: What Categories Reveal about the Mind*. The University of Chicago Press, Chicago and London.
Lefschetz, S. (1937): "On the fixed point formula," *Annals of Mathematics 38*(4), 819–822.
Lefschetz, S. (1942): *Algebraic Topology*, Vol. 27. Amer. Math. Soc. Colloquium Publications, New York.
Luo, Q. (2001): "KKM and Nash equilibria type theorems in topological ordered spaces," *Journal of Mathematical Analysis and Applications 264*, 262–269.
Mas-Colell, A. (1974): "A note on a theorem of F. Browder," *Mathematical Programming 6*, 229–233.

Mas-Colell, A. (1985): *The Theory of General Economic Equilibrium: A Differentiable Approach*. Cambridge University Press, Cambridge.

Mas-Colell, A. (1986): "The price equilibrium existence problem in topological vector spaces," *Econometrica* 54(5), 1039–1053.

Mas-Colell, A., Whinston, M. D., and Green, J. R. (1995): *Microeconomic Theory*. Oxford University Press, New York.

McKenzie, L. W. (1981): "The classical theorem on existence of competitive equilibrium," *Econometrica* 49(4), 819–841.

McLennan, A. (1991): "Approximation of contractible-valued correspondences by functions," *Journal of Mathematical Economics* 20(6), 591–598.

Mehta, G. and Tarafdar, E. (1987): "Infinite-dimensional Gale–Nikaido–Debreu theorem and a fixed-point theorem of Tarafdar," *Journal of Economic Theory* 41, 333–339.

Michael, E. (1956): "Continuous selections I," *Annals of Mathematics* 63(2), 361–382.

Morishima, M. (1950): *Dougakuteki Keizai Rironon (Dynamic Economic Theory)*. Koubundou, Tokyo. English translation: *Dynamic Economic Theory*, Cambridge University Press, New York, 1996.

Nakaoka, M. (1977): *Fudo-ten Teiri to Sono Syuhen (Fixed-point Theorems and Related Problems)*. Iwanami Syoten, Tokyo.

Nash, J. (1950): "Equilibrium states in N-person games," *Proceedings of the National Academy of Sciences of the U.S.A.* 36.

Nash, J. (1951): "Non-cooperative games," *Annals of Mathematics* 54, 289–295.

Neuefeind, W. (1980): "Notes on existence of equilibrium proofs and the boundary behavior of supply," *Econometrica* 48(7), 1831–1837.

Nikaido, H. (1956a): "On the classical multilateral exchange problem," *Metroeconomica* 8, 135–145. A supplementary note: *Metroeconomica* 8, 209–210, 1957.

Nikaido, H. (1956b): "On the existence of competitive equilibrium for infinitely many commodities," Tech. Report, Dept. of Econ. No. 34, Stanford University.

Nikaido, H. (1957): "Existence of equilibrium based on the Walras' law," ISER Discussion Paper No. 2, Institution of Social and Economic Research.

Nikaido, H. (1959): "Coincidence and some systems of inequalities," *Journal of The Mathematical Society of Japan* 11(4), 354–373.

Nikaido, H. (1968): *Convex Structures and Economic Theory*. Academic Press, New York.

Nishimura, K. and Friedman, J. (1981): "Existence of Nash equilibrium in n-person games without quasi-concavity," *International Economic Review* 22, 637–648.

Park, S. (2001): "New topological versions of the Fan–Browder fixed point theorem," *Nonlinear Analysis* 47, 595–606.

Park, S. (2004): "New versions of the Fan–Browder fixed point theorem and existence of economic equilibria," *Fixed Point Theory and Applications* 2, 149–158.

Park, S. (2005): "Comments on collectively fixed points in generalized convex spaces," *Applied Mathematics Letters 18*, 431–437.
Park, S. and Kim, H. (1996): "Coincidence theorems for admissible multifunctions on generalized convex spaces," *Journal of Mathematical Analysis and Applications 197*, 173–187.
Putnam, H. (1981): *Reason, Truth and History*. Cambridge University Press, New York.
Putnam, H. (1983): *Realism and Reason*, Vol. 3 of *Philosophical Papers*. Cambridge University Press, New York.
Putnam, H. (1990): *Realism with a Human Face*. Harvard University Press, Cambridge, Massachusetts.
Putnam, H. (1995): *Pragmatism*. Blackwell, Cambridge.
Putnam, H. (2002): *The Collapse of the Fact/Value Dichotomy and Other Essays*. Harvard University Press, Cambridge, Massachusetts.
Putnam, H. (2004): *Ethics without Ontology*. Harvard University Press, Cambridge, Massachusetts.
Quine, W. V. O. (1953): *From a Logical Point of View: 9 Logico-Philosophical Essays, Second Edition, Revised 1961*. Harvard University Press.
Quine, W. V. O. (1960): *Word and Object*. M.I.T. Press.
Rotman, J. J. (1988): *An Introduction to Algebraic Topology*. Springer-Verlag, New York.
Rubinstein, A. (2006): *Lecture Notes in Microeconomic Theory: The Economic Agent*. Princeton University Press, Princeton and Oxford.
Samuelson, P. A. (1947): *Foundation of Economic Analysis*. Harvard University Press, Cambridge, Mass.
Samuelson, P. A. (1952): "Economic theory and mathematics — an appraisal," *Papers and Proceedings, American Economic Review XLII(2)*.
Scarf, H. (1967): "The core of an n-person game," *Econometrica 35*, 50–69.
Scarf, H. (1973): *The Computation of Economic Equilibria*. Yale University Press, New Haven.
Scarf, H. (1982): "The computation of equilibrium prices: An exposition," in *Handbook of Mathematical Economics*, Volume II (K. A., J. and M. I., D. ed.), North-Holland, New York.
Schaefer, H. H. (1971): *Topological Vector Spaces*. Springer-Verlag, New York/ Berlin.
Shafer, W. and Sonnenschein, H. F. (1975): "Equilibrium in abstract economies without ordered preferences," *Journal of Mathematical Economics 2*, 345–348.
Shapley, L. (1973): "On balanced games without side payments," in *Mathematical Programming* (Hu, T. and Robinson, S. ed.), pp. 261–290, Academic Press, New York.
Shoenfield, J. R. (1967): *Mathematical Logic*. A. K. Peters, Ltd., Massachusetts.
Smith, A. (1776): *The Wealth of Nations*.
Spanier, E. H. (1948): "Cohomology theory for general spaces," *The Annals of Mathematics, 2nd Ser. 49(2)*, 407–427.

Steenrod, N. E. (1936): "Universal homology groups," *Amer. J. Math* **58**, 661–701.
Takahashi, W. (1970): "A convexity in metric space and nonexpansive mappings, I," *Kōdai Math. Sem. Rep.* **22**, 142–149.
Takahashi, W. (1988): *Hi-senkei Kansu Kaiseki-gaku: Fudo-ten Teiri to Sono Syuhen (Non-linear Functional Analysis: Fixed-point Theorems and Related Problems)*. Kindai Kagaku-sha, Tokyo.
Tan, K.-K. and Yuan, X.-Z. (1994): "Existence of equilibrium for abstract economies," *Journal of Mathematical Economics* **23**, 243–251.
Tarafdar, E. (1977): "On nonlinear variational inequalities," *Proc. Amer. Math. Soc.* **67**, 95–98.
Toussaint, S. (1984): "On the existence of equilibria in economies with infinitely many commodities and without ordered preferences," *Journal of Economic Theory* **33**, 98–115.
Tukey, J. W. (1940): *Convergence and Uniformity in Topology*. Princeton University Press.
Urai, K. (1998): "Incomplete markets and temporary equilibria, II: Firms' objectives," in *Ippan Kinkou Riron no Shin Tenkai (New Developments in the Theory of General Equilibrium)* (Kuga, K. ed.), Chapter 8, pp. 233–262, Taga Publishing House, Tokyo.
Urai, K. (2000): "Fixed point theorems and the existence of economic equilibria based on conditions for local directions of mappings," *Advances in Mathematical Economics* **2**, 87–118.
Urai, K. (2002a): "Why there isn't a complete description of the human society: The rationality and individuals," Discussion Paper No. 02-04, Faculty of Economics and Osaka School of International Public Policy, Osaka University.
Urai, K. (2002b): "Why there isn't a complete description of the human society II: The society and value," Discussion Paper No. 02-05, Faculty of Economics and Osaka School of International Public Policy, Osaka University.
Urai, K. (2002c): "Why there isn't a complete description of the human society I: The individual and rationality," Kokyuroku No. 1264, Research Institute for Mathematical Sciences, Kyoto University.
Urai, K. (2006): "Social recognition and economic equilibrium," Discussion Paper No. 06-30, Faculty of Economics and Osaka School of International Public Policy, Osaka University.
Urai, K. and Hayashi, T. (1997): "A generalization of continuity and convexity conditions for correspondences in the economic equilibrium theory," Discussion Paper, Faculty of Economics and Osaka School of International Public Policy, Osaka University.
Urai, K. and Hayashi, T. (2000): "A generalization of continuity and convexity conditions for correspondences in the economic equilibrium theory," *The Japanese Economic Review* **51**(4), 583–595.
Urai, K. and Yokota, K. (2005): "Generalized dual system structure and fixed point theorems for multi-valued mappings," in Kokyuroku, Research Institute for Mathematical Sciences, Kyoto University.

Urai, K. and Yoshimachi, A. (2004): "Fixed point theorems in Hausdorff topological vector spaces and economic equilibrium theory," *Advances in Mathematical Economics* 6, 149–165.

Uzawa, H. (1962): "Walras' existence theorem and Brouwer's fixed point theorem," *Economic Studies Quarterly* 13, 59–62.

Vietoris, L. (1927): "Über den höheren Zusammenhang kompakter Räume und eine Klasse von zusammenhangstreuen Abbildungen," *Mathematische Annalen* 97, 454–472.

Vind, K. (1983): "Equilibrium with coordination," *Journal of Mathematical Economics* 12, 275–285.

von Neumann, J. (1937): "Über ein ökonomisches Gleichungssystem und eine Verallgemeinerung des Brower'schen Fixpunktsatzes," *Ergebnisse eines mathematischen Kolloquiums viii*. English translation: "A model of general economic equilibrium," *Review of Economic Studies*, xiii, (1945–6), pp. 1–9.

Walras, L. (1874): *Eléments d'économie politique pure*. Corbaz, Lausanne. English translation: *Elements of Pure Economics*, George Allen and Unwin, London, 1926.

Watson, B. (1968): *The Complete Works of Chuang Tzu*. Columbia University Press, New York.

Weber, M. (1904): "Die "Objektivität" Sozialwissenschaftlicher und Sozialpolitischer Erkenntnis," *Archiv für Sozialwissenschaft und Sozialpolitik* 19, 22–87. English translation: "Objectivity in social science," in *The Methodology of the Social Sciences* (Edward A. Shils and Henry A. Finch, tr. & ed.).

Wittgenstein, L. (1919): *Tractatus Logico-Philosophicus*. Routledge & Kegan Paul, London. First published in Annalen der Naturphilosophie 1921. English edition first published 1922 by Kegan Paul. Routledge & Kegan Paul edition published 1974.

Wittgenstein, L. (1953): *Philosophische Untersuchungen*. Blackwell, Oxford.

Yannelis, N. (1985): "On a market equilibrium theorem with an infinite number of commodities," *J. Math. Anal. Appl.* 108, 595–599.

Yannelis, N. (1991): "The core of an economy without ordered preferences," in *Equilibrium Theory in Infinite Dimensional Spaces* (Kahn, M. A. and Yannelis, N. C. ed.), pp. 102–123, Springer-Verlag, Berlin.

Yannelis, N. and Prabhakar, N. (1983): "Existence of maximal elements and equilibria in linear topological spaces," *Journal of Mathematical Economics* 12, 233–245.

Index

0-neighborhood base, 23
$B_n^c(\mathfrak{M})$, 134
$C(A)$, 40
$C(Z)$, 41
$C_n^c(\mathfrak{M})$, 134
$C_v^q(X, A, \mathfrak{M})$, 176
$C_n^v(\mathfrak{M})$, 135
$C_q^v(\mathfrak{M})$, 175
$C_q(K)$, 129
E', 23
E^*, 23
$H_n^c(\mathfrak{M})$, 134
$H_n^v(X)$, 136
$H_n^v(\mathfrak{M})$, 136
$H_q^v(X, A; \mathfrak{M})$, 175
$I(f)$, 181
$K(c)$, 150
$K \times \{0, 1, I\}$, 151
$K \times \{0, 1\}$, 137
N, 14, 212
P-adic expansion, 257
Q, 218
R, 19, 218
Sd_q, 150
$St(N; \mathfrak{N})$, 132
$X^v(\mathfrak{M}) \cap W$, 136, 143
$X_q^v(\mathfrak{M}) \cap W$, 145
Z, 218
$Z_n^c(\mathfrak{M})$, 134
$Z_n^v(\mathfrak{M})$, 136

$[\langle \sigma^q \rangle, \langle \sigma^{q-1} \rangle]$, 129
$\mathbf{Cover}(X)$, 132
Δ^A, 39
Δ^n, 127
$\mathbf{Tops}$, 128
$\mathbf{Vert}(K)$, 128
$\langle e^1 \cdots e^{n+1} \rangle$, 127
$\langle e^{i_0} \cdots e^{i_k} \rangle$, 127
$\langle v_0, \ldots, \hat{v}_i, \ldots, v_q \rangle$, 129
$\langle x^{i_0} \cdots x^{i_k} \rangle$, 127
$\bar{A}$, 40
$\bigcup \mathcal{U}$, 14
$\exists$, 203
$\forall$, 204
$\mathfrak{M}$-morphism, 126
$\mathbf{Obj}(\mathfrak{S})$, 125
$\ulcorner \theta \urcorner$, 223
$\hom(X, Y)$, 125
$\leftrightarrow$, 204
$\rightarrow$, 204
$\wedge$, 204
$\neg$, 203
$\vee$, 203
diam, 146
ω, 273
$\vdash$, 208
$\preccurlyeq$, 131
$\preccurlyeq^*$, 132
$\mathcal{E}$, 106
$\mathcal{F}(X)$, 40

285

$\mathscr{P}(A)$, 14
$\natural A$, 19, 39
$\tilde{H}_0(X)$, 142
$\tilde{Z}_0(X)$, 142
$\varphi_x[t]$, 207
$e^1 \cdots e^{n+1}$, 127
e^a, 39
e^i, 127
$e^{i_0} \cdots e^{i_k}$, 127
f_A, 40
$x^{i_0} \cdots x^{i_k}$, 127
x_i, 59
$x_{\hat{i}}$, 59
(x,y), 14

absolute neighborhood retract, 146
abstract economy, 60
 equilibrium for —, 60
abstract simplicial complex, 128
abstract summation, 129
acyclic, 9, 142
 method of — models, 11, 160
addition
 vector space, 22
affine function, 197
affine independent, 127
affine subspace, 22
Alaoglu
 Theorem of —, 117
algebraic dual, 23
allocation, 105
 attainable, 105
 feasible, 105
alternating tensor, 177
ANR, 146
antisymmetric, 15
atomic formula, 204
attainable
 allocation, 105
auxiliary base set, 125
axiom
 —of choice, 16, 214
 —of extensionality, 210
 —of foundation, 214
 —of infinity, 212

 —of paring, 210
 —of power set, 211
 —of replacement, 213
 —of separation, 211
 —of union, 211

balanced, 184
 — game, 187
barycenter (abstract sense), 198
barycentric operator, 198
barycentric subdivision, 150
base
 0-neighborhood —, 23
 — of $K \times \{0, 1, I\}$, 151
 — of $K \times \{0, 1\}$, 137
 for a topology, 18
 neighborhood —, 23
basis, 22
 vector space
  standard, 127
better set correspondence, 58
bijective function, 15
binary relation, 15
bound, 130
boundary, 130
boundary operator, 129
bounded variable, 205
Brouwer, 27
 fixed point theorem, 27
Browder type
 mapping, 34, 48
Browder-type (wide sense), 65
Burali–Forti Paradox, 273

canonical bilinear form, 5
cardinal, 275
cardinality, 15, 214, 275
carrier, 159
Cartesian product, 14, 213
 of the family of sets, 15
category, 126
category of topological spaces, 128
Cauchy net, 266
Čech boundary, 134
Čech cycle, 134

Index

Čech homology group, 135
   n-th —, 134
Čech simplex, 133
chain, 129
chain complex, 130
chain equivalence, 131
chain equivalent, 131
chain homomorphism, 131
chain homotopic, 131
chain homotopy, 131
chain map, 131
choice function, 16, 215
circled, 264
class, 125
class $\mathcal{B}$ mapping, 168
class $\mathcal{D}$ mapping, 172
class $\mathcal{K}$ mapping, 170
closed, 16
   correspondence, 28
closed formula, 205
closed subcomplex, 136
closed term, 205
closedness (condition), 68
closure, 16
cluster point, 17
coalition, 187
coalition structure, 194
cochain group, 176
cofinal, 17
Coincidence Theorem, 189
compact, 20
   space, 20
Compactness Theorem, 208
complete, 266
completeness, 207, 209
complex, 8
   in R^n, 127
   underlying space, 127
composition of morphisms, 125
consistency, 206, 208
   introspective—, 224
   logical—, 224
consistent, 208
constant, 203
   individual —, 203

constraint correspondence, 60, 76
contiguous, 134
continuous
   correspondence, 29
continuous function, 16
Continuum Hypothesis, 275
contravariant functor, 126
convergence
   net, 17
convex
   — subset (abstract), 40
   — subset of vector space, 23
   structure, 2, 40
      C1 (condition), 40
      C2 (condition), 40
      C3 (condition), 40
convex combination operator, 40
convex function, 196
convex hull, 40
convex linear combination, 127
convex morphism, 197
coordinate correspondence, 64
correspondence, 15
   Browder type, 34, 48
   closed, 28
   continuous, 29
   Kakutani type, 50
   lower semicontinuous, 29
   upper semicontinuous, 29
   upper semicontinuously
      differentiable, 81
countable set, 15
covariant functor, 126
covering, 19
   binary —, 132
   finite, 19
   intersection —, 132
cycle, 130

De Morgan's laws, 14
decomposition, 19
Deduction Theorem, 209
differentiable
   upper semicontinuously —, 81
differentiable (preference), 46

dimension of complex, 127
dimension of polyhedron, 127
direct limit, 126
direct product, 14
directed set, 15
direction, 42, 44
directional form, 177
domain, 14
dual space
  algebraic, 23
  topological, 23
dual system, 5, 43
  generalized, 43
  structure, 6, 43
    V1 (condition), 43
    V2 (condition), 43
    V3 (condition), 43
    V4 (condition), 44
    V4* (condition), 62
    V5 (condition), 81
  topological, 44
duality
  structure, 5, 6

Eaves Theorem, 187
economy, 106
Eilenberg–Montgomery theorem, 11, 164
empiricism, 202, 243
  Two dogmas of —, 243
epistemic
  values, 244
equilibrium, 243
  — of economy $\mathscr{E}$, 106
  market, 244
  Nash, 244
  physical and moral, 243
equivalence class, 18
equivalence relation, 15
essential element, 126, 144
Euclidean space, 19
eventually in, 17
expansion
  P-adic—, 259
extension, 15
externality, 70

face
  simplex in R^n, 127
  standard n-simplex, 127
fact, 247
Fan–Browder
  fixed point theorem, 27
Fan–Browder fixed point theorem, 54
feasible
  allocation, 105
finite character, 217
finite intersection property, 20
first separation theorem, 24, 270
first-order predicate calculus, 13
fixed point, 27
fixed point theorem
  Brouwer, 27
  Fan–Browder, 27
  Kakutani, 51
  Kakutani–Fan–Glicksberg, 51
fixed-point index, 179
fixed-point set, 27
fixed-point-free extension, 32
formula, 125, 202
free variable, 205
frequently in, 18
function, 15, 203, 213
functor
  contravariant, 126
  covariant, 126

game
  non-cooperative, 57
gauge function, 270
generalized sequence, 17
generic structure, 125
Gödel Completeness Theorem, 208
Gödel Lemma, 227, 231
Gödel Second Incompleteness Theorem, 226
graded
  vector space, 10

Hamel basis, 22
Hausdorff
  space, 20
Heine–Borel covering theorem, 20

Index

homeomorphic, 20
homeomorphism, 20
homology group, 9
  for pair, 175
homology theory, 8, 128
  Čech type, 8
  simplicial, 129
homotopy, 131
homotopy equivalence, 131
homotopy type, 131
hyperplane, 24

image, 15
  —of set, 17
incidence number, 129
index, 181
induced orientation, 127
inference, 203
injective function, 15
Instantiation, 205
integer, 218
interior, 16
introspective consistency, 224
inverse image, 15
  —of set, 16
inverse limit, 126
isomorphic, 126
isomorphism, 126
isotone, 15

K1-type (wide-sense), 66
Kakutani
  fixed point theorem, 51
Kakutani type
  mapping, 50
Kakutani–Fan–Glicksberg
  fixed point theorem, 51
Kakutani-type (wide sense), 66
KKM Theorem, 183
KKMS Theorem, 183

L-majorized mapping, 190
law of the excluded middle, 225
least element, 215
Lefschetz Fixed-Point Theorem, 175
Lefschetz number, 167, 175

linear form, 23
linear functional, 23
linear mapping, 22
linear subspace, 22
linearly independent, 22
local connectedness, 11
local-intersection property, 27
locally connected, 146
logical consistency, 224
logical positivism, 202, 243
logical sign, 203
lower section, 14
lower semicontinuous
  correspondence, 29
  function, 196
Löwenheim–Skolem Theorem, 209

Markov–Kakutani fixed-point
  theorem, 197
mathematical objects, 125, 202
mathematical relation, 125, 202
mathematical structure, see
  structure, 125
maximal element, 59
metatheory, 209
method of acyclic models, 160
methodological individualism, 201
minimum wealth condition, 106
model, 203, 208
Modus Ponens, 205
monoid, 198
monotone, 15
morphism, 125

Nash equilibrium, 60
  generalized —, 60
natural number, 14, 212
neighborhood, 16
neighborhood base, 23
neighborhood retract, 146
  absolute —, 146
nerve, 133
nerve of covering, 8, 128
nest, 217
net, 17

non-cooperative game, 57
  strategic form —, 57
non-side-payment game, 187
norm covering, 146
norm refinement, 146
normal
  space, 21
normal refinement, 148
NR, 146

objects
  mathematical, 125, 202
occurrence
  bounded —, 205
  free —, 204
one to one function, 15
onto function, 15
open, 16
open mapping, 265
open-lowersection property, 27
order preserving, 15
ordered pair, 14, 211
ordinal, 273
orientation
  induced, 127
  simplex, 127
  standard, 127
oriented simplex, 127, 128
oriented simplicial complex, 128

partial realization, 146
partition, 19
partition of unity, 21, 261
payoff function, 59
polyhedron, 127
power set, 14, 211
predicate, 203
predicate logic, 207
  — with equality, 207
preordered set, 15
preordering, 15
principal base set, 125
prismatic chain homotopy, 154
product, 14
product convex structure, 69
product correspondence, 64

product ordering, 16
product relation, 16
product simplicial complex, 137, 151
product topology, 18
projection, 18, 133
propositional logic, 205, 206
  completeness of —, 207
  consistency of —, 206

Quine corner convention, 223
Quine, W.V.O., 243
quotient set, 19
quotient space, 266
quotient topology, 19

radial, 264
range, 14
rational number, 218
rationality, 245
real number, 218
realization, 146
  geometric—, 133
  partial—, 146
recursive set, 223
reduced homology, 142
reductio ad absurdum, 225
refinement, 19
reflexive binary relation, 15
regular, 61
regular simplex, 161
relation, 14, 215
  mathematical, 125, 202
restriction, 15
retract, 146
  absolute neighborhood —, 146
  neighborhood —, 146
revealed preference relation, 96
rules of inference, 203
Russell Paradox, 14

scalar multiplication, 22
Scarf Theorem, 187
second separation theorem, 24
section, 14
  lower, 14
  upper, 14

Index

self-reliance, 250
semantics, 203
semi-norm, 270
semigroup, 198
sentence, 205
separate (points in the dual), 91
separation
  first theorem, 24, 270
  hyperplane, 24
    strict, 24
  second theorem, 24
sequence, 17
set, 13
sign of a simplex, 161
simplex
  in R^n, 127
    k-face, 127
  orientation, 127
  standard, 127
simplicial complex, 127
simplicial homology theory, 129
skeleton, 150
Smith, Adam, 243
social coalitional equilibrium, 194
social system with coordination, 193
species of structures, 125
Sperner Lemma, 161
Sperner Theorem, 185
standard
  n-simplex, 127
  basis, 127
  orientation, 127
standard n-simplex, 127
  face, 127
standard orientation, 127
star refinement, 132
strategic form, 57
strategy profile, 57
structure, 125
  base set
    auxiliary, 125
    principal, 125
  convex, 2, 40
  dual system, 6, 43
  duality, 5, 6
  generic —, 125

species of —, 125
    axiom of —, 125
  typical characterization, 125
subbase
  for a topology, 18
subcovering, 19
subnet, 17
subset, 210
support, 129
supporting direction, 44
supporting element, 44
surjective function, 15
symmetric, 15
system of duality, 5

Tarski Truth Definition Theorem, 231
tautology, 206
term, 125, 202
theory, 203
theory of sets, 13
top
  — of $K \times \{0, 1, I\}$, 152
  — of $K \times \{0, 1\}$, 137
topological dual, 23, 44
topological space, 16
topological vector space, 23, 263
topologist's sine curve, 147
topology, 16
  product, 18
  usual —, 19
totally ordered
  —set, 217
transfinite induction, 216
transitive, 213
transitive binary relation, 15
Tuckey's Lemma, 217
typical characterization, 125

underlying space
  of complex, 127
union, 211
upper bound, 217
upper section, 14
upper semicontinuous
  correspondence, 29
  function, 196

upper semicontinuously (USC)
  differentiable correspondence, 81
Urysohn's Lemma, 260

value, 247
variable, 203
  bounded —, 205
  free —, 205
  individual —, 203
variational inequality problem, 191
vector space, 22
  graded, 10
  topology, 23, 263
vertex assignment, 161
vertex set, 128
Vietoris
  theorem, 11
Vietoris chain, 135

Vietoris cycle, 136
Vietoris homology group, 136
Vietoris mapping, 143
Vietoris simplex, 135
Vietoris–Begle mapping, 143
view of the world, 75, 244

weak topology, 23
wealth function, 105
weighted sum function, 40
welfare economics
  first fundamental theorem of —, 109
  second fundamental theorem of —, 109

Zermelo–Fraenkel set theory, 13